SUPER DESTROYERS

357

SUPER DESTROYERS

From the Torpedo Boat Era to the Dominant Surface Warship of Today

ROBERT C STERN

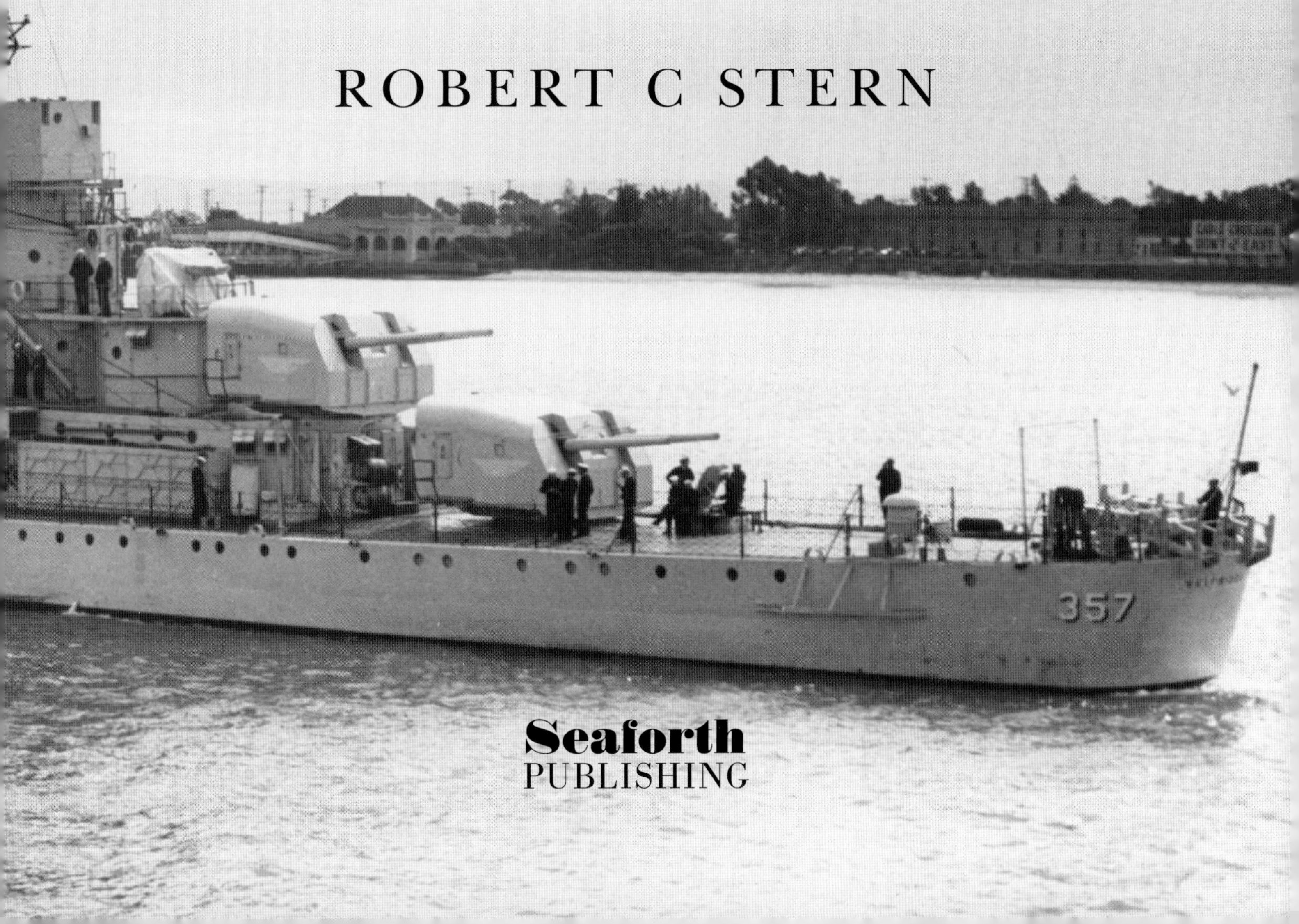

Seaforth
PUBLISHING

Dedication

To Lionel Leventhal, without whose assistance and guidance I never would have started down this road in the first place.

To Dick.

This edition first published in Great Britain in 2024 by
Seaforth Publishing,
An imprint of Pen & Sword Books Ltd,
George House, Beevor Street, Barnsley S71 1HN

www.seaforthpublishing.com
Email: info@seaforthpublishing.com

British Library Cataloguing in Publication Data
A catalogue record for this book is available from the British Library

978-1-5267-7745-4 (Hardback)
978-1-5267-7746-1 (ePub)
978-1-5267-7747-8 (Kindle)

Pen & Sword Books Limited incorporates the imprints of Atlas, Archaeology, Aviation, Discovery, Family History, Fiction, History, Maritime, Military, Military Classics, Politics, Select, Transport, True Crime, Air World, Frontline Publishing, Leo Cooper, Remember When, Seaforth Publishing, The Praetorian Press, Wharncliffe Local History, Wharncliffe Transport, Wharncliffe True Crime and White Owl.

Typeset and designed by Mousemat Design Limited
Printed and bound in India by Parksons Graphics Pvt Ltd

Frontispiece: As built, the US Porter class destroyers resembled small light cruisers even more than did the British Tribals, as can be seen in this pre-war image of USS *Selfridge* (DD 357). The four large twin 5in/38 main-battery mounts – fully enclosed for the first time – the quadruple 1.1in/75 mount fore-and-aft (here covered in canvas), the two huge, box-like Mk35 single-purpose gun directors – the after one mounted on a prominent superstructure – and the two heavy tripod masts, all reinforced the appearance of a much larger ship. (NHHC)

Contents

Acknowledgements

Having written as many books as I have had the good fortune and stubbornness to create, I have come to depend heavily on a support community of relatives, colleagues, and friends without whom all this would never have been possible. As my writing career is winding down, it is especially important that I try to thank as many of these individuals as possible:

My wife Beth, ever-patient, who has read over my manuscripts, often in very rough form, and has given me myriad helpful suggestions; John Jordan, for his sharing of his knowledge of the French Navy of this period; Enrico Cernuschi, for his generosity with his time and the sharing of rare photographs, and for his great knowledge of the Regia Marina; Fulvio Venturi and Conrad Waters, for having made available a number of rare photographs; Randy Stone, for his knowledge of American and Japanese destroyers, particularly their actions in the South Pacific; and Vince O'Hara, as ever, for his gracious assistance in providing help when my resources seemed to reach a dead end.

A few images used in this book are uncredited. This, of course, does not mean they appeared out of thin air, but rather that my sometimes disorganised system for storing them, particularly before I was able to digitise them and move them onto disk storage, failed to preserve their source. To anyone whose name is omitted here who helped with images used in this book, I offer my most sincere apologies and thanks.

The following abbreviations are used in crediting photographs:

DoD	US Department of Defense
LoC	US Library of Congress
MoD	Ministry of Defence (United Kingdom)
MoDJ	Ministry of Defense (Japan)
MoDR	Ministry of Defence (Russia)
NARA	US National Archives & Records Administration
NHHC	US Naval History & Heritage Command
SLV	State Library of Victoria (Australia)
USMM	Ufficio Storico Marina Militare (Italy)
USN	US Navy

Preface

While writing this book I have become less, rather than more, clear on the meaning of the naval ship type chosen for its title. My starting point was my pair of recent books on the impact of the between-the-wars naval disarmament treaties on various ship types, and the well-known and well-liked booklet *Super Destroyers* published by Conway's in 1978.[1] My idea was to extend the treatment of the first two books to cover the development of the large destroyers that emerged between the wars, largely as a result of the impact of those treaties. But it became clear that the period from the late 1920s through the early 1930s that saw the development of the classes that Antony Preston identified as the first 'super destroyers' was far from unique. If the term – never a formal naval type designator – means a ship representing a quantum leap in size and capability over the previous generation, then there may have been other periods when boats were designed that could be so-labelled. It is not my intent in this book to be definitive. Rather, each such leap will be pointed out as a milestone along the path of development of the destroyer type from the earliest torpedo-carrying steam launches to the current examples, which, along with aircraft carriers, are the major surface combatant types in today's navies.

Introduction

Before there could be 'super destroyers', there had to be 'destroyers'. The first ships with that latter description were laid down in 1892 for the Royal Navy. At that time, though, they would not have been called destroyers; that type descriptor did not come into general usage for another ten or more years. That designation was a contraction of the full type descriptor by which they were originally known: 'torpedo boat destroyers'. In that full name lies the story of the origin of the type.

A torpedo is a type of warm-water fish with a fearsome reputation, specifically, one of several species of electric rays known to cause numbing shocks to anyone incautious enough to step on or handle one without protection. This ability to cause harm from a mere touch led to an early association of the fish's name with the idea of naval explosive devices, which were conjured by inventors almost from the time that knowledge of gunpowder came to Europe from the Far East in the thirteenth century. It is known that the 'spar torpedo' – an explosive device at the end of a wooden boom attached to the bow of a small boat – was employed by the Royal Navy in the sixteenth century. Floating mines, called 'torpedoes' at the time, date to the seventeenth century. The famous quote attributed to the Union Rear Admiral David Farragut, 'Damn the torpedoes, full speed ahead', referred to anchored mines in the bay at Mobile, Alabama, during the American Civil War.

As a drifting or anchored mine, a torpedo was a useful defensive weapon, but with the exception of the spar torpedo, which was as dangerous for the operator as for the target, none of the various attempts at developing a self-propelled ('automobile' or 'locomotive') torpedo were notably successful until the British engineer Robert Whitehead produced his first prototype *minenschiff*, improving on ideas originally proposed in 1866 by the Italo-Croatian Giovanni Luppis. Working for the Austro-Hungarian Navy, Whitehead developed a cylindrical device powered by compressed air that could be launched from an underwater tube and propel itself towards a target; upon striking an object, its internal explosive charge would detonate. This first model was small, slow, and erratic (because it lacked any means of self-stabilisation), but it had the basic ingredients of a modern torpedo, and it attracted interest from a number of European navies. Whitehead continually improved his design, borrowing good ideas wherever he found them. For example, he added a pendulum and pressure-sensitive valve that allowed it to maintain a steady depth, and later counter-rotating propellers that helped it maintain a steady course. Speed and range both increased to the point that, by 1890, the Whitehead torpedoes being sold to the Royal Navy had a maximum speed of thirty knots and a maximum range of 800yds.

An interesting summary of the state of torpedo technology appeared in the *New York Times* in September 1890.[1] The anonymous author of this short article expressed some surprise that the US Navy had reached an agreement that would allow licensed manufacture of Whitehead torpedoes in the United States. The source of the author's surprise was that, despite the fact that the navy had tested no fewer than six different American torpedo designs over the preceding few years, named as the Weeks, Howell, Sims, Sims-Edison, Patrick, and Lay-Haight torpedoes, and had ordered small numbers of two of the designs – three of the Patrick and thirty of the Howell – it had turned to a foreign design to produce in large numbers.[2] The navy's response was that the Whitehead torpedo was 'not perfect, but is the only practical automobile torpedo now in use . . .' It was described as being able to maintain a true course and depth at a steady speed between twenty-five and thirty knots for the first few hundred feet of its run. Thereafter, as the pressure in its compressed air flask dropped, the torpedo's speed and stability decreased commensurately.

The torpedo was obviously a weapon of tremendous potential, carrying a large warhead capable of disabling even the largest of warships, but also was one that, unlike improved naval guns, required new types of platforms for delivery. The first four chapters of this book will cover the evolution of torpedo delivery craft through the development of the super destroyer. The remainder will be devoted to the story of the super destroyer type and its use in action.

Ships vs. Boats: Early in the process of writing this book, I was faced with the issue of whether to refer to the vessels being discussed as 'ships' or 'boats'. The distinction between the two is to an extent size-related, with small vessels being normally called boats and larger ones ships, but where the line is drawn between the two has never been exact. Some of it has to do with the language involved and the naval traditions of the nation(s) using that language. For example, in the US Navy, submarines are always called boats, regardless of their size. No similar tradition exists for torpedo craft. Small patrol torpedo (PT) craft have always been called boats, as were larger torpedo boats, but once destroyers were being built at displacements of 1000 tons and larger, the usages were less consistent, but, in general tended towards the use of the term ships. While I would have much preferred using a single term throughout the book, it would be seriously inappropriate to refer to the 2000-ton destroyers of the Second World War as boats, much less the much larger destroyers built since; torpedo boats and their lesser brethren will be referred to as boats in the book, and destroyers and the warships that are now being built that are destroyer-sized, regardless of their designation, will be termed ships. Problems remain, such as with those nations which before and after the First World War built standard destroyers, often exceeding 1000 tons, but continued to designate them torpedo boats. An example of this was France, which built *torpilleurs d'escadre*, some of which exceeded 1700 tons, or Italy, which did a mass redesignation of old destroyers as torpedo boats on 1 October 1929. To the extent possible, I will attempt to maintain consistent usage.

A Note on Nomenclature and Units, etc.: Names of geographical locations mentioned in this book are those that would have been used by an educated English speaker of the first half of the twentieth century. Where those differ from the current name or spelling of a place, the current usage is given in parentheses when first encountered. Ranks and rates for the sailors of navies other than the US Navy, excepting only the Royal Navy, are translated to the closest USN equivalent.

The names of nations, particularly those in central Europe, changed frequently during the period being discussed in this book. For the most part, I have used the names that a contemporary English speaker would have used, with the exception of the Austro-Hungarian Empire; I have, for the sake of brevity, referred to that state as Austria whenever it is mentioned.

One of the many designs tried out by the world's navies in the early days of 'automobile torpedo' development was the American-developed Howell Mk 1 torpedo, which used a 131 lb flywheel spun up to over 10,000rpm to provide power to a pair of contra-rotating propellers that gave it a range of 400yds at 25kt. Howell torpedoes had several advantages over their contemporaries, including the ultimately successful Whitehead design, such as inherent gyroscopic stability brought about by the spinning flywheel, but the design was doomed by the lack of growth potential in terms of speed or range. This example, described as the last of the Howell torpedoes, is seen in its loading cradle on the foredeck of USS *Stiletto* (WTB 1) in 1896.

Ships in many of today's navies are crewed and commanded by both men and women, so the use of male pronouns to refer to officers and ratings may seem anachronistic and even sexist, but during the time period covered by all but the last pages of this book, navies and their ships were an exclusively male domain and I hope I will be excused this pronoun usage.

Mention should be made of the fact that, as the reader probably has already noticed, the author is American; for the most part, this should make no difference to the British reader. However, there is one point of terminology on which the Americans and British differ that should be pointed out lest confusion result. To the Americans, the highest continuous deck in a ship is the main deck; to the British, it is the upper deck. To the Americans, the next lower deck is the second deck; to the British, this is the main deck. The author uses American terminology throughout, except that quotes from British sources retain British terminology.

When first referenced, US Navy ships are identified by their hull number (eg, USS *Fletcher* (DD 445)) in which the letters designate hull type (DD – destroyer) and the number is a one-up counter of hulls of that type as ordered.[3] For Royal Navy ships of the treaty era, pennant numbers are given when they are first mentioned (eg, HMS *Zulu* (F18)).[4] Warship prefix designators, such as 'USS', 'HMS' or 'RN', are also used only the first time a ship is mentioned. Some nations, such as Imperial Japan and Nazi Germany, used no such designator and, in those cases, none is used in this book.

Dimensions of ships are given in feet (12in/304.8mm/abbreviated 'ft') and inches (25.375mm/abbreviated 'in'). Distances are given in feet and yards (3ft/0.9144m/abbreviated 'yds') and nautical miles (2025.37yds/1.853km/abbreviated 'nm'). These are the units used by most seamen in the twentieth century and remain in use in America and, to a lesser extent, Great Britain. Gun calibres are given in the system used by the nation to which a ship belonged. Radar wavelengths are given in metric units. Weights are given in those units used by the nation whose weapon or craft is being described. For the US and UK, this was the English system of pounds and ounces; for the rest of the world, this was the metric system (1kg = 2.205 lb; 1 lb = 16oz = 453.6g). One must be careful when speaking of tons. This tripped up experts at the time and can confuse modern readers as well. When the American and British navies spoke of tons, they meant 'long tons', which equalled 2240 lb; the rest of the world meant 'metric tonnes', which equalled 1000kg or 2204.6 lb. (To confuse matters even more, there was also the 'short ton', used in commerce, which equalled 2000 lb, but was never used in regard to ship characteristics.)

The displacements given for ships in the specifications that follow are the best the author could find. They are generally based on the official figures reported by the building nation and, as such, may have been over- or (more likely) under-reported for various reasons. Even if they were generally correct for a ship at its completion, they would rapidly become incorrect as equipment and weapons were added or removed, and crew size changed. Displacements should be used for comparative purposes only.

To aid any readers unfamiliar with the conversion of gun calibres between 'pounders' (abbreviated 'pdr'), inches and metric measures, the following crib is provided. The reader should always remember that nations felt no obligation to hew exactly to the precise diameters with which their guns were labelled. A good example was the German 15cm gun introduced in destroyers with the S113 class of 1918; this gun, which in its various models was produced from 1898 through the end of the Second World War, always had an actual bore diameter of 14.91cm (5.87in). Also, some conversions from old-style 'pounder' designations to inch or metric calibres were far from exact; the ones given here are simply the most common and should be taken as representative rather than necessarily exact for any given gun:

50cal	= 12.7mm
0.79in	= 20mm/2cm
1in	= 25.4mm/2.54cm
1pdr/1.46in	= 37mm/3.7cm
2pdr/1.575in	= 40mm/4cm
3pdr/1.85in	= 47mm/4.7cm
1.97in	= 50mm/5cm
6pdr/2.244in	= 57mm/5.7cm
9pdr/2.56in	= 65mm/6.5cm
12pdr/3in	= 76.2mm/7.62cm
3.45in	= 88mm/8.8cm
3.9in	= 100mm/10cm
4in	= 102mm/10.2cm
4.1in	= 105mm/10.5cm
4.7in	= 120mm/12cm
5in	= 127mm/12.7cm
5.1in	= 130mm/13cm
5.3in	= 135mm/13.5cm
5.46in	= 138.6mm/13.86cm
5.5in	= 140mm/14cm
5.9in	= 150mm/15cm
6in	= 152mm/15.2cm
64pdr/6.3in	= 160mm/16cm
7in	= 177.8mm/17.8cm
9in	= 228.6mm/22.9cm
300pdr/10in	= 254mm/25.4cm
11in	= 280mm/28cm

List of abbreviations/acronyms used in this text:

AA	Anti-Aircraft
AAC	Anti-Aircraft Common (US Navy term for the anti-aircraft form of HE projectiles)
AP	Armour-Piercing
APC	Armour-Piercing Capped
AR	Action Report

ASW	Anti-Submarine Warfare
BLR	Breech-loading rifle (a rifled naval artillery piece loaded from the breech – cf., MLR)
BuC&R	Bureau of Construction and Repair (the US Navy department responsible for the named activities from 1862 until 1940)
BuEng	Bureau of Steam Engineering (the US Navy department responsible for power systems)
BuOrd	Bureau of Ordnance (the US Navy department responsible for guns and gunnery)
BuShips	Bureau of Ships (formed in June 1940 by the merger of BuC&R and BuEng)
CAP	Combat Air Patrol (US Navy term for defensive fighter cover over a task force)
CID	Committee of Imperial Defence (a standing committee of members of the British Cabinet, heads of the military services and Dominion prime ministers, intended to advise the prime minister on military strategy; in existence from 1902 until 1939)
CDN	Conseil de la Défense Nationale (National Defence Council – French Cabinet-level standing committee tasked with establishing military policy and priorities in the inter-war period)
CinC	Commander-in-Chief (primarily a US Navy usage; pronounced 'sink'; in British usage it is often hyphenated, as 'C-in-C')
CO	Commanding Officer
CNO	US Navy Chief of Naval Operations
CPC/CPBC	Common Pointed Capped/ Common Pointed Ballistic Capped (Typical QF shell used against armoured targets, a common shell given an armoured cap and, optionally, a ballistic cap, cf., SAPC/SAPBC)
CPO	Chief Petty Officer
CS	Cruiser Squadron (Royal Navy formation of heavy/armoured cruisers)
DCT	Director Control Tower
DNC	Director of Naval Construction (the Royal Navy's head of the ship design bureau, generally a civilian marine architect)
DNO	Director of Naval Ordnance (the RN's lead gunnery officer on Admiralty staff)
DP	Dual-Purpose (referring to guns that could be fired at aircraft and surface targets and the directors that controlled them)
EMM	État-Major de la Marine (General Staff of the Marine Nationale)
FDO	Fighter Direction Officer (US Navy term for an officer directing the airborne defence of an area by fighter aircraft)
FY	Fiscal Year (in US terminology, the budgetary year in which money was allocated for a ship; often does not coincide with the calendar year; expressed in the form 'FYyyyy', as in 'FY1928')
GM	Metacentric Height (a measure, in feet, of the height of a ship's metacentre – the point where a vertical line from the ship's centre of buoyancy intersects the ship's inclined centreline)
GMT	Greenwich Mean Time
GRT	Gross Register Tons (a measure of a cargo ship's carrying capacity, only indirectly related to displacement)
HA	High-Angle (generic descriptor for anti-aircraft-related equipment)
HACS	High-Angle Control System (Royal Navy anti-aircraft fire control direction system)
HADT	High-Angle Director Tower (Royal Navy anti-aircraft fire control position)
HC	High-Capacity (synonymous with HE)
HE	High-Explosive (synonymous with HC)
HFDF	High-Frequency Direction Finding ('Huff-Duff')
HMAS	His Majesty's Australian Ship (Royal Australian Navy warship designator)
HMS	His Majesty's Ship (Royal Navy warship designator)
HSF	Hochseeflotte (High Seas Fleet – the battlefleet of the Imperial German Navy (Kaiserliche Marine) from 1907–1918)
HTS	High-Tensile Steel (homogenous armour-grade steel highly resistant to deformation under stress, widely used as a construction steel for warships after the First World War)
IFF	Identification Friend or Foe (an electronic interrogation system that allowed a radar scope to display whether a contact was friendly or not)
ihp	indicated horsepower (the theoretical energy output of reciprocating steam engines, as opposed to the measured energy output of rotating shafts, cf., shp)
IJN	Imperial Japanese Navy
JMSDF	Japanese Maritime Self-Defense Force (post-WWII successor to the IJN)
KM	Kriegsmarine (the German navy after 1935)[5]
LCS	Light Cruiser Squadron (Royal Navy formation of light cruisers)
LD	laid down
MLR	Muzzle-loading rifle (a rifled naval artillery piece loaded from the muzzle – cf., BLR); also sometimes rendered as RML for 'rifled muzzle-loader'

MN	Marine Nationale (the French navy)
NCM	Naval Conference (Ministerial) Committee (a British Cabinet-level committee established in April 1934 to monitor naval matters)
NIACC	Naval Inter-Allied Control Commission (board established by the Treaty of Versailles to oversee the post-First World War German Navy)
NMS	Warship designator for the Royal Romanian Navy
NY	Navy Yard
oa	overall (used to specify a ship's length, in this case indicating that the measurement is between the farthest extremities at the ship's ends)
OpNav	US Navy Office of the CNO
PLAN	People's Liberation Army Navy (Post-1949 Chinese navy)
PM	The British Prime Minister
PO	Petty Officer
pp	between perpendiculars (used to specify a ship's length, in this case indicating that the measurement is between a ship's foremost and aftermost vertical frames)
PPI	Plan Position Indicator ('God's Eye View' 360° radar display)
psi	pounds per square inch – a unit of gas pressure (1psi = 6894.76 pascals)
QF	Quick-firing
RA	Regia Aeronautica (Royal Italian Air Force – from 1923–1946)
RAF	Royal Air Force
RAN	Royal Australian Navy
RCN	Royal Canadian Navy
RDF	Radio Direction-Finder (originally simply a passive system for identifying the direction from which a radio signal originated; sometimes used by the British as a synonym for radar)
RM	Regia Marina (Royal Italian Navy – from 1861–1946)
RML	See MLR
RN	Regia Nave (Royal Ship – warship prefix designator used by the Regia Marina)
RN	Royal Navy (United Kingdom)
rpg	rounds per gun
rpm	rounds per minute
SAM	Surface-to-Air Missile
SAP	Semi-Armour-Piercing
SAPC/SAPBC	Semi-Armour-Piercing Capped/ Semi-Armour-Piercing Ballistic Capped (later name for CPC/CPBC)
SecNav	Secretary of the Navy (US Cabinet-level civilian in charge of the Navy, equivalent of British First Lord of the Admiralty)
shp	shaft horsepower (the measured horsepower of the rotating shaft of a steam turbine or internal combustion engine after any reduction gearing, cf., ihp)
SMS	Seiner Majestät Schiff (His Majesty's Ship – Kaiserliche Marine warship designator)
STCN	Service Technique des Constructions Navales (design bureau of the Marine Nationale)
STS	Special-Treatment Steel (a high nickel content armour-grade high-tensile-strength weldable steel used by the US Navy for vertical bulkheads and protective decks)
TBD	Torpedo Boat Destroyer
TF	Task Force (USN designation for a large ad hoc force given a specific task)
TG	Task Group (USN designation for a subdivision of a TF)
TNT	Trinitrotoluene (an explosive widely used in bomb and torpedo warheads and artillery shells, replacing Lyddite (trinitrophenol) as the preferred explosive at the beginning of the twentieth century due to its chemical stability and its ability to withstand shock)
TS	Transmitting Station (RN term for the gunnery control station below decks; Plotting Room in USN parlance)
UK	United Kingdom
UNREP	Underway Replenishment
USAAF	United States Army Air Force, successor to the United States Army Air Corps (USAAC)
USN	United States Navy
USS	United States Ship (USN warship designator)
VLS	Vertical Launch System
VT	Variable-Timing (The designation for the rounds fitted with the proximity fuse used in heavy AA guns, such as the US Navy's 5in/38)
VTE	Vertical-Triple-Expansion (an advanced form of compound steam engine)
wl	waterline (used to specify a ship's length, in this case indicating that the measurement is between the points fore and aft where the hull meets the water at designed displacement)
W/T	Wireless Telegraphy (Royal Navy term for radio, particularly Morse as opposed to voice radio)
XO	Executive Officer

Chapter 1

Steam launches, torpedo boats, and competing ideas: 1876–1885

WITH THE INVENTION OF the modern torpedo, a whole new realm of tactical possibility became available to naval planners. At the end of the nineteenth century, the universal utilisation of steam propulsion meant a commander leading ships into battle no longer had to be concerned with his position relative to the wind (or most other environmental factors) except as to how they might affect visibility or the working of guns. Of much greater concern would be the number and quality of ships he commanded compared to the enemy's, and factors such as the training of his crews, the fuel and munitions state of his ships, and his and the enemy's positions relative to supporting forces or safe harbours. In terms of the quality of his force compared to his enemy's, his main concern would be the speed of his ships and the number and size of the guns carried by his ships compared to those on the enemy's ships. Until the introduction of effective torpedoes into the world's navies, there was little else, other than perhaps the threat of ramming, that a tactical commander needed to fear in a deep-water engagement.

The introduction of the Whitehead torpedo changed everything effectively overnight. The world's major navies very rapidly began fitting torpedo launching devices – initially fixed tubes and cradles, followed soon by aimable torpedo tubes – to their ocean-going warships. The first ships to be so fitted were cruisers, because of their speed and their generally more independent role in a battlefleet. (The American newspaper report cited in the Introduction specifically mentioned the construction of new cruisers as being the driving factor in the US Navy's decision to license Whitehead torpedo technology.[1])

Dedicated delivery platforms

Even before the automobile torpedo was used for the first time in battle, albeit unsuccessfully, at Pacocha Bay in southern Peru in May 1877, it had occurred to smart people in multiple navies that the erratic, short-ranged, and slow torpedoes of the day did not pose much of a threat to enemy ships if fired in small numbers at gun-battle ranges. What was needed was a means to fire a larger number of torpedoes from multiple points much closer to the target. There were a number of ways this could be accomplished. The first and simplest of these was to mount a torpedo or two on the steam launches already carried by larger warships and send them out to launch their weapons from a point closer to the enemy. This led directly to the realisation that the large warship was an unnecessary part of this set-up. The ship that transported torpedo-bearing steam launches close to an enemy did not have to be an expensive and difficult-to-replace large warship; any ship large enough to carry one or (preferably) more steam launches within tactical range of a target would do just as well, as long as it could approach the target safely. This fitted well with a second conclusion reached by many 'experts' at this time, which was that torpedoes were a weapon best employed at night. In darkness, in the days before night-piercing radar, it would be possible for a large ship to approach undetected within range of the target before putting her torpedo-armed launches in the water. A further advantage of nocturnal attack was that targets were more likely to be anchored or steaming slowly.

By the time of the Battle of Pacocha, these ideas had coalesced into concrete form in multiple navies. One nexus for this development was the somewhat-mislabelled Russo-Turkish War. The war, which lasted from April 1877 until the end of January 1878, saw the Ottoman Empire opposed by Russia and several semi-independent Balkan states, including Serbia and Romania, and an emerging Bulgarian principality. It was primarily a land war, but saw a number of major naval milestones during its brief course.

One of those occurred on the Danube on 26 May 1877, three days before the Battle of Pacocha halfway around the world.[2] The Romanian spar torpedo boat NMS *Rindunica*, which had been built at the Yarrow & Co (Poplar, East London) yard two years earlier, was 46ft long and displaced approximately 10 tons.[3] When delivered in 1875, *Rindunica* was unarmed; it was only after the Romanians signed a pact with the Russians at the beginning of hostilities on 4 April 1877 that she was equipped with a spar torpedo of Russian make and began operating on the river under joint command. (The joint command was 'joint' in name only; the boat was effectively commanded by Russian Lt F V Dubasov. A second Russian officer, a Romanian 'co-commander', a Romanian machinist, and a river pilot rounded out the five-man crew. The Russians went so far as to give *Rindunica* a Russian name – *Tsarevich* – for the duration of the river campaign.) Three other similarly armed small torpedo launches with similar jointly-commanded crews – *Kseniya* (Lt Shestakov), *Dzhigit* (Ens Persin), and *Tsarevna* (Ens Bal) – made up the Danube torpedo flotilla commanded by Lt Dubasov.[4]

On the night in question, *Rindunica* along with her three flotilla-mates was operating on the Danube River near Macin. That night was chosen because it offered a bright moon just short of full for illumination despite an overcast sky, but there

An artist's rendering of the Battle at Macin, 26 May 1877, between Romanian/Russian steam launches armed with spar torpedoes and the Turkish monitor *Seyfi*. This image, which appears to be basically accurate, shows *Seyfi*, to the left, firing her main-battery gun at full depression above the attacking steam launches, with *Kseniya* in the foreground about to deliver her attack, and *Rindunica* beyond her trying to back away, having delivered her attack. *Tsarevna*, on the far side of *Rindunica*, is attempting to come alongside to her aid, and *Dzhigit*, her spar torpedo extending out in front of her, enters the picture from the right.

was a strong breeze and steady rain to muffle the sound of the small boats' steam engines.[5] Starting from the Romanian river port of Galati, where the allies had established a base in a suburb at Tighina, the flotilla headed upstream to Braila, and then turned east into the channel leading to the Turkish-held port of Macin, passing the wreck of the 2540 ton Ottoman coast defence turret ship *Luft-i Djelil*, which had ventured too close to Russian positions near Braila on 10 May and had succumbed to mortar fire.[6] Since that loss, the three remaining Turkish ships in the area had stayed close to Macin. They were the large monitor *Seyfi*, the smaller ironclad river monitor *Feth-ül Islam*, and the paddle-steamer *Kiliç Ali*. *Seyfi* was actually a rather formidable vessel for its time, one of a pair of sisters designed and built at the Tersâne-i Âmire yard at Constantinople, 144ft long, displacing 404 tons, protected by 3in wrought iron belt and battery armour, and with a pair of turret-mounted Krupp 12cm/22 RK (*ringkanone* – built-up gun) MLRs.[7] The three Ottoman ships were at anchor, deployed across the Macin channel, with *Seyfi* in the middle and somewhat in advance of the other two.

The Russo-Romanian flotilla steamed up the Danube in line-abreast formation, with *Rindunica*, with Lt Dubasov aboard in command of the flotilla, at the starboard end of the line and slightly ahead of the other three. The Macin channel was too narrow for this formation, so the two boats on the port side of it – *Dzhigit* and *Tsarevna* – took station behind the leading pair. By the time the Turks sighted the approaching torpedo launches and could react, there was only time for *Seyfi* to get off three main-battery rounds, all of which missed high. Dubasov ordered *Rindunica* to attack with her spar torpedo, which she drove into *Seyfi*'s port side aft. This caused a massive explosion with water and debris blown up and outward, a good deal of it landing on *Rindunica*. This, of course, was one of the great problems with spar torpedoes, that any practical length of spar left the attacking ship vulnerable to the effects of the detonation of the torpedo. Dubasov ordered *Rindunica* to back away at best speed, while he and the others in the crew manned the steam pump and tossed overboard what debris they could.

Seyfi rolled sharply away and back, but settled on an even keel, down by the stern, but showing no signs of sinking. She even began firing her main battery again, but was still unable to depress her guns enough to hit the launches. Seeing this, Dubasov ordered Shestakov in *Kseniya* on his port side to attack. This he did with similar effect, detonating his torpedo further forward on *Seyfi*'s port side, in way of her turret, resulting in a similar large explosion, a similar deluge of water and debris on the launch and a similar reaction by the Turkish monitor, except that this time, as she rolled back to an even keel, she was obviously sinking. *Kseniya* was not able to back away quickly enough and became entangled in the debris around the sinking monitor. Ens Bal in *Tsarevna* saw *Kseniya* in distress and came alongside the damaged launch to help pull her free. *Tsarevna*'s crew also helped bail out *Kseniya*, necessary because her steam pump was one of the casualties of the torpedo explosion. That left only *Dzhigit*; despite orders from Dubasov that any remaining Turkish ships were to be attacked by any launch still carrying an unexpended torpedo, Ens Persin exercised discretion and withdrew with the other three launches. All four arrived back at Tighina, having lost none of their men, with the distinction of being the first to have made a successful torpedo attack of any kind and to have survived to tell the tale. The best available estimate is that approximately twenty Ottoman sailors were lost in *Seyfi*, about forty percent of her normal complement. The success of this attack denied the Ottoman forces in the Danube access to the delta and the Black Sea; they had no option but to retreat further upstream towards Hirsova, which rendered them effectively irrelevant for the rest of this brief conflict.

The first 'real' torpedo success

The success of the Russian Danube squadron and the sinking of *Seyfi* perhaps should have validated the spar torpedo as an effective naval weapon, but it did nothing of the sort, in large part because of what happened eight months later on the other side of the Black Sea. Despite having their Danube squadron bottled up, the Ottoman navy enjoyed considerable success in the open waters of the Black Sea in the early days of the war, not so much because of the quality of their ships or their leadership, but because the Russian Black Sea fleet, largely destroyed during or dismantled after the Crimean War (1853–6), had not yet been rebuilt to anything resembling its former strength. This left the Turks in control of the sea and able to block any Russian movement by water. Further, the Ottomans had hopes that the large Muslim populations in the Russian Caucasian provinces that bordered the eastern Black Sea – Abkhazia, Chechnya, and Ossetia – might be inspired to rebel against their masters by a show of force by the Turks.[8] This led them to concentrate their naval efforts on the eastern shore.

Seen near the end of his career, as a vice admiral, probably in 1904, the innovative Russian naval officer Stepan Osipovich Makarov had, as a senior lieutenant in 1876, helped pioneer the use of 'automobile torpedoes' dropped from steam launches, which would be carried to the vicinity of their target by the tender *Velikiy Knyaz Konstantin*, under his command. (National Archives of Norway)

The Russians, lacking any significant naval force in the Black Sea, were forced to improvise. Even before the war began, the oceanographer and inventor, Sr Lt Stepan Osipovich Makarov, one of those aforementioned smart people thinking about the potential of torpedoes, was concerned about this state of affairs. In December 1876, Makarov proposed to the Russian naval staff that a large, fast steamer be converted into a tender for torpedo-carrying steam launches, and the first purpose-built torpedo boats.[9] By the end of January 1877, a fast (ten knots) iron screw steamer owned by the Maritime Company of Odessa, *Velikiy Knyaz Konstantin* (Grand Duke Constantine), was taken over for conversion into a torpedo boat tender/carrier.[10] For her new role, she had been modified with four sets of davits, each capable of handling a steam launch armed with a torpedo, and four small MLRs. The four boats assigned to *Konstantin* were an assorted lot of small steam launches varying in displacement between 3.5 tons and 7 tons, and in construction; three of the four – *Navarin*, *Sinop*, and *Miner* – were wooden-hulled, while the fourth – *Chesma* – had a bronze hull. At this point each was armed with a towed mine. Fittingly enough, *Konstantin* was commanded by Makarov.

Convinced that his idea would prove to be a force multiplier, Makarov trained his crew and the crews of the launches (and two somewhat larger torpedo boats which could be towed behind *Konstantin*) to the point where, at least in daylight, the four launches could be lowered and rehoisted in seven minutes. By May, *Konstantin* was declared ready for operations and she was put to work immediately, attacking a Turkish squadron anchored at Bat'umi (Georgia). This attack and four others that followed in June and August, were utter failures, primarily because their chosen weapon, the towed mine, was unreliable and difficult to handle. On more than one occasion the tow line to the mine became entangled in the launch's propeller or, in one case, the target's paddle wheel, and became more of a danger to the attacker than to the target. The five attacks managed to sink only one ship; one of the two larger torpedo boats that *Konstantin* had towed to Sulina (Romania) on 9 June 1877, apparently ran up on Turkish defensive booms, puncturing her hull and sinking rapidly.

Makarov was well aware of these problems with towed mines even before the campaign began.[11] He knew that the Russians had obtained a licence to manufacture Whitehead torpedoes the year before, but he also knew that before the outbreak of war nothing had been done to establish an assembly facility, so that Russian-made torpedoes would not be available anytime soon. They had, however, purchased six of the current model torpedo from the Whitehead factory in

Sadly, a poor-quality image, but an interesting one nonetheless, that shows the torpedo launch *Chesma* posing in front of the torpedo launch tender *Velikiy Knyaz Konstantin* in 1878. The relative size of the launch and the tender can be judged, even in this (quite possibly manufactured) image, by looking at the size of the crewmen in *Chesma*, and then comparing *Chesma* to what appear to be a pair of similar launches close alongside the tender in the background. (Drashpil Collection – NHHC)

Austria to serve as manufacturing models and test examples. With the coming of war, the Navy Minister came under intense pressure to release some or all of these six for use against the Ottomans, not least from Makarov, who had considerable influence in St Petersburg. In June 1877 two of the six were allocated for his use.

As was the case with every other navy considering the first use of automobile torpedoes, Makarov faced the problem of how to launch this new weapon. He decided on perhaps the simplest solution, given his circumstances; two of his steam launches would each tow a raft fitted with an above-water tube from which the torpedo would be launched; but during the first practice drills a storm arose, sinking one of the rafts, causing the remaining torpedo to be withdrawn. However, Makarov persisted and he was once again in possession of two Whitehead torpedoes by the end of 1877. Two of his steam launches had improved, though still very much improvised, mechanisms for launching their torpedoes. *Chesma* had a single underwater launching tube affixed below her keel, while *Sinop* had an above-water tube at her bow.[12] On 27 December, these boats attacked a Turkish squadron at Bat'umi, targeting the large ironclad *Mahmudieh*. Because of the darkness of the overcast night, neither launch noticed that they were attacking bow-on rather than from a broadside position, and both torpedoes missed. One ran up on the beach, where it was found by Ottoman soldiers the next day; the other struck *Mahmudieh*'s anchor chain, broke apart and detonated upon sinking to the bottom of the harbour, causing a spectacular, albeit inconsequential, explosion. Fortunately, both launches were able to return to *Konstantin* without damage.

Makarov had enough political sway at the capital to overcome the reluctance of the theatre commander to allocate any more of the precious few Whitehead torpedoes. With two more torpedoes in hand, he was able to mount another attack against the same target on 26 January 1878.[13] Again, *Chesma* and *Sinop* carried single torpedoes mounted as before. *Konstantin* carried the two launches to within 5nm of Bat'umi harbour, put them in the water, and then stood off to await their return. Approaching the roadstead after midnight at their best speed, approximately six knots, the officer commanding *Chesma*, Lt I M Zatzarennyi, picked out the most visible ship, at the near end of the line of anchored Ottoman vessels. This turned out to be an unarmoured steamer, *Intibah*.[14] Approaching to less than 100yds, both launches fired at *Intibah*'s exposed side; this time both torpedoes hit, causing the steamer to list rapidly and sink within minutes, taking more than twenty of her crew with her.

The target may not have been as important as Makarov may have wished, but the event was nevertheless momentous, vindicating his persistent advocacy of the weapon and its delivery method. For the first time, a ship had been sunk by automobile torpedoes, and the attackers had escaped intact. From this point forward, no naval commander would be able to move ships or anchor them in a harbour without taking the possibility of torpedo attack into account.

Looking for a better delivery system

Torpedo technology continued to improve rapidly and the options for delivering torpedoes against an enemy multiplied equally rapidly. The method pioneered by Makarov and the Russians, using small, slow, short-ranged steam launches carried to a point near the site of the intended attack, clearly had some advantages, particularly when carried out at night, but there were also limitations that could be overcome by improving naval technology. The most obvious limitation of the steam launches used by Makarov at Bat'umi was their slow speed, which made them vulnerable should they be detected, and small size, which limited them to carrying only one torpedo. A larger torpedo boat would be able to accommodate a larger engine, making it faster, and would be more seaworthy and longer-ranging, and also offered the option of carrying more torpedoes (and possibly even reloads). The obvious disadvantages, that larger boats would be more visible, even at night, and more expensive, making their construction in large numbers more difficult to justify, were at least partially offset by the greater survivability offered by greater speed and size.

Clearly this was something of a zero-sum game, as is often the case in warship design. There was, at the end of the 1870s, no one obvious right answer to the question of what the ideal torpedo delivery system would look like. Therefore, it is no surprise that each of the world's major navies tried out different, often multiple, design ideas, groping for a successful formula. What is generally considered to be the first purpose-built torpedo boat was laid down at the British shipbuilders Thornycroft & Co (Chiswick, West London) for the Norwegian Navy; her specifications were:

KNM *Rap* (as built): 1873[15]

Displacement:	7 tons
Length (oa):	57ft
Beam (wl):	7ft 6in
Draft (max):	3ft
Powerplant:	90ihp; inverted double cylinder direct-acting engine; one locomotive-type boiler
Speed:	14kt (design); 14.96kt (actual)
Protection (side):	0.2in steel (in way of command and engineering spaces)
Armament (intended):	1 spar torpedo

All indications are that *Rap* was unarmed when delivered in 1873 and may well have remained so for six years, although it is possible she briefly carried a towed mine as an experimental installation. What is certain is that in 1879 she was fitted with

This heavily retouched image purports to show KNM *Rap*, generally considered the first purpose-built torpedo boat. Built for the Norwegian navy by Thornycroft & Co (Chiswick), this image does show her general appearance at the time of her completion in 1873. At that time, she carried no armament, not even the spar torpedo she was intended to carry.

a pair of launching cradles for Whitehead-type automobile torpedoes, but by this time that would have made her far from unique.

The British Royal Navy, as the world's largest and unquestionably pre-eminent navy at the time, had the most to fear from the emergence of an inexpensive weapon such as the automobile torpedo, which could put its investment in large ships at risk. It therefore had the greatest impetus to investigate thoroughly all contemporary developments of torpedo delivery systems, if only to understand better how to counter them. This was especially true seeing the number of orders for torpedo boats being received by Thornycroft and its most serious competitor, Yarrow, from smaller European navies, which, besides the Norwegian, included those of Sweden, Denmark, Austria, the Netherlands, Italy, Russia, and, most ominously for the British, France. Of these nations, the British would at this time been most concerned with the Russians and French; despite the fact that the French and British had been allied against Russia only a few years before during the Crimean War, a vast majority of Britons, including most in the Royal Navy, were much more worried about possible threats from the French Marine Nationale (MN) with its bases on the English Channel, Atlantic and Mediterranean and its long history of hostility towards the British, than they were about the potentially more powerful, but much more distant, Russians. If any nation were to find in the torpedo a practical and affordable means to neutralise Britain's dominant fleet in some possible future conflict, the British feared it would be the French.

Thus the Royal Navy, beginning as early as 1872, felt obliged to explore the methods of torpedo warfare available to its potential enemies. To this end, in May 1873, the Admiralty established a Torpedo Committee to define the methods by which automobile torpedoes could be launched.[16] This resulted in the definition of four different basic approaches to torpedo delivery which the Royal Navy should consider.

(From this point forward in the book, the word 'torpedo' will be understood to refer to the 'automobile torpedo' in general and more specifically to those made from, or developed to compete with, Whitehead's designs. For all practical purposes, by 1880 spar torpedoes and towed mines had ceased to be considered primary weapons. As late as 1884, the French used small steam launches armed with spar torpedoes to attack the Chinese screw frigate *Yuyuen* in Shipu Bay (Taiwan); one launch succeeded in setting off its torpedo against *Yuyuen*'s hull, sinking the unarmoured target, but being sunk itself in the process.)

The four basic approaches defined by the Royal Navy's 1872 Torpedo Committee were:

Torpedo carriage by large vessels: This was the most obvious expedient and the simplest to implement, as torpedoes could be mounted on existing large warships with little difficulty. The weapons and the equipment to handle them were relatively small and light compared to large guns, and could be fitted in unused spaces between other equipment. From the end of the 1870s, torpedo tubes were regularly part of the designs of larger Royal Navy warships. (The Colossus class turret ships were built with a pair of trainable 14in torpedo tubes mounted above the waterline; unlike torpedoes fired from cradles or fixed tubes, these torpedoes could be fired at a target within a range of angles fore and aft of athwartships without turning the ship. Underwater torpedo tubes in large ships, in fixed mountings, but better protected from enemy gunfire, would come along towards the middle of the 1880s.)

Torpedo carriage by ships' boats: This was essentially what the Russians demonstrated in 1877–8. This approach had the advantage of simplicity and low cost, but the great disadvantages of slow speed, total lack of protection for the crew, and instability in all but the calmest waters (the torpedo and its handling gear could weigh as much as the boat). While this approach was studied by the Royal Navy, it seemed to offer no advantages over the next, which was pursued with considerable interest.

Torpedo carriage by purpose-built small boats: These boats, which were often, at least at first, not much bigger than ships' boats, were nonetheless much better suited for the purpose of torpedo delivery. The first of these boats built for the Royal Navy, by Thornycroft, had the following specifications:

HMTB *51–62*: 1878[17]

Displacement:	10.75 tons
Length (oa):	60ft 6in
Beam (wl):	7ft 6in
Powerplant:	120–150ihp; one locomotive-type boiler
Speed:	15–16kt (design)
Armament:	2 × 14in Whitehead-type torpedoes (2 × 1) in Thornycroft-designed launching frames

These boats were not much larger than the ship's boats they were intended to replace, distinguished only by being somewhat more seaworthy with decking over most of the boat's length. Nonetheless, they lacked the range or speed to be seen as warships that could be sent out on their own into open water from a harbour for coast defence, nor would they be able to accompany the fleet any distance to attack enemy forces at sea or in port. It was clear to the Royal Navy that these small torpedo boats would need to be carried to a point near their intended target, much as Makarov had done at Bat'umi.

The Royal Navy could not improve on Makarov's solution to the problem of delivering small torpedo boats to a distant target. It is somewhat ironic that it was a 'war scare' in 1878,

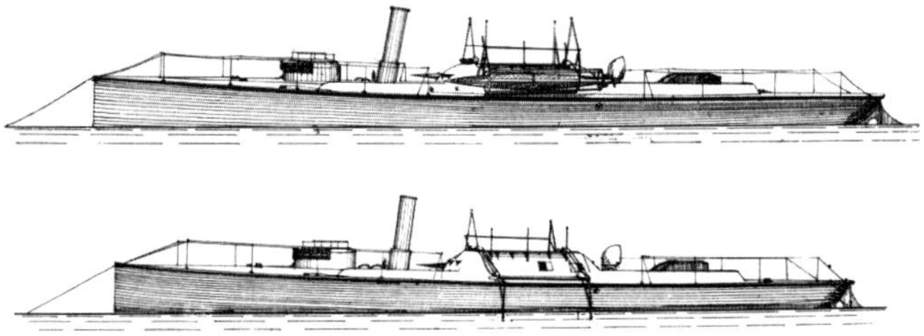

A drawing showing a Royal Navy torpedo boat from the Thornycroft series ordered in 1878 – HMTB *51–62* – taken from C W Sleeman's *Torpedoes and Torpedo Warfare*, published two years later. These drawings are of particular interest because they show the operation of Thornycroft's patented launching frame, that suspended an 'automobile torpedo' above the boat's main deck until ready to launch and then lowered it into the water away from the boat's side. (Sleeman)

a threat of war with Russia, more imagined than real, that spurred Parliament to lavish sums of money on the Royal Navy and spurred the navy to act on its plans for deploying the small torpedo boats then in the works. A large merchant ship, SS *British Crown*, then under construction at Harland & Wolff (Belfast, Northern Ireland), was taken over on the ways by the Royal Navy early in 1878. The modifications necessary to allow her to act as a 'torpedo depot ship and floating factory' were relatively minor; her midships superstructure was not built up beyond one deck, and cradles were fitted fore and aft of the small bridge so that six small torpedo boats could be carried on deck. A substantial mainmast aft of the bridge carried a large boat crane which could lift the torpedo boats one at a time to or from the water alongside the tender. (Compared to the davits that raised and lowered the steam launches for Makarov's *Konstantin*, this was a far slower means of deploying the carried boats, but it must be remembered that these torpedo boats were between two and three time heavier than the Russians' steam launches and were almost twice as long, so their handling by davits would have been problematic.) When completed in 1878 and commissioned as HMS *Hecla*, she proved to be very wet ship, faring poorly in all but the calmest seas, but she was, nonetheless, successful at proving the concept of a torpedo boat carrier to the Royal Navy's satisfaction. To that end, in 1888, a purpose-built torpedo boat carrier, HMS *Vulcan*, was laid down at Portsmouth Dockyard.

Torpedo carriage by purpose-built ships: Even in 1875, well before the first of the small torpedo boats were ordered, it was obvious to the Admiralty that torpedo craft of the proposed dimensions would not be sufficiently seaworthy to operate outside of sheltered waters. A larger, sturdier ship would be

Classed as a torpedo depot ship, HMS *Hecla*, completed in 1878, served as a prototype for a torpedo boat carrier similar in concept to Makarov's *Velikiy Knyaz Konstantin*. A purpose-built torpedo depot ship followed two years later, but the idea was soon shown to be impractical, as torpedo boats rapidly grew too big to carry aboard another ship. (NHHC)

required to operate alongside the fleet or independently in open water. The as yet unanswered question was how large would such a ship need to be?

No sooner had the 1872 Torpedo Committee report been presented than the Royal Navy ordered a craft from the Pembroke Dockyard designed to carry and launch automobile torpedoes.

HMS *Vesuvius*: 1873[18]

Displacement:	245 tons
Length (pp):	90ft
Beam (wl):	22ft
Draft (max):	8ft 6in
Powerplant:	350ihp; two-shaft compound engines
Speed:	9.7kt
Armament:	1 × 16in Whitehead-type torpedo in underwater bow tube; 9 reloads

Laid down in March 1873, HMS *Vesuvius* was innovative in a number of ways; indeed, perhaps too innovative. In the expectation that she would be employed in nocturnal attacks, every effort was made to make *Vesuvius* stealthy. She was given a deep hull with a low freeboard, which would place her large reciprocating steam engine mostly under the waterline, in theory muffling its considerable noise. To prevent the production of much smoke, her boilers were designed to burn coke rather than coal, and to suppress any visible smoke or sparks that might be produced, the ship was designed to exhaust the results of that combustion through vents below the waterline.

Launched in March 1874, *Vesuvius* was towed to Portsmouth for completion. She was never tested in her original designed configuration. Early trials showed that the underwater exhaust vents made raising steam difficult, so a tall funnel was added, and her slow speed led to her quickly being relegated to a training role. The belief that high speed relative to other contemporary warships was a necessary characteristic of a torpedo-carrying vessel seems to have sprung up early in this time period and rapidly became an unshakable article of faith.

It is not hard to see why this would be the case. Because the torpedoes of the time were slow and short-ranged, torpedo carriers had to get very close to a target – within a few hundred yards at most – to have any chance of success. That meant a

Shown soon after completion with her tall funnel, HMS *Vesuvius*, the Royal Navy's first purpose-built torpedo vessel, laid down in 1873, is seen followed closely by a barge. Her slow speed and low freeboard led very quickly to her relegation to secondary duties. (NHHC)

torpedo carrier had to be quite stealthy to approach undetected at night, a very difficult achievement given the noise made by the reciprocating steam engines of the day and the likelihood that flame or sparks would emerge from the boat's funnel. The only other option that gave a torpedo carrier much chance of success, or even survival, was high speed. That this second option became the one that dominated design thinking was promoted by persistent lobbying by specialist firms such as Thornycroft and Yarrow, and by the naval officers who supported their designs; this led to the Royal Navy ordering a larger torpedo boat – similar to TB *51*, but larger in all dimensions – from Thornycroft even before the smaller ones.[19]

HMTB *1* (HMS *Lightning*): 1876[20]

Displacement:	32.5 tons
Length (oa):	87ft
Beam (wl):	5ft 2in
Powerplant:	460ihp; one-shaft compound engine
Speed:	19kt
Armament:	2 × 14in Whitehead-type torpedoes (2 × 1) in Thornycroft-designed launching frames

With three times the displacement of the TB 51 class boats, it is easy to see why *Lightning* and her successors quickly were dubbed 'first-class torpedo boats' and the smaller ones given the generic type descriptor 'second-class torpedo boats'. Despite this distinction, *Lightning* proved to be only a fair sea boat and had limited range. The torpedo launching frames proved unable to operate with the ship making speed and were replaced within a few years by a single above-water bow-mounted torpedo tube with a pair of reloads on trolleys carried amidships. Despite the limitations evident in testing *Lightning*, nineteen similar boats from Thornycroft, Yarrow, and other makers were ordered, starting in 1878.

There would be a steady increase in size, speed and offensive armament of torpedo boats, all made possible by more than two decades of equally steady improvement in engine and weapons technology, none of which could have been confidently foreseen by even the most far-sighted of smart people in the mid-1870s; it was at that time far from obvious that torpedo boats of either class would provide a seagoing torpedo carrier capable of accompanying a fleet or attacking enemy targets any appreciable distance from a safe harbour. Nor was it obvious what a ship that could accomplish these tasks would look like. Therefore multiple candidate designs were considered and a surprising number were actually built by the Royal Navy.

Not long after agreeing to buy *Lightning* from Thornycroft, the Admiralty ordered a totally new type of ship designed to attack ships at anchor with torpedoes. Intended as a much larger and faster follow-on to *Vesuvius*, HMS *Polyphemus* was designed to survive the run-in to a defended harbour before she brought her considerable offensive power to bear. Besides a fixed bow torpedo tube, she was equipped with four fixed broadside torpedo tubes, each with multiple reloads, the first broadside tubes designed into a British warship. Additionally, she had a spur-ram bow to allow her to ram enemy targets after harbour defences had been breached and torpedoes fired away. The original plans drawn up by the Royal Navy's Director of Naval Construction (DNC) Nathaniel Barnaby in 1874 called for a hull that was a prolate spheroid – shaped like an American football – with

The Thornycroft-built HMTB *2* was completed in 1878. As completed, she was fitted with HMTB *1 Lightning*'s final armament set-up, which comprised a single, trainable torpedo tube at her bow, with reloads on trolleys on both sides abaft the funnel. HMTB *2* and her eleven close sisters differed visually from *Lightning* in lacking a covered cabin aft. (NHHC)

only the most minimal upperworks, which were designed to detach and serve as life rafts in need.[21] The parts of the main hull that projected above the waterline were to be protected by 2⅜in of iron armour over in shell plating. By the end of the next year, the requirements had changed. The design now was largely in the hands of Barnaby's assistant (and future DNC) Philip Watts, and had morphed into a much larger vessel with a more prominent ram bow and a permanent, but still extremely low, above-water hull structure. As such, she was laid down at Chatham Dockyard in September 1878. Progress on her construction was slow; she was not completed until September 1882.

HMS *Polyphemus*: 1882[22]

Displacement:	2640 tons
Length (pp):	240ft
Beam (wl):	40ft
Draft (max):	20ft 6in
Powerplant:	7000ihp; two-shaft compound engines
Speed:	18kt (forced draft)
Protection:	3in steel deck
Armament:	5 × torpedo tubes (5 × 1); 18 × 14in Whitehead-type torpedoes; 12 × 1in Nordenfelt automatic rifles (6 × 2)

The Nordenfelt guns, six pairs of heavy machine guns, were mounted as a defence against torpedo boats, which, it was presumed, would be encountered on the run-in towards the target. Each dual mount was fitted in a cylindrical gunhouse topped by a domed cover.

By the time *Polyphemus* entered service, the technical issues involved in mounting a torpedo attack on anchored ships had evolved. Aware that torpedo boat attacks of the type Makarov had mounted at Bat'umi were a real possibility at any time, harbour defences became an urgent priority. While emplacing guns to protect a harbour entrance was time-consuming and costly, other harbour protection measures were far simpler and less expensive to implement; these included rigging booms across harbour entrances from which torpedo nets could be hung, laying minefields across all but specified channels to the entrance (and regularly changing those channels to prevent them being detected) and stationing local defensive torpedo craft, such as steam launches or second-class torpedo boats, in position to sortie against attacking forces. In 1885 the Royal Navy conducted an exercise at Berehaven, on the south coast of Ireland, intended to replicate an attack on the Russian main naval base at Kronstadt. War with Russia again threatened and the Royal Navy worried about its ability to strike at the Russians' Baltic Fleet should it retreat into its well-defended island base near St Petersburg.

The 1885 Berehaven Exercise was held in an inlet off the main harbour in which was constructed a set of harbour defences intended to replicate those at Kronstadt.[23] Two layers of floating booms – an inner set of light wooden spars and an outer line of heavier booms held in place by a five-inch steel hawser – were held five yards apart by light spars, with rope nets, designed to entangle the propellers of any ship that managed to breach or override the outer boom line, hung from the inner booms. The rest of the defences were all notional, existing only in the imagination of the judges and participants. There was an elaborately designed imaginary minefield, three-layers deep, outside the booms. Farthest out this had four rows of small 'observation mines', meant to signal the presence of intruding enemy ships; lining the shore would have been several rows of shallow electrically-detonated mines; and closest to the booms were more rows of shallow contact mines meant to stop small boats. All this was backed up by imaginary field guns and machine guns emplaced ashore, and by a pretend squadron of torpedo boats acting in a defensive role. Only the booms and nets were real.

Designated a 'torpedo ram', HMS *Polyphemus*, seen here as completed in 1882, was an innovative attempt to solve the problem of how to attack a well-defended harbour with torpedoes. To achieve this mission, DNC Nathaniel Barnaby designed a small – 2640-ton – and fast – 18kt maximum speed – low freeboard protected cruiser whose curved armoured deck extended well out beyond the sides of her hull, giving her something of the appearance of a primitive submarine. Her main armament was five submerged torpedo tubes, not visible in this image, and six double 1in Nordenfelt automatic guns, each in an armoured cylinder with a domed cover. (NHHC)

Technology was evolving so rapidly that *Polyphemus* was obsolescent long before she was old. Her one brief moment of glory came in 1885, when she was the centrepiece of the Berehaven Exercise, seen here, at which she charged a set of booms and nets constructed to resemble those at the Russian base at Kronstadt. She crashed through and over the booms and nets with no evident damage, but the exercise was conducted without any defending guns or mines, which likely would have caused serious damage. (NHHC)

Two different types of attack on these defences were attempted in order to evaluate their effectiveness. On the first night, 29 June, a swarm of torpedo boats and steam launches approached the booms with the aim of planting charges to blow them up. The 'judges' of the exercise ruled that the minefields and the shore-based defensive fire would have been sufficient to deter the planting of charges against the booms by the steam launches, and that the mines and defending torpedo boats would have kept the larger torpedo boats far enough away. The next day, *Polyphemus* had her turn. The rules were changed to presume that a channel had been swept clear of mines to allow the big ship a straight run-in on the booms. Starting from 2nm away, *Polyphemus* built up to 17kt, her best practical speed in open water, successfully dodged attacks by six torpedo boats, which launched a total of ten torpedoes at her, all of which missed, after which she continued on at full speed on to and through the booms, which she penetrated with no noticeable reduction in speed or damage to the ship.

This exercise was well-publicised and the image of *Polyphemus* breaking the booms widely distributed, but the lessons to be drawn from the Berehaven Exercise were clear to the Admiralty and quite different from the story told to the public. The Admiralty understood there was little future in attacking well-defended enemy harbours and no more ships of the torpedo ram type were ordered for the Royal Navy.[24] *Polyphemus*, despite her many good qualities, simply had too little freeboard to be a practical seagoing asset, and adding more to her freeboard to improve seaworthiness would only degrade those other qualities. The solution to the 'problem' of torpedo delivery would have to be found elsewhere. Even though first-class torpedo boats continued to grow in size and capability – by 1884 reaching a displacement of over sixty tons, a length of 113ft, and the ability to steam at a top speed of twenty knots – they were still too small and short ranging to act as a fleet element. Even before the Berehaven Exercise, the navies were exploring three other possible next steps.

It was the Germans who took one of those steps, a step they had in fact started taking before *Polyphemus* was laid down. This was a small cruiser with considerably strengthened torpedo armament. The logic of this design concept was that, even if torpedo boats were not yet capable of accompanying a fleet as it sortied across long stretches of open water, they nevertheless were likely to be encountered when that fleet approached an enemy shore, so a relatively fast and well-armed

Considered by many to be the first 'torpedo cruisers', the avisos of the Imperial German Navy's Blitz class, represented by SMS *Blitz* seen here as built in 1883, were only slightly slower and less well-armed than *Polyphemus* and significantly more seaworthy, offering a more useful naval vessel with little sacrifice in military qualities. (NHHC)

ship that could defend the fleet against enemy torpedo boats and also make torpedo attacks had considerable intuitive appeal.

A ship described as an 'aviso' less than half the size of *Polyphemus* had been ordered by the Germans from Thames Iron Works (Blackwall, London) in 1875. SMS *Zieten*, otherwise very much a traditional aviso – meaning a small cruising ship intended for scouting and message running – even down to her classic clipper-like lines, stood out because she was initially armed only with two submerged 15in torpedo tubes, one each bow and stern; she would be the last German warship ordered from a foreign yard. (Later, she had various gun armaments added.) It is not clear that *Zieten* was ever intended as anything other than an experimental vessel; she was employed primarily as a tender for small torpedo boats.

The importance of *Zieten* was not so much in what she was or did, but in the line of ships that followed in her design lineage and how other navies, particularly the French MN and the Royal Navy, reacted to those descendants. The immediate follow-ons to *Zieten* were a pair of larger avisos designed in 1879 and laid down in 1881 with more conventional cruiser characteristics, but with enough thought being given to torpedo fit as to cause at least some observers to consider them the first of a new ship type called 'torpedo cruisers'.[25]

SMS *Blitz*: 1883[26]

Displacement:	1463 tons
Length (wl):	228ft 2in
Beam (wl):	28ft 1in
Draft (max):	15ft 2in
Powerplant:	2800ihp (normal draft); two-shaft horizontal compound engines
Speed:	16kt (normal draft)
Armament:	1 × torpedo tube (bow); 3 × 35cm torpedoes; 1 × 12.5cm BLR; 4 × 8.7cm BLR (4 × 1)

The Royal Navy felt compelled to respond to reports of the construction of *Blitz* and her sister *Pfeil*. Additional impetus no doubt came from reports of the French beginning construction of a class of four ships of similar concept and characteristics in early 1883.[27] There were serious disagreements on the British side as to how to respond to these moves, although there was no question that a response was necessary. Nearing the end of his tenure as DNC, Barnaby proposed an improved *Polyphemus*, while Thornycroft put forward the idea of an enlarged, lightly-armoured torpedo boat.[28] The Admiralty decided that the British reply would be a much

more direct response, a pair of what were originally described as torpedo cruisers – reclassified as 3rd class cruisers soon after their completion – about half the size of *Polyphemus*, which were laid down in 1884 with the following characteristics:

HMS *Scout*: 1885[29]

Displacement:	1580 tons
Length (pp):	220ft
Beam (wl):	34ft 3in
Draft (max):	14ft 6in
Powerplant:	3200ihp (forced draft); two-shaft two-cylinder horizontal direct-acting compound engines; four boilers
Speed:	17kt (forced draft)
Protection:	in steel protective deck
Armament:	1⅜in × torpedo tube (bow); 4 × trainable torpedo cradles (2 each broadside); 4 × 5in BLR (4 × 1); 8 × Hotchkiss 3pdr/40 QF (8 × 1); 2 × Gardner .45cal MG (2 × 1)

The single torpedo tube in the two Scouts was above water at the bow; the four other torpedoes were carried in trainable cradles that fired through broadside ports fore and aft. The main-battery guns were on small main deck sponsons, the secondary battery fired through second deck embrasures. There was so much confidence that this design would succeed that six of a follow-on class of eight similar ships, the Archer class, had been laid down before *Scout* was completed in August 1885. This was unfortunate, as the ships proved to be disappointing in service, being poor sea boats, particularly the follow-on Archers, which carried heavier armament, and proved to be unstable and wet, and too slow to engage effectively their intended foe, enemy torpedo boats, which by this time were exceeding twenty-three knots top speed.

HMS *Fearless*, sister to HMS *Scout*, represented the Royal Navy's answer to the German Blitz class. She was very similar in size to *Blitz*, but much more heavily armed and proved to be a poor sea boat with questionable stability, and her even more heavily-armed successors suffered even more from this defect. (NHHC)

In 1883, at the same time that the British were planning the Scout class, and the French had started work on their own torpedo cruisers of the Condor class, the MN started construction on a class of eight much smaller ocean-going torpedo cruisers, which they called *aviso-torpilleurs*. In appearance, they were much more like enlarged torpedo boats than reduced cruisers, with minimal upperworks atop curved sides. It was mainly in size and armament that they differed from the torpedo boats of the day:

MN *Bombe*: 1887[30]

Displacement:	369 tons; later ships in class 430 tons
Length (pp):	194ft 3in
Beam (wl):	19ft 7in
Draft (max):	10ft 5in
Powerplant:	1800ihp; two-shaft vertical compound engine; four locomotive-type boilers
Speed:	19kt
Protection:	½in steel conning tower
Armament:	2 × fixed torpedo tubes (2 × 1); 4 × Hotchkiss 3pdr/40 QF (4 × 1); 3 × Hotchkiss 1pdr five-barrel revolver (3 × 1)

The eight ships of the Bombe class were, in many ways, ahead of their time; well-armed and, for ships of their size, well-protected (by the armoured conning tower and coal bunkers up the sides above a 'cellular' protective deck), but they were plagued by difficulties with their locomotive-type (firetube) boilers, that prevented many from being delivered until 1888–90, by which time their design innovations had been overtaken by other developments.

Nonetheless, from across the Channel, the construction of the Bombe class was the source of considerable concern. Despite still holding out a great deal of hope for *Scout* and her already-ordered follow-ons, the Royal Navy felt compelled, in 1885, to respond in kind to the perceived threat from the MN. Proposed by Barnaby not long after *Scout* had been laid down was a craft intended to be the smallest possible ship that could carry out the duties envisaged for *Scout*, namely providing defence of the fleet against torpedo boat attack and carrying out torpedo attacks on enemy ships at sea or in port. Derived from the Archers' design, it was variously described as a 'torpedo boat catcher' or 'torpedo gunboat', entering service with the Royal Navy under the latter type designator. The specifications of the first Royal Navy torpedo gunboat, laid down in November 1885, were as follows:

Heading off in a different direction from the British or Germans, the Marine Nationale responded to the 'problem' of torpedo attacks with what they called an *aviso-torpilleur* (dispatch-torpedo boat), which was essentially an enlarged torpedo boat rather than being a shrunken cruiser. The first of the type, MN *Bombe*, is seen here soon after completion in 1887, displaying the pronounced tumblehome characteristic of many French torpedo boats. (NHHC)

HMS *Rattlesnake*: 1887[31]

Displacement:	550 tons
Length (pp):	200ft
Beam (wl):	23ft
Draft (max):	10ft 4in
Powerplant:	2700ihp (forced draft); two-shaft triple-expansion engine; two locomotive-type boilers
Speed:	19kt (forced draft)
Protection:	¾in steel protective deck[32]
Armament:	2 × torpedo tubes (1 each bow & stern); 2 × trainable torpedo cradles (1 each broadside); 8 × 14in torpedoes; 1 × 5in BLR; 6 × Hotchkiss 3pdr/40 QF (6 × 1)

Rattlesnake was also greatly anticipated, and in the face of yet another Russian war scare – this one in 1885 – three near-sisters were ordered and laid down before she was even launched.[33] When she joined the fleet in May 1887, she

The British felt compelled to respond to the French innovations, coming up with their own torpedo gunboats, such as HMS *Grasshopper*, seen here soon after completion in 1888. As had been the case with the Scout class that came barely a year earlier, these ships were designed in haste and proved to be ill-conceived, being slow and possessing poor seakeeping qualities. (NHHC)

appeared at first to meet the high expectations that awaited her. Even before her commissioning, the following was reported to Parliament by the Admiralty on 17 March 1887 and recorded as 'DECISION OF THE BOARD UPON SEA-GOING TORPEDO-VESSELS AND SECOND CLASS TORPEDO-BOATS':

> The Board have carefully considered the types and dimensions of the torpedo-boats, which experience and experiment have shown to be best adapted for sea-service or harbour-defence. The practical tests during recent evolutions imposed upon first-class torpedo-boats, that is, boats intended for open-sea service, and

In an attempt to improve the seaworthiness of the successors to the Grasshoppers, the Royal Navy added a forecastle deck, such as seen here in HMS *Gleaner* in 1891. This did reduce their wetness forward to an extent, but added weight, making these boats almost one-third heavier than their predecessors, which resulted in no gain in speed despite a 30 per cent increase in power.

The mid 1880s was a time of tremendous innovation in design of torpedo craft, as evidenced by the Austrian SMS *Adler*, designed and built by Yarrow's, seen here before delivery in 1885. She had two torpedo tubes side-by-side at her bow and pedestal mounts, sided, between her conning tower and funnels, for 37mm QF guns, not yet fitted at the time of this photograph. (NHHC)

upwards of 100 feet in length, have shown that they could not, in actual warfare, stand the strain of the daily wear and tear to which they might be exposed. The discomfort and hardships which their crews have to endure in rough weather, the loss of speed to which, after steaming a certain number of hours at full pressure, they are subject, owing to the tendency of their necessarily-contracted boiler tubes to choke, disqualify boats of these dimensions from taking the prominent part in warfare in the open sea which their adherents claimed for them. The Board have therefore decided in future to build for seagoing purposes vessels of larger dimensions, capable of acting both on the

> offensive and defensive in torpedo warfare in the open sea. These vessels to be of the 'Rattlesnake' type, capable of maintaining for many hours the initial speed of the ordinary torpedo-boat, armed with quick-firing guns, and another gun of a larger calibre, and affording adequate accommodation and protection to the complement they carry for sea-going purposes.
>
> The Board have decided upon adopting a new type of second-class torpedo-boat, capable of being lifted on board ships of a certain displacement, and which would be carried by the ship for service in the place of the larger or first-class torpedo-boats which now are supposed to accompany a squadron.
>
> The 125-feet torpedo-boats now building or built have, in most instances, been allotted to the defence of the great military ports and coaling stations, where they will remain.[34]

There was a clear preference in the Admiralty and Parliament for the *Rattlesnake*-type ship; the unanswered question, in March 1887, when the above report was made, was whether *Rattlesnake* was in fact the type of ship she was expected to be. The disappointing answer, it was soon discovered, was that she was not, nor, in most cases, were her larger successors;

Another futuristic-looking torpedo boat design from this period was the French *torpilleur N°63* of the Normand 33 metre type, seen in 1884. This boat was from the first tranche, characterised by a straight, vertical bow, which were more successful than the second group that had a ram bow that threw up a large bow wave at speed. (NHHC)

three more classes of follow-on torpedo gunboats, a total of twenty-nine ships, were built to the end of 1895. The first ones were found to be poor sea boats, due to shallow draft, narrow hulls and the lack of forecastle in the earliest class; recommendations from the fleet included adding a forecastle deck with a turtleback to aid in handling rough weather.[35] The follow-on classes were larger, beamier and had forecastles and, in the case of the Halcyon class, the final class of torpedo gunboats, a poop deck as well, but, without being re-boilered, they rarely achieved the speed thought necessary for the tasks they were designed to accomplish. As the realisation sank in that *Rattlesnake* was not the ship that had been expected, more proposals were advanced that would address the known needs of the Royal Navy.

It should be noted, simply as an aside, that the emergence of the torpedo boat threat to the world's fleets drove one other significant technical innovation – the electrification of warships. By the mid-1870s it was obvious that torpedo craft operating at night presented a new and serious threat to warships, which needed a means to illuminate the waters around them in order to find the small and elusive enemy. (Starshells would not come into regular use until the First World War.) Flame-powered signalling lamps with concave mirrors to direct the light had been in use for centuries, but they could not produce enough light to illuminate targets at a useful distance. The principle of the electric arc lamp, fully capable of producing enough light to enable a useful naval searchlight, was well-known by 1870, lacking only a shipborne source of sufficient electricity to power it. The Royal Navy first began installing electricity generators in warships in the middle of that decade; the turret ship HMS *Inflexible*, launched in 1876, was completed with multiple direct-current generators for powering electric lights in the engineering spaces, exhaust fans, and searchlights.

The first American torpedo boat was built as a private venture by Herreshoff Manufacturing Co of Bristol, RI. Already well-reputed for building fast yachts and steam launches, Herreshoffs built a 94ft boat on speculation, completed in 1887. This boat was purchased by the US Navy and commissioned as USS *Stiletto* (WTB 1). Various armament combinations were tried out, including the bow torpedo tube, seen here launching a Howell torpedo in 1890, and the trainable deck-mounted torpedo tube visible further aft. (NHHC)

The first American purpose-built torpedo boat was ordered from Herreshoffs in 1886, USS *Cushing* (TB 1), although she did not enter service for four years. Like many of her contemporaries, she had her turtleback forecastle carried all the way aft in an attempt to protect the crew spaces from inundation at speed. (LoC)

These images show USS *Porter* (TB 6), taken probably in 1898 at Newport, RI. A quarter larger than *Cushing*, the Porters were designed with much more 'ship-like' hull shape, with a tall forecastle and higher vertical sides carried all the way aft, making them much drier in weather. (LoC)

Chapter 2

Torpedo boat destroyers to lead destroyers: 1884–1915

IN 1884 THE BUILDER J Samuel White (East Cowes), which had been building torpedo boats of both classes for the Royal Navy, began design work on a torpedo boat that was significantly larger than its contemporaries; this was done as a private venture, because the Admiralty had expressed little interest in the type. In their own descriptions of the boat, which they dubbed HMS *Swift* in their internal documents, she was designated variously as a 'torpedo boat catcher' or a 'torpedo boat destroyer'. In their several initial approaches to the Admiralty, they offered *Swift* in two different armament configurations, one as a dedicated 'torpedo boat destroyer' with primarily gun armament and minimal torpedo capability, and the other as a type of 'super torpedo boat' with a balanced gun and torpedo outfit equal to or better than that carried by any contemporary torpedo boat. Despite the lack of official support, construction of the putative HMS *Swift* began in 1884 with the following specifications:

There are some who state that MN *Bombe* (seen in the last chapter) should be considered the first prototype 'destroyer' despite never laying claim to the distinction. Others think it more appropriate that honour should go to 'HMS *Swift*', built as a private venture by J S White, which was dubbed a 'torpedo boat destroyer' in one of the possible configurations it was offered to the Royal Navy. In the end, the navy purchased the boat configured as a standard torpedo boat and named her HMTB *81*, seen here sometime after 1888. (NHHC)

HMS *Swift* (HMTB *81*): 1885[1]

Displacement:	137 tons
Length (pp):	150ft
Beam (wl):	17ft 6in
Draft (max):	9ft 6in
Powerplant:	1330ihp; one-shaft three-cylinder compound engine
Speed (trials):	23.75kt
Armament: (as completed)	1 × 14in fixed torpedo tube (bow); 2 × 18in trainable torpedo tubes (2 × 1, sided); 4 × Hotchkiss 3pdr/40 QF (4 × 1)

The same 1885 war scare that led to the premature decision to order the follow-ons to *Rattlesnake* caused the Admiralty to purchase *Swift* before she had run any of her builder's trials. At that point she was unarmed and she easily made her designed speed, which gave her a huge tactical advantage over the contemporary torpedo gunboats or torpedo cruisers, but it did not ease the problem the Royal Navy had in deciding exactly how to use this new boat or even what to call her. Many months passed before the decision was made to arm her as in the above specifications, with a balanced weapons fit, and to name her in the numbered torpedo boat sequence – HMTB *81*. Nonetheless, she stood out as being unique in the Royal Navy's fleet, far bigger than any other torpedo boat and

much faster than the similarly armed torpedo gunboats.

Her very uniqueness in some ways was an obstacle to any hope of immediately following up on her design. HMTB *81* had no ardent supporters in the fleet, as opposed to torpedo gunboats and the small cruisers that followed after the Archers. The British would continue to build large first-class torpedo boats, which gradually increased in size and capability, but none were directly inspired by the undisputed success of HMTB *81*, which served well with the Royal Navy through the First World War before being sold off in 1921.

Torpedo boat to torpedo boat destroyer

It was ultimately the danger posed to the fleet by the ever larger and more capable first-class torpedo boats that finally led to the inescapable conclusion that a ship was needed that could protect the fleet from the threat they represented. Advances in the design and development of torpedo boats included the introduction of a 'turtle back' forecastle, first used by Yarrow in 1885 in the Austrian torpedo boat SMS *Falke*, and then in the Royal Navy's HMTB *80* completed in 1887. The turtleback was supposed to improve a boat's entry into rough seas, keeping water away from the bridge and any forward-mounted armament, but in fact it caused a boat to dig her bow into waves and threw water back onto the bridge, exactly the opposite of the intended result. Nonetheless, this became a feature of British torpedo craft and those of many other nations for years to come.

The short road from *Swift* and *Falke* to the first true torpedo boat destroyer would have been relatively uneventful if the British had had their way. Advances in technology, both that of torpedoes, which were improving steadily in speed, range, and accuracy, and of ships' engines, which continued to gain in efficiency, providing more power and economy in units of any given size and weight, seemed to be addressing any issues facing their designers. They were content continuing to build improved classes of torpedo gunboats and first-class torpedo boats. Rather, they would have been content to do so had not the French upset those plans.

The strategic issues facing the MN at the end of the nineteenth century were in many ways similar to those of the Royal Navy, but the way the French and British viewed their places in the world and the roles their navies played in supporting those positions could hardly have been more different. To the British, an island nation, the navy represented the primary means of protecting national sovereignty against foreign enemies. The French, although the nation had extensive coastlines on an ocean, two seas, and a channel, always saw themselves first as a land power. Despite this, by the end of the nineteenth century the French had amassed a colonial empire second only to the British, stretching around the world. This left the MN with global responsibilities, but not necessarily the means to address them. Add to this a geopolitical situation that, on both sides of the English Channel, looked confusing.

For centuries, the British and French had been rivals, often at war with each other, because the British believed their safety would be best assured by preventing any European land power from gaining full ascendency over the continent and thus the freedom to turn against Great Britain without worrying about their backs. In practical terms, this often found the British lined up against the French up to the time of Napoleon. But, after Waterloo, and more importantly the Congress of Vienna, the power dynamic on the continent shifted in favour of Russia and Prussia, away from Austria and France, not necessarily to the benefit of the British. It even found the British and French allied against Russia in the Crimean War in the 1850s, but that did not result in any warmth or trust between the two nations facing each other across the Channel.

Despite having unrestricted access to the Atlantic and Mediterranean from excellent ports such as Brest, La Rochelle, and Toulon, and useful colonial harbours at Mers-el-Kébir, Casablanca, and Dakar, the French were in an economically and politically weakened state in the second half of the nineteenth century, and could not afford to build a fleet capable of challenging the Royal Navy's command of the seas and effective control of the waters around France – at least in conventional terms. But the French were ever open to new and unconventional ideas and responded to their reduced circumstances after the defeat of Napoleon with a military philosophy – called the Jeune École (Young School) – that advocated the use of small, expendable vessels mounting new weapons to counter the Royal Navy's preponderance of heavy warships. One of the new weapons the Jeune École adopted with the greatest enthusiasm was the automobile torpedo, and the MN pursued the development of torpedo craft of all types, much to the continued annoyance of the Royal Navy across the Channel.

Thus it was that the MN placed an order on 9 August 1884 with Chantiers et Ateliers A Normand at Le Havre for three torpedo boats of a size greater than any similar boat except *Swift* (HMTB *81*), the main difference being, from the Admiralty's point of view, that they were intended to accompany French naval forces into open waters against enemy forces, including, potentially, even the Royal Navy. The specifications of the nine ships of the class were as follows:

MN *Balny*: 1886[2]

Displacement (normal):	58 tons[3]
Length (wl):	133ft 10in
Beam (wl):	10ft 11in
Draft (max):	3ft 9in
Powerplant:	580ihp; one-shaft compound engine; one locomotive-type boiler
Speed (actual):	19kt
Armament (as completed):	2 × 14in fixed torpedo tubes (bow); 2 × Hotchkiss 1pdr five-barrel revolver (2 × 1)

The first of the most direct antecedents of the destroyer were the large, ocean-going torpedo boats (*torpilleurs de haute mer*) developed by the Marine Nationale starting in 1885. The first of the type, MN *Balny*, seen here at Algiers (al-Jaza'ir), was too small to perform the tasks expected of her, but later classes were enlarged and proved more successful. (NHHC)

In service, the large torpedo boats of the Balny class proved disappointing, showing themselves to be too small to perform the duties for which they were intended, but that did not lessen their impact on the thinking of the Admiralty across the Channel, which was as impressed by their type designation – *torpilleurs de haute mer* (seagoing torpedo boats) – which fully expressed the threat of their intent, as by any actual capabilities.

Realising the potential of the type and understanding why the Balny class had failed to fulfil that promise, the MN quickly followed that class with the much larger Ouragan class, which had its own set of problems due to using an experimental quadruple-expansion engine. Frustrated by this failure, the French looked across the Channel and ordered their next *torpilleur de haute mer* from Thornycroft, hoping to take advantage of that firm's reputation for exploiting new technology. This ship had the following specifications:

MN *Coureur*: 1888[4]

Displacement (normal):	101 tons
Length (wl):	147ft 7in
Beam (wl):	14ft 3in
Draft (max):	4ft 5in
Powerplant:	1500ihp; two-shaft compound engines; two Thornycroft watertube boilers
Speed (actual):	23.5kt
Armament (as completed):	2 × 14in fixed torpedo tubes (bow); 2 × Hotchkiss 3pdr/40 QF (2 × 1)

Coureur was innovative in a number of ways. The most important innovation from the French point of view was the Thornycroft watertube boilers, which offered considerable power increase at a saving in weight and fuel efficiency. She was of light all-steel construction, which proved to be *too* light, causing trouble from the date of her delivery in 1889. The MN built two near-sisters at Forges et Chantiers de la Méditerranée at Graville-Le Havre; these were more stoutly built and incorporated triple-expansion engines, which added sixteen tons to their displacement and, despite increasing power by 250ihp, which netted the same speed, this led to poorer handling in rough water.

Despite any problems the French may have had deploying these three ships, and a follow-on set of three even faster boats authorised in April 1891– one of which, MN *Chevalier*, was briefly the fastest boat in the world, reaching 27.6 knots on trials – it was these boats which rivetted the attention of the incoming Third Sea Lord, RAdm John A 'Jackie' Fisher, in February 1892, whose job it would be to direct ship procurement for the Royal Navy. Impatient and impulsive as always, Fisher reacted quickly and decisively to the threat posed by the big, fast torpedo boats being built across the Channel. This, at least, is the popular story regarding the origin of the torpedo boat destroyer as a type.[5] What is clear is that the Royal Navy's 1892–3 Estimates included money for five (or six) significantly larger torpedo boats with two different possible configurations of intended armament (depending on whether they were to be fitted out as 'super torpedo boats' or 'torpedo boat catchers'), to be carried at a sustained speed of twenty-six knots, two to three knots faster than the first-class torpedo boats then being ordered.[6] To achieve the intended combat capability, speed, and endurance, it was accepted that these boats would have to be approximately twice the displacement of contemporary torpedo boats and would carry larger guns (one 12pdr and multiple 6pdrs as opposed to multiple 3pdrs).

The decision to build large torpedo boats to combat smaller torpedo craft was not without controversy. As seen above, only five years earlier the Royal Navy had been promoting torpedo gunboats of the *Rattlesnake*-type as the

Another interesting, but unsuccessful, experiment by the Marine Nationale was the ordering of a *torpilleur de haute mer* from a British yard in order to test Thornycroft's watertube boiler, which promised more power per unit weight. The result was MN *Coureur*, seen here, probably at Toulon, in 1892. She was fast, but was very lightly constructed, as was the case with all British torpedo boats at this time, making her unsuited for work in the open ocean. (NHHC)

best defence for the fleet, relegating what were then large (125ft) torpedo boats to harbour defence. There were those in Parliament who well remembered the Admiralty's enthusiasm for these torpedo gunboats and their subsequent disenchantment, and were sceptical this time around. During debate on the Naval Estimates on 14 March 1892 Capt G E Price, representing Devonport, raised the following points – the 'noble Lord' to whom he refers was First Lord of the Admiralty George Hamilton, equivalent to the American Secretary of the Navy, the political head of the Royal Navy:

> As regards torpedo vessels, a good deal has been said about them in the Press and elsewhere. It is a well-known thing that France and many other countries in Europe have far more torpedo vessels than this country. But the noble Lord says the object for which they require torpedo boats is not the same as that for which we require them. I do not follow that argument; but, at all events, he goes on and says we intend to beat these torpedo-boats by vessels of a larger type – what are called torpedo-boat catchers, a well-known Admiralty name for them, which, I think, is a misnomer. But these torpedo vessels, I think, have gone to an extreme in their size. What is the object of building these vessels so large? . . . Then about their armament. I think it is a mistake to have these comparatively large, quick-firing guns. We must sacrifice two or three things. We must sacrifice a large draught of water for one thing, and the large number of small guns which we might have in their place; and consequently we shall be spending a great deal more money. For the money which is to be paid for the five ships that you are going to lay down you might have seven or eight smaller vessels, which would do the work equally well.[7]

There is no record of the First Lord responding to Capt Price's questions in any manner and the Estimate was approved with the monies for Fisher's destroyers intact.

As would be the wont of many other Royal Navy Controllers, Fisher was willing to leave most design details to the yards doing the building, so as to allow them the opportunity to demonstrate their best thinking, particularly regarding engineering and hydrodynamics. In this case orders along with general specifications went out immediately for two ships each to the experienced Yarrow and Thornycroft yards and a similar order was given somewhat later to the Laird Brothers yard at Birkenhead (near Liverpool) which had far less experience building small craft for the Royal Navy. The first four ships were all laid down in July 1892; the two from Lairds a year later. Yarrow rushed their pair to completion, turning over HMS *Havock* in January 1894 and *Hornet* in July; the former had locomotive boilers while the latter was completed with more efficient watertube boilers. Thornycroft took significantly more time to compete their pair; HMS *Daring* commissioning in February 1895 and *Decoy* four months later, but they were faster and were generally considered superior to the Yarrow-built ships. *Daring*'s specifications as built are as follows:

HMS *Daring*: 1895[8]

Displacement:	265 tons
Length (wl):	185ft
Beam (wl):	19ft
Draft (max):	7ft
Powerplant:	4644ihp (trials); two-shaft four-cylinder compound engines; three Thornycroft watertube boilers
Speed:	29.25kt (trials); 27.5kt (actual)
Armament (as completed):	1 × 18in fixed torpedo tube (bow); 2 × 18in trainable torpedo tubes (1 × 2; centreline aft; opposite directions); 1 × 12pdr/40 QF; 3 × Hotchkiss 6pdr/40 QF (3 × 1)

The trials speed of 29.25 knots given above for *Daring* must be taken in context, but regardless, it made her, at least briefly, the new fastest ship in the world, surpassing the mark set by MN *Chevalier* four years earlier. The reader must bear in mind that *Daring*'s record speed, as had also been the case with *Chevalier*'s, was achieved under the most favourable possible conditions, being run at light displacement and in shallow and calm waters, which all tend artificially to enhance a ship's performance. (The effect of shallow water on a ship at various speeds is complex, but, as a general rule, '[a]s the speed increases, the excess resistance increases up to a maximum at a certain critical velocity, and then diminishes. With still further increase of speed, the resistance in shallow water ultimately becomes, and remains, less than that in deep water at the same speed.'[9]) The best actual sustained high speed *Daring* could maintain was a still quite impressive 27.5 knots.

Why, the astute reader might ask, go through the highly-organised theatre of running trials under such conditions in order to record artificially high top speeds that did not represent actual military capabilities? It turns out there were many reasons for doing so. The Royal Navy, like many others, looked to private yards to advance the state of the art in many naval technologies, particularly when it came to steam engineering, and often offered monetary incentives for them to produce ships that exceeded required speed or endurance; they accepted the concomitant risk that yards would be tempted to build ships of dangerously light construction, assuming that the builders' sense of enlightened self-interest and careful testing would prevent egregious excess.

There was also an inescapable element of national pride involved. There was considerable prestige associated with being known to have built the fastest warship in the world. This competition was never 'official', there being no record keeping authority to verify reports, but that did not in any way diminish the seriousness with which nations took these claims. (The British, at this time, were concerned about the possibility that the French, in particular, might be building more or faster torpedo boats.[10]) In much the same way, nations competed for the 'unofficial' claim of having the fastest trans-Atlantic steamer, the so-called 'Blue Riband'.

The greatest differentiation within this initial order of destroyers was in their engineering setups. This was exactly the kind of technological experimentation the Admiralty hoped to foster by pushing detailed design responsibility out to the individual yards. As noted above, *Havock*, the first completed, had locomotive-type firetube boilers, which was a 'mature' technology with known limitations; *Hornet*, the second in the water, used the newer, watertube boilers which generated more power per unit and made more efficient use

Briefly heralded as the fastest warship in the world, MN *Chevalier*, a *torpilleur de haute mer* built by Normand, is seen making a reported speed of 27.22kt on trials on 13 October 1893. Her rated service speed of 24.5kt still qualified her as one of the fastest ships in any navy. (NHHC)

Thornycroft was one of the two other yards contracted to build the Royal Navy's first torpedo boat destroyers. Because their two ships used an experimental four-cylinder triple-expansion engine along with watertube boilers, they took longer to build than Yarrow's, but proved to be better built and faster. One of those first two from Thornycroft, HMS *Decoy*, is seen here in 1895, soon after completion. (NHHC)

of fuel, but the type had a long list of teething problems, mainly related to the joins of the watertubes to the boiler tanks, which had a tendency to leak if not perfectly soldered. (Other problems included corrosion and deposit build-up inside the watertubes, which had to be regularly cleaned and pressure-tested to assure efficiency.) The two Thornycroft ships were completed with that yard's patented watertube boilers and Thornycroft's unique four-cylinder triple-expansion engines, an engineering plant that was heavier and larger than that provided by the two other yards, but also provided more power and a higher sustained speed. Lairds was no doubt chosen as the third yard in this initial set, despite their relative lack of experience building smaller ships, because they had an established relationship with the French Normand yard, which allowed them to offer licence-built Normand boilers in their ships, those being the watertube boilers with the best reputation for reliability.

At the time of their ordering, the Admiralty was undecided between the two armament options available for these ships. One was the 'super torpedo boat' option described in the above specifications, with three torpedo tubes and the mix of one 12pdr and three 6pdr guns. Soon after entering service, most of these ships had their after twin torpedo tubes replaced by another pair of 6pdr guns, one on each side aft, confirming their role as destroyers, rather than just oversized torpedo boats. As will be seen below, in not much time the need to make this choice would become unnecessary, as it became possible to make the successors of these initial destroyers both bigger and faster, adequately armed with both guns and torpedoes.

Along the way, this process irrevocably differentiated this new torpedo boat destroyer from the first-class torpedo boat it was intended to destroy. If this new type of torpedo-carrying boat could both deliver torpedo attacks and protect the fleet from an enemy's torpedo boats, and do so while accompanying the fleet over long distances, then there was little need for the continued production of large torpedo boats useful only in coastal waters. The first-class torpedo boat soon disappeared from the building plans of the world's navies; a need could still be found for smaller, very fast torpedo vessels, smaller even than second-class torpedo boats, such as the coastal motor boats which were developed during the First World War.[11]

More destroyers, bigger and faster

Having pushed the state of the art in torpedo boat development with the first six torpedo boat destroyers ordered in 1892–3, Great Britain could not rest content with this accomplishment; it was necessary to follow these first destroyers with substantial numbers of bigger and faster follow-ons, because it was correctly seen as inevitable that other nations, putative friend or potential foe, would rapidly follow the

The Royal Navy followed the six initial '26-knotters' with six '27-knotters' ordered in 1893–94 from Thornycroft and Yarrow. HMS *Hasty*, seen here in 1896, was a Yarrow-built ship; lightly built and fast, the Yarrow-built ships did not stand up as well to hard use as those from other builders. Note the lack of a bow torpedo tube and the presence of the 12pdr gun atop her conning tower forward, an indication of the gradual emergence of guns rather than torpedoes as the primary armament of torpedo boat destroyers. The two closely spaced funnels indicates that the boilers were aligned in opposite directions, with their 'working faces' widely separated. (NHHC)

Royal Navy's lead. It was British policy, officially codified in 1889, that the Royal Navy would maintain a level of strength equal to at least that of her two closest rivals.[12]

Fisher did not wait for the first six destroyers to be launched, much less tested, to place more orders. He asked for more speed and standardised the armament with a full suite of guns and torpedoes. The first six of this group, ordered from the experienced firms Yarrow and Thornycroft, would have three torpedo tubes and four guns – one 12pdr and three 6pdrs, like the first destroyers – but feedback from the commanders of these first ships was negative on the effects of the bow torpedo tube, which threw up a heavy spray at all but the slowest speed, so that all the remaining ships in this group were completed without a bow torpedo tube. Later, most would be fitted with two additional single 6pdr mounts.

Beyond the initial six ordered in this group, collectively known as the '27-knotters' because the Admiralty specifications called for one knot of greater speed over the first batch of destroyers, thirty ships were ordered from twelve yards in 1893–4. Besides Lairds (later Cammell Laird), these included a number of yards that would become regular builders of Royal Navy destroyers in the twentieth century, such as Armstrong (later Vickers-Armstrong), Hawthorn Leslie, Fairfield, Palmers, and White. All were delivered with similar specifications; those of the three Palmers-built ships are given as an example:

HMS *Porcupine*: 1896[13]

Displacement:	275 tons
Length (pp):	200ft
Beam (wl):	19ft 9in
Draft (max):	8ft
Powerplant:	3900ihp; two-shaft triple expansion engines; three watertube boilers
Speed:	27.84kt (trials)
Armament (as completed):	2 × 18in trainable torpedo tubes (2 × 1); 1 × 12pdr/40 QF; 3 × Hotchkiss 6pdr/40 QF (3 × 1)

All the thirty-six ships in this group entered service between 1895 and 1901 and seem to have been well liked in the fleet. As before, Yarrow's ships were found to be fast, but very lightly built; the ships from Thornycroft, Palmers, and Hawthorn Leslie were considered the most robust, those from Palmers being particularly noteworthy as fine sea boats. In order to aid the less-experienced shipbuilders, the Admiralty shared Yarrow's and Thornycroft's design drawings among all the participating yards. Yarrow protested vehemently, stating that this was a leak of proprietary information. So loud (and public)

As more torpedo boat destroyers were built for the Royal Navy, and as more of the 'follow up order 27-knotters' came into service, such as the Fairfield-built HMS *Handy* seen here, the destroyermen began to influence the layout of the ships, particularly when it came to command spaces. The thinly-protected conning tower was never a popular site for the captain and the other officers from which to conn the ship in most situations, being cramped and having poor visibility; gradually the command functions migrated to the gun platform above the conning tower, and minimal amenities were provided there including adding canvas screening around the railing to provide some wind and weather protection and setting up a somewhat protected chart table for the navigator, seen projecting forward from the railing. It was assumed that the officers would retreat to the conning tower when shooting started. (NHHC)

were Yarrow's complaints that the Royal Navy refused to offer the firm any more shipbuilding contracts until 1901.

The Royal Navy continued to order destroyers at a rapid pace, prodded to a great extent by the reaction of other nations to their massive programme of destroyer construction. By the time the 1897–8 programme was ordered, no fewer than ninety-two destroyers had been authorised for the Royal Navy. The last fifty, starting with those funded by the 1894–5 Estimates, were known as the '30-knotters', because the Admiralty issued requirements to the participating yards calling for a trials speed three knots higher than the preceding classes. Beyond the speed requirement, the Admiralty set the ships' complements at sixty-three officers and men and the armament as two trainable torpedo tubes and six guns.[14] The destroyers built at the Palmers (Jarrow, Tyneside) yard were considered the best sea boats of the generally lightly built 30-knotters. The six Palmers-built ships of the 1895–6 tranche had the following specifications:

HMS *Flying Fish*: 1898[15]

Displacement (light):	390 tons[16]
Length (pp):	215ft
Beam (wl):	20ft 9in
Draft (max):	9ft 9in
Powerplant:	6200ihp; two-shaft triple-expansion engines; four watertube boilers
Speed:	30.371kt (trials)
Armament (as completed):	2 × 18in trainable torpedo tubes (2 × 1; centreline); 1 × 12pdr/40 QF; 5 × Hotchkiss 6pdr/40 QF (5 × 1)

Of note is the increase in displacement in these ships over the preceding classes, although dimensions increased very little. This was because, in order to reach the thirty-knot contract speed, these ships needed half as much power again as the 27-knotters, while dimensions increased by only approximately 10 percent. None of the 30-knotters ever reached their contract speed in service, rarely achieving twenty-five knots with the fleet. Many of them were criticised for their light construction and the cramped arrangement of their stokeholds. Nonetheless, almost all of the 30-knotters served through the First World War, generally as coastal defence units.

Three out-of-sequence 'Specials' were ordered by the Admiralty as part of the 1896–7 Estimates. Believing that even more speed could be wrung out of the basic design of the Royal Navy destroyer, three yards – Thornycroft, Lairds and Thomson (Clydebank) – were given contracts each to produce a single ship, otherwise similar in characteristics to the 30-knotters, capable of a trials speed of thirty-three knots. The immediate impetus for this programme was the news, received in October 1895, of the trials in France of the *torpilleur de haute mer* MN *Forban*, reported to have reached 31.029 knots on trials in full load condition.[17] There is no question that this news, particularly since these trials had been run in far less artificial conditions than normally used by the Royal Navy, caused considerable alarm in Whitehall, despite the fact that *Forban* was only about half the size of the contemporary 27-knotters. As noted above, national pride, aligned with prudent

The '30-knotters' from Palmer's, such as HMS *Flying Fish* seen here, had the best reputation for reliability among the fifty ships to this general design ordered between 1894 and 1899. Note the generally enlarged gun platform over the conning tower, now with noticeable wing platforms on either side to accommodate more easily the officers as well as the gun crew. (NHHC)

concern about a potentially superior enemy weapon across the English Channel, prompted Admiralty action.

In this case, it seems that the limits of reciprocating engine technology had been reached. None of the three ships, despite pushing the power produced to 9250ihp in the case of Laird's HMS *Express*, were able to reach the speed target despite repeated attempts.[18] However, technology had an answer; in 1897, Charles Parsons demonstrated his steam-turbine-powered launch at the Spithead Navy Review, showing the world the potential of this new engine type. To its credit, the Admiralty reacted quickly, ordering a turbine-powered destroyer from Parsons Marine Steam Turbine Co (Wallsend, Tyneside) on 12 January 1898.[19] Because Parsons was an engine maker, not a shipyard, the ship's hull was subcontracted out to Hawthorn Leslie. The specifications of this ship are as follows:

HMS *Viper*: 1900[20]

Displacement (light):	344 tons
Length (pp):	210ft 3.5in
Beam (wl):	21ft 0.25in
Powerplant:	12,500shp; four-shaft Parsons turbines; four watertube boilers
Speed:	36.838kt (trials)
Armament (as completed):	2 × 18in trainable torpedo tubes (2 × 1; centreline); 1 × 12pdr/40 QF; 5 × Hotchkiss 6pdr/40 QF (5 × 1)

The most notable aspect of these specifications, when compared to those of *Flying Fish*, is the great increase in power and speed. This was purely a function of the adoption of turbines – high-pressure and astern turbines on the inner shafts, low-pressure on the outer – which gave two immediate important benefits: greatly enhanced reliability at high speed – the longer reciprocating engines ran at high-speed, the greater the risk of mechanical failure – and much reduced vibration throughout the ship, particularly astern. However, trials very quickly demonstrated the great downside of direct-drive turbine engines; they operated efficiently only at high speed.

Inevitably, the limits of reciprocating engines were reached with the '30-knotters' – the Admiralty ordered three '33-knotters' but none made their designed speed – which led two builders, the engine-maker Charles Parsons (in conjunction with the Hawthorn Leslie yard) and the shipyard Armstrong-Whitworth, to build turbine-powered destroyers on speculation. Both ships were purchased by the Royal Navy before completion. This undated image shows the Parsons-Armstrong-Whitworth candidate, HMS *Viper*. Nothing in her external appearance indicated her novel powerplant.

Early attempts to run turbine-powered ships at slower speeds demonstrated that, compared to reciprocating engines, turbines lost efficiency in direct proportion to any decrease from high speed. There were technical solutions to this problem, but they were, at this point in time, costly, heavy, and wasteful of space, all of which made them impractical for use in a small ship like a destroyer.[21] The optimal solution, gearing much like an automobile's transmission that allows the engine to run at high speed, but turn a shaft at a much lower speed, was obvious, but the ability to produce the gears of the necessary precision and robustness was a decade in the future.

There was a great difference of opinion in the fleet over how successful this first turbine-powered destroyer had been compared to her more conventional contemporaries. Starting in 1902, and continuing for the next two years, the Admiralty ordered from their regular shipyards a more-or-less standard design of large destroyer, all named after rivers. Their design was influenced by a German design – SMS *S90* – a boat ordered from the Schichau yard at Elbing (now Elblag, Poland) in 1898. There was not much about *S90* or her eleven sisters to arouse any interest, except for one design feature. Because the Marine Amt was worried about the ability of their destroyers to operate in the notoriously rough waters of the North Sea, and having had experience with the turtleback forecastle of their earlier torpedo boats, which caused a constant cascade of water over the bridge in most weather conditions, they asked for a raised forecastle deck forward of the small bridge. While far from perfect, this arrangement proved to be a vast improvement over the standard British design, and became one of the design features required by the Admiralty in all River class destroyers.

The other big change in this class of destroyers was the abandonment of the artificial trials speed requirement. Instead, the Admiralty asked that these ships be more stoutly-constructed and be able to maintain a top speed of 25.5 knots in all but the worst weather conditions, which would allow them to maintain a speed six to seven knots faster than the eighteen-knot top speed of the battle fleet. That would be enough of a margin to permit them to manoeuvre around the fleet as needed.[22] This led to larger ships with lower official speeds, which caused some considerable criticism in the popular press, but resulted in ships that were quite well received in the fleet. The specifications of a typical Yarrow-built ship of the first group of the River class are as follows:

HMS *Usk*: 1904[23]

Displacement (light):	550 tons
Length (pp):	225ft
Beam (wl):	23ft 6in
Draft (max):	10ft
Powerplant:	7500ihp; two-shaft triple-expansion engines; four watertube boilers
Speed:	25.5kt
Armament (as completed):	2 × 18in trainable torpedo tubes (2 × 1; centreline); 1 × 12pdr/40 QF; 5 × Hotchkiss 6pdr/40 QF (5 × 1)

Hawthorn Leslie was given a contract to build two ships of this first group, one with reciprocating engines and one with Parsons turbines. The comparative trials simply bore out what was already known, that turbines, in their then-current state, were more reliable, especially at high speed, and caused much less vibration in the ship, but were far less economical to operate at cruising speed. During the Royal Navy's 1904 manoeuvres, the turbine-powered HMS *Eden* was tested in conditions approximating war service against her conventionally-powered sisters and found to be sufficiently superior in performance as to cause the Royal Navy's Committee on Designs to recommend the adoption of turbines for all future destroyers.[24] Adding another variable into the issue of how to power destroyers was the emerging question of the use of oil fuel in addition to or in the place of coal. Even with the greater endurance allowed by reciprocating engines, range remained an ongoing issue if destroyers were to accompany the fleet. With conventionally-powered destroyers using coal for fuel, the range calculations for the River class destroyers varied between 1600 and 1800nm. Oil fuel promised to deliver

Despite the success of the experimental turbine-powered *Viper* and *Cobra*, the Royal Navy concentrated on other features of destroyer design in an effort to match the seaworthiness exhibited by destroyers being built by potential enemies such as France and Germany. This led to the River class destroyers, such as HMS *Usk*, seen here, probably in 1904. These ships displaced more than 40 per cent more than the '30-knotters' without any increase in size or armament and actually had a lower top speed; what accounted for the greater mass was a long, high forecastle deck and a separate bridge built atop an enclosed pilothouse aft of the gun platform. The bridge was still open and still protected only by a canvas-covered railing, but now could be manned without interfering with the gun crew. The navigator's chart table, seen at the near bridge wing, is still open, but has a more elaborate hood offering some protection from the elements. The muzzle of the 6pdr gun sited in the sponson abreast her bridge can just be seen extending out from under its canvas cover, in the inset in the forecastle that allowed it to fire almost directly forward. (SLV – Green Collection)

approximately twenty-five percent more energy per unit weight, which would at least partially address the endurance issue of turbines; additionally, oil fuel was easier to store in small, awkward bunker spaces and required far fewer fire room personnel. The downsides to the adoption of oil fuel came down to two main issues. One was that coal bunkers provided some degree of protection against small calibre gunfire above the waterline in unarmoured ships like destroyers; also coal was compressible, which allowed it to absorb a good deal of explosive shock below the waterline, offering some protection against torpedoes or mines. Oil, being a liquid, was incompressible; any shock to the outer shell of an oil tank would simply be transmitted inboard. The other factor was cost and availability. The United Kingdom had plentiful, inexpensive, high quality coal; it had no oil to speak of. Any oil burned in Royal Navy ships would have to be purchased abroad and imported by sea, a potentially critical vulnerability in wartime. Nonetheless, at the same time that the Royal Navy was deciding to adopt turbines, the decision was made to adopt oil fuel for destroyers.[25]

The next great 'milestone' in destroyer development was indisputably due to 'Jackie' Fisher, who returned to Whitehall as First Sea Lord on 20 October 1904, having spent most of the preceding seven years away from the Admiralty. He arrived with a fully thought out programme of innovative ship designs that he wanted implemented immediately, designs that would revolutionise naval warfare, making virtually all existing warships obsolete. The core of this programme was the idea that the Royal Navy's battle fleet of the future would comprise all-big-gun battleships and armoured cruisers (these latter the future battlecruisers of the Invincible class) and large, fast destroyers capable of escorting the capital ships at high speed.[26] For the specifications of these new destroyers, Fisher had set a high bar. They were to maintain a top speed of thirty-three knots for eight hours, a cruising speed of at least sixteen knots

When 'Jackie' Fisher returned to the Admiralty from the Mediterranean in 1904, now as First Sea Lord, he brought with him a full set of ideas for new ships that would revolutionize the Royal Navy and force all other navies into a long stern chase to catch up. Besides the dreadnought battleship and battlecruiser, he proposed the construction of a class of destroyers that were bigger, faster, and better armed than any before. These would become the first Tribal class. This image shows HMS *Maori*, one of the final tranche of five ships ordered in the 1907–8 program. The torpedoes used by the Royal Navy could now reach 7000yds at moderate speed, so the guns carried by torpedo boat destroyers had to be able to disable torpedo boats at well beyond that range; hence the adoption of the 4in/40 BL Mark VIII. Note that the vestigial conning tower has now disappeared, replaced by a low, broad gun platform just ahead of the enlarged pilothouse with its open bridge above. The chart table on port side of the bridge is still open to the weather, but looks more substantial. (NHHC)

with a range of 1500nm, use only oil fuel, hold provisions for four days, and carry a weapons suite including two or more 12pdrs of a new, higher-velocity 18cwt model. Five bids were accepted for the first tranche in 1905–6, no two of which were alike in any particulars. Normal displacements ranged from a low of 850 tons to a high of 882 tons. A further two were ordered a year later, both almost 100 tons heavier because Fisher asked the two selected yards – Thornycroft and J S White – to include all the 'best features' of the first five ships. A year after that, a further five of this class were ordered; the ships of this class were called the 'Tribals', because they were all given names of indigenous warrior groups within the British Empire, or war-like bands from myth or history renowned for their bravery. This last group of five were bigger still; the specifications for HMS *Maori*, an average-sized example from this third group, were as follows:

HMS *Maori*: 1910[27]

Displacement (normal):	1026 tons
Length (pp):	279ft 11.75in
Beam (wl):	27ft 1in
Draft (max):	10ft
Powerplant:	15,500shp; three-shaft Parsons turbines; six Yarrow watertube boilers
Speed:	33kt
Armament: (as completed)	2 × 18in trainable torpedo tubes (2 × 1; centreline); 2 × 4in/40 BL Mark VIII (2 × 1)

Even a quick comparison of the specifications of *Maori* and *Usk* showed that to gain 7.5 knots in top speed required doubling the power produced by the ship's engineering plant, which in turn required very nearly doubling the ship's displacement and increasing its overall length by more than twenty percent. As with the second batch of Tribals, *Maori* mounted a pair of 4in/40 Mark VIIIs in place of the three 12pdrs carried by the first five. This increase in gun size in Royal Navy destroyers was not simply a case of 'Fisher being Fisher,' always looking to make ships bigger and stronger. Starting in 1904 and gaining momentum over the next several years, a revolution in torpedo design was occurring, replacing compressed air as the means of propulsion with 'heated air', meaning the exhaust generated by burning a fuel (generally kerosene at first, later methyl alcohol) in a combustion chamber. (The idea for 'heated air' had originated when it was noted that releasing the high-pressure air stored in the compressed air flasks caused icing in the torpedoes' rotors through a physical process known as adiabatic cooling. Engineers tried various methods to warm the released air, noting that the more the air was heated, the greater the energy that was delivered to the rotors.) Burning fuel in a combustion chamber led to the opposite, heat build-up which could cause its own set of problems. The British and others very quickly devised methods of spraying water into the air stream at the point of ignition, which provided the twin benefits of cooling the chamber and adding energetic steam to the exhaust, further increasing the power derived from the combustion of the air-fuel mixture. Where in 1904, 2000yds had been considered the maximum effective range of a torpedo, by 1908 the British were putting torpedoes with these 'wet-heater' motors into production – the 18in Mark VII soon followed by the 21in Mark I 'Short' – which could reach 7000yds or more at thirty knots. This greatly increased the range at which torpedo boats could launch an effective attack, and would need to be engaged to prevent such an attack, making the 12pdr look like too light a weapon to disrupt torpedo attacks.

At normal displacements ranging from 850 tons to 1090 tons, the Tribals were quite possibly the first 'super destroyers', intended to do everything a destroyer should and some of the tasks of a small cruiser, but that did not make them popular in the fleet. They were considered inadequate sea boats, despite most of them having high forecastle decks, and had extremely limited endurance, which left them unable to carry out the fleet escort task for which they were intended.[28] There is no question that their problems stemmed from attempting to force too much performance into too small a ship. With a larger hull, they would have been better sea boats and would have had room for more fuel stowage, but that would have added tonnage, potentially making them slower. This is the 'classic' zero-sum game of warship design. They fought well throughout the First World War, assigned to the Dover Patrol, where their short range and suboptimal seakeeping was less of a handicap.

Even before the first Tribal was ordered, Fisher was pressing for yet more, specifically, more speed and firepower. He asked the DNC – at the time this post was held by Philip Watts – to design a ship that would accompany the fleet of battlecruisers he envisaged as the core of the future Royal Navy.[29] During his time at Portsmouth, just prior to taking over as First Sea Lord, Fisher had put considerable work into the specifications of his proposed ships, working closely with the naval constructor W H Gard, which enabled him to present Watts a list of desiderata. For the large destroyer, these included a top speed of thirty-six knots, to be achieved by a powerplant generating 19,000shp, armament of two 4in guns and two torpedo tubes, all in a displacement of approximately 900 tons. Fisher, who was given to coming up with clever names for his ship projects, called this one HMS *Uncatchable*. Watts' initial reaction to this request was not recorded, but it is known that he informed Fisher that no such ship could be built on that displacement.

Undeterred, Fisher pressed forward with all his customary drive. He made known to Watts that all the particulars of this vessel were negotiable except the thirty-six knot top speed; he wanted this ship to have a similar margin of speed over his proposed fleet of twenty-five knot battlecruisers as the Rivers had over the old battle fleet.[30] By his calculations, that meant that this new ship needed an eleven-knot margin to manoeuvre safely in all conditions. Watts responded in January 1905 with a sketch design for a ship of 1140 tons and 19,000shp that could make a top speed of 33.5 knots. When Fisher insisted that he still required the missing 2.5 knots, a revised sketch was produced calling for 1350 tons and 29,000shp. The constant accretion of requirements only conspired to add to the ship's size and weight. The Third Sea Lord insisted that a ship this size needed to be better armed, so two more 4in guns were added to the design, as was a radio room, just before bids were requested.

Even more so than in the case of the Tribals, the bids varied widely in size and cost, because this was moving into unexplored territory. After much debate, the Admiralty ruled that five of the bids were technically compliant and that the one in the middle in terms of cost and size, the bid from Cammell Laird, had the greatest chance of success. This ship was laid down in December 1906, to be named HMS *Swift*, with the following specifications:

HMS *Swift*: 1910[31]

Displacement (normal):	2170 tons
Length (pp):	345ft 0.375in
Beam (wl):	34ft 2in
Draft (max):	10ft 6in
Powerplant:	30,000shp; four-shaft Parsons turbines; twelve Laird watertube boilers
Speed:	35kt
Armament: (as completed):	2 × 18in trainable torpedo tubes (2 × 1; centreline); 4 × 4in/40 BL Mark VIII (4 × 1)

Swift was delivered late, not starting trials until September 1909, and never made her contract speed of thirty-six knots; the Admiralty was inclined to be lenient with the yard because of the highly experimental nature of the ship. After many trials with different shapes and arrangements of propellers, the best speed she ever achieved was 35.037 knots on a light displacement.[32] *Swift* was well received in the fleet; she was found to be stable in rough seas, but she lacked any flare forward, which made her a wet ship, and her endurance was far below that which had been desired. The main complaint against her was that, in the end, she was only one ship, while, for the same investment of time and money, many more of a more reasonably sized ship could have been acquired.

At the same time as the Tribals and *Swift* were being built, Fisher had pushed through the large-scale production of a class of much smaller boats, averaging 230 tons, half the size of the Rivers, known as the Cricket class coastal destroyers. Turbine-powered and oil-fired, the Crickets were lightly built and handy and proved of great value in the coastal defence role but, like the Tribals, were ill-suited to accompany the fleet in open water.

From the beginning, Fisher's programme of destroyer production met opposition in Parliament. He had taken the posting of First Sea Lord on the promise of reducing naval expenditures; some in Parliament mistook this for taking a conservative approach to shipbuilding.[33] The shipbuilding programme he submitted for his first year at the head of the naval service raised more than a few eyebrows among the lawmakers, particularly at the rapid rate at which he proposed building his new, untried designs. One member, a Mr. Buchanan (Perthshire East), addressed the issue on 6 March 1905, noting that three new designs – the Tribals, Crickets and *Swift* – were to be funded at the same time, to the exclusion of ships of proven design:

> Another feature of the programme was that two new forms of destroyers were to be built – five ocean destroyers and twelve coastal destroyers, and also one destroyer of a new type. It was quite possible that here the Admiralty might make the same financial mistake as was made in the case of scouts. There were still eight scouts on the Estimates this year. From the accounts in the newspapers regarding the trials the scouts could not be regarded as an unmixed success. It was urged at the time the proposal to build scouts was brought forward that it would be unwise to order eight of them right away. Costly experiments should at first be made on a limited scale. It was possible that in regard to the new destroyers there had also been a want of thinking out the subject on the part of the Admiralty.[34]

Speaking for the Admiralty at this debate, the Financial Secretary to the Admiralty, a post held by a member of Parliament, in this instance by Ernest G Pretyman, member for Suffolk Woodbridge, did not respond to Buchanan's point in any manner, as the latter had rambled on to other issues before Pretyman could reply. As the Naval Estimate put forward by Fisher for 1905–6 reduced the prior year's request by £3.5M, there was little objection from the Commons about the fine points.

A number of members, who declared themselves still confused about the reasons for the change from the 30-knotters to the 'officially' slower Rivers, which had been approved only four years earlier, were now thrown into complete befuddlement by the apparent abandonment of the Rivers design for three all-new ones. A Mr Kearley from Devonport expressed this confusion explicitly on 13 March of that same year:

> As regarded torpedo boats, it was stated that the Board of Admiralty had decided to combine speed and sea-keeping power. Last session, he [Kearley] called attention to the danger of the policy the Admiralty were adopting in diminishing the speed of torpedo boats, whereas every other nation of naval standing was

After the frenzy of the Fisher era, the Royal Navy returned to something resembling a more normal pattern of destroyer development with the Beagle class ordered in the 1908–9 program. While bigger than most of the Tribals, the Beagles, such as HMS *Foxhound*, seen here soon after completion, probably in 1911, sacrificed six knots of nominal top speed for stouter construction and greater stability. They reverted to coal-burning boilers, in a concession to the conservative faction in the Admiralty, a change in no way discernible externally. (NHHC)

> increasing their torpedo boat speed. The Japanese boats which were built in England had a speed of thirty-one or thirty-two knots, whereas the speed of our boats was reduced from thirty knots to twenty-five.[35].. Last session the hon. Gentleman [Pretyman] denied that there was any comparison to be drawn between our torpedo boats and those of Japan; yet the Japanese boats travelled thousands of miles of ocean and arrived safely in the Far East, and were engaged in the early part of the war against the Russian Fleet with a considerable amount of success. Last session he suggested that it would be possible to have in torpedo boats sea-keeping power combined with considerable speed; and the hon. Gentleman replied that the most eminent experts declared that they could not design a boat of greater strength and greater speed.[36]

Pretyman replied that he only meant that such a ship could not be designed at the same cost as the River class ships.[37] When the House reconvened that evening to continue the debate, Pretyman addressed both the issue of the supposed inferiority of the 'slower' 25.5 knot Rivers compared to their 30-knotter predecessors, and the need for the new classes being requested in Fisher's proposed Estimates:

> His hon. and gallant friend raised the question as to the two classes of *destroyers*, one for ocean work and the other for work in narrow waters. It was not true to say that the new twenty-five-and-a-half 'knotters' had proved inferior to the old thirty 'knotters.' In anything like dirty weather the twenty-five-and-a-half 'knotters' had shown their superiority. As he had stated to the Committee, they had failed to produce a better design which would combine greater speed with greater stability at a limited cost. After all cost was a convenient standard and index. The evolution of these two classes was the result of the old battle of opposing factors – navigability and invisibility. It was not true to say that they had abandoned the policy of reducing the size of the destroyers; but they could not get in one class maximum speed and maximum invisibility both in narrow waters and in the open. Therefore, instead of trying to build one class of destroyer, which would

> attain a maximum speed in all waters, they had designed two classes, each of which advanced to some extent, separately, as regarded speed and invisibility, so as severally to meet requirements in narrow waters and in the open sea.[38]

Following the success of the Beagle class, the Royal Navy opted to return to oil fuel, which allowed the same qualities that made the Beagles so successful to be delivered in a ship that displaced almost 20 per cent less. The resulting Acorn class became the basis for the standard Admiralty design for the remaining years before the First World War. HMAS *Swan* (61), seen here soon after the war, was one of six ships of the follow-on Acheron class, nearly identical to the Acorns, built for the Royal Australian Navy. Of interest is the half-shield fitted to the forward 4in gun. This light shield, providing protection from weather and small-arms fire only, stopped at the waistline, leaving the gun crew's lower extremities curiously unprotected. (SLV – Green Collection)

Neither the Tribals nor Crickets had any direct successors. The former were too expensive and too short-legged to be considered seriously for production in large numbers and the Crickets was understood to be really just large torpedo boats and not true destroyers in any sense. (Although thirty-six Crickets were built, only the first twelve were given names; the remaining twenty-four were numbered as torpedo boats.) *Swift* was always considered experimental, and after the difficulties experienced with her introduction into service, there was no attempt to develop her design. Regardless, by 1907, the time when the Admiralty was ready to seek funding for the next destroyers to be ordered, the strategic picture had changed dramatically for Great Britain, compared to just three years earlier. Fisher had always been far more concerned about Germany as a potential enemy than France, a conviction that only increased over the years he was First Sea Lord.[39] The difference, in terms of naval requirements, between a potential war with France and one with Germany was dramatic. The need for coastal defence against torpedo craft was considered to be minor, compared to France, but the Germans were known to expect their torpedo craft to accompany their fleet into the North Sea, which required seakeeping ability in excess of that offered the Royal Navy by either the Tribals or the Crickets.[40]

Faced with this problem, the Admiralty turned, in 1908, to a design that had been shelved in 1904 when Fisher arrived at Whitehall. This was for an improved-River class destroyer that became known as the Beagle class of 1908. As the design was revisited after a lapse of four years, a number of dramatic changes were requested that made it difficult to trace many aspects of the Beagles' lineage back to the Rivers. For one thing, so great was the Admiralty's concern that these new destroyers be able to maintain their station with the fleet in typical North Sea weather conditions, that their displacement was increased by more than 400 tons over that of the Rivers, to a size not much smaller than the Tribals. This concern was largely due to unconfirmed – and incorrect – reports that the Germans were building a torpedo boat the size of a River class destroyer – *G137* – capable of maintaining thirty-three knots in open water.[41] This raised the alarming possibility, given the already well-reputed ability of German torpedo boats to manoeuvre in open water due to high freeboard forward and set-back bridges, that the British fleet might encounter torpedo craft in the North Sea both faster and more seaworthy than their own.

The increase in size was also, in part, due to the Beagle class reverting to coal fuel. This was due to a strong reaction, both in the fleet and in Parliament, against Fisher's move to oil fuel, when Great Britain had cheap and abundant coal, and when the supply of oil fuel was, at the same time, far more costly and less secure. (The first secure British access to oil supplies came in 1909 with the establishment of the Anglo-Persian Oil Company to exploit an oil discovery in the Abadan region at the head of the Persian Gulf, although it was only four years further on that the Royal Navy, under a new First Lord of the Admiralty, Winston Churchill, took a controlling interest in the company.) In 1908, the anti-oil factions in the Royal Navy and the government were still strong enough that the requirements for the next class of destroyers included coal fuel and a sustainable speed of twenty-seven knots. In order to achieve that speed with coal-fired boilers, which of necessity were larger and required more crewmen to service than oil-fired ones while delivering less steam, the number of boilers had to be increased to five. That, and the increased armament, now including, for the first time, 21in Mark I 'Short' wet-heater torpedoes, which together with their mounts were half again heavier than the torpedo batteries carried by the Rivers, accounted for the dramatic increase in size over the Rivers. The specifications for a typical Beagle class destroyer are as follows:

HMS *Beagle*: 1910[42]

Displacement (normal):	945 tons
Length (pp):	270ft
Beam (wl):	27ft
Draft (max):	8ft 6in
Powerplant:	14,300shp; three-shaft Parsons turbines; five Yarrow watertube boilers
Speed:	27kt
Armament : (as completed)	2 × 21in trainable torpedo tubes (2 × 1; centreline); 4 × Mark I 21in 'Short' torpedoes; 1 × 4in/40 BL Mk VIII; 3 × 12pdr/40 QF (3 × 1)

The Admiralty anticipated that the Beagles would be successful, despite their coal-fired boilers. So confident were they in their success that, well in advance of any of them even being launched, they returned to Parliament asking for twenty follow-ons. These were to differ from the Beagles in only two ways. For one thing, they would be oil fuelled, which would allow a reduction in crew from ninety-six to seventy-two and in boilers from five to four; the smaller, more powerful oil-fired boilers with their reduced manpower requirements allowed the ships to be reduced in length by thirty feet and almost 200 tons in displacement. This in turn allowed a somewhat increased armament to be carried at the same speed by a powerplant that produced less power. Indeed so certain was the Admiralty of this design that for the first time a standard design was produced by the navy and supplied to all participating yards, which were allowed to deviate only in the details of the powerplant, which only had to meet power, weight, and size requirements. This class, the Acorn class ordered as part of the 1909–10 programme, had the following specifications:

HMS *Acorn*: 1911[43]

Displacement (normal):	760 tons
Length (pp):	240ft
Beam (wl):	25ft 3in
Draft (max):	8ft 6in
Powerplant:	13,500shp; three-shaft Parsons turbines; four Yarrow watertube boilers
Speed:	27kt
Armament: (as completed)	2 × 21in trainable torpedo tubes (2 × 1; centreline); 2 × 4in/40 BL Mk VIII (2 × 1); 2 × 12pdr/40 QF (2 × 1)

The Beagles, and to an even greater extent the Acorns, lived up to their expectations, and became the standard Royal Navy destroyer design for the years up to and most of the way through the First World War. Succeeding classes increased in size and top speed gradually and the 12pdr guns were eventually dropped in favour of more 4in mounts and the adoption of smaller calibre anti-aircraft machine guns.

International competition, of a sort

This story has been told, so far, as if destroyer development was something that happened almost entirely in Great Britain. In part that is because the Royal Navy did most often lead that development, and when they were pushed by developments in other nations, those have been duly noted. Nevertheless, it is worthwhile to trace briefly the highlights of the development of destroyers going on in the world's other major navies.

Unlike the 'jump' from building pre-dreadnought capital ships to building all-big-gun battleships and cruisers, instigated by Jackie Fisher in 1905, the move from building first-class torpedo boats to building destroyers did not involve a major reset in shipbuilding technology or design philosophy, so most major navies did not hesitate to follow the Royal Navy's lead.

In fact, a number of them, such as the French with their *torpilleurs de haute mer* and the Germans, had already started down a similar path of design evolution. The following sections cover the development of destroyers in the world's major navies up to the beginning of the First World War.

France – The nation probably most invested in the development of torpedo craft as offensive and defensive weapons was France, which, through the teachings of its Jeune École, had bought into the concept of using light, torpedo-carrying vessels as a means of extending the power of its relatively small navy. The French, at the end of the nineteenth century, quite realistically reasoned that Germany was their most likely opponent in any future war, but that a war with Germany would, like the Franco-Prussian War of 1870, not be primarily a naval war unless the British joined on the German side. The problem, from the French point of view, was that they had no way of being certain of what Great Britain would do in the event of a future war with Germany. The British had stayed strictly neutral in 1870, and although the two nations had been showing signs of increasing co-operation since that would culminate in the signing of the Entente Cordiale in 1904, even then that agreement was not initially a military alliance. The French believed they had every reason to worry about a British blockade of their long and vulnerable coastline, and therefore needed to have the means to counter that threat. A large force of expendable torpedo craft was an affordable means to address that threat and fit well with the Jeune École mindset.

From the beginning of the development of torpedo-carrying craft, the MN was never far from the forefront in building fast, seaworthy craft, and often was in the lead, challenging British domination of their shared maritime frontier. As described above, the French pioneered the development of larger torpedo boats designed to operate

The first French destroyer (*contre-torpilleur*) was easy to distinguish from the preceding *torpilleurs de haute mer* by its size, but little else. More than 40ft longer and displacing more than three times as much, MN *Durandal*, seen here probably in 1905, had a raised turtleback forecastle carried aft as a flying deck on which the crew served the ship's weapons. She and her three sisters served as the prototypes for more than fifty similarly sized '300-tonne' destroyers. The extremely widely separated funnels indicate that the two Normand watertube boilers faced in opposite directions, their working faces sited together amidships. (NHHC)

beyond coastal waters – the so-called *torpilleurs de haute mer.* By the early-1890s, the MN was building large and fast torpedo boats that drove the Royal Navy to lay down the first dedicated torpedo boat destroyers starting in 1892.

It is often difficult to tell when the progression of a nation's development of torpedo craft crosses that invisible line that separates a particularly large and fast torpedo boat from a true destroyer – it has something to do with the proportion of armament given over to quick-firing guns of a large calibre, as opposed to torpedo armament. In the case of the MN, that was less difficult than most because the size of the first French destroyers, the Durandal class, was almost three times that of the preceding *torpilleurs de haute mer*, making them easily distinguishable:

MN *Durandal*: 1899[44]

Displacement:	296 tons
Length (oa):	188ft 8in
Beam (wl):	20ft 8in
Draft (max):	10ft 5in
Powerplant:	4800ihp; two-shaft triple-expansion engines; two Normand watertube boilers
Speed (actual):	26kt
Armament:	2 × 381mm trainable torpedo tubes (2 × 1; centreline); 4 × 381mm torpedoes; 1 × 9pdr/50 Mod 1891; 6 × Hotchkiss 3pdr/40 QF (6 × 1)

The Durandals were given the type designation *contre-torpilleurs*, which the MN used to designate the destroyer type, yet compared to the first Royal Navy destroyers laid down six years earlier, they were decidedly slower and less-well-armed, and they compared even less well against their nearer British contemporaries. The French seemed well pleased with the type and built a total of fifty-five ships of roughly the same 300-tonne design before developing a larger type.[45] The real motivations for transitioning in 1907–8 to a ship of 50 percent larger displacement were practical issues regarding stability. To a greater extent than the British, at least in the early years of the development of the type, the French expected their destroyers to operate in open water, and they found the 300-tonne ships, particularly the later classes, suffered stability problems deriving from too much topweight carried on too small a hull. This was particularly true when the 381mm torpedo was replaced by a 450mm model starting with the Claymore class in 1905. With the Spahi class laid down beginning in 1908, the French were building larger destroyers with better stability and speed, although they were still smaller, slower, and less well-armed than their British contemporaries, and most were still powered by reciprocating engines. Some of these transitional 450-tonne *contre-torpilleurs* had experimental turbine or mixed powerplants.

Reacting more quickly this time, the MN placed orders under the 1908 and 1910 programmes for yet larger destroyers, with the following characteristics:

Realising in 1907 that a limit had been reached in what could be achieved in a hull displacing 300 tons, the Marine Nationale moved quickly in steps up to an '800-tonne' standard, as exemplified by MN *Fourche* of the Bouclier class, seen here most likely in 1911, soon after completion. Armed with a weapons suite similar to its British contemporaries and capable of similar speed, *Fourche* displayed a much more conventional hull form, with a high, raised forecastle and little, if any, tumblehome in the rest of her hull. (NHHC)

MN *Bouclier*: 1912[46]

Displacement (normal):	745 tons
Length (oa):	250ft
Beam (wl):	25ft 6in
Draft (max):	10ft 2in
Powerplant:	13,000shp; two-shaft turbines; four boilers
Speed:	30kt
Armament:	4 × 450mm trainable torpedo tubes (2 × 2; sided); 6 × 450mm torpedoes; 2 × 100mm/45 Mod 1893 (2 × 1); 4 × 9pdr/50 Mod 1902 (4 × 1)

The eighteen destroyers of this '800-tonne' class and a nearly identical follow-on class, and the two destroyers built to one more similar class, were the last French destroyers completed before wartime restrictions shut down warship construction. The 800-tonne type was a conscious effort to make a destroyer that could operate with the fleet in bad weather and match up against the Royal Navy's Tribals, but they were seen as lacking in hull strength, engineering reliability, and overall fighting power compared to their presumptive opponents. The French naval design bureau – the STCN (Service Techniques des Constructions Navales) – drew up specifications for another, significantly larger, class to come after the 800-tonne *contre-torpilleurs*, but these were never started, much less completed, due to the demands of war.[47]

Four large, well-armed destroyers were ordered by Argentina from the French yard Ateliers et Chantiers de Bretagne, Nantes, in 1910.[48] They were requisitioned by the MN as they neared completion when war broke out in 1914. The French found them to be useful, but difficult to maintain due to their non-standard machinery. Desperately needing more destroyers and unable to build more at home due to wartime materiel shortages, the French turned to the only possible source of new-construction warships – Japan. The French government ordered a dozen small destroyers of a proven design, the Kaba class, in December 1916. The Japanese reacted with admirable haste, allocating the twelve hulls to their four government yards and two private firms (Kawasaki, Kobe, and Mitsubishi, Nagasaki); all twelve were delivered complete at Port Said on 26 October 1917. Other than logistics problems caused by their armament, which was of Japanese manufacture and required unique munitions, these ships were a welcome addition to the French fleet and served well into the late 1930s.

Germany – The path to destroyer development for Germany differed from most other nations for several reasons. One reason was that Germany as a nation that could maintain a national navy did not exist until after the Franco-Prussian War of 1870, and another was that even after Germany achieved

A transitional design built by Thornycroft for the Kaiserliche Marine, SMS *D10* was designated a *divisionsboot* rather than a destroyer (*zerstörer*), although she was smaller than previous division leader designs. She had both a broad gun platform built atop a conning tower, where two of her five 5cm/40 guns were mounted side-by-side, and, aft of that, a bridge atop a pilothouse for the division commander. (NHHC)

The home-grown destroyers built by the Germans after testing the British-built *D10* were somewhat smaller, carried less armament, and looked a great deal different, as can be seen from even a quick glance at this image of SMS *S95*, dating from approximately 1905. This ship, and her twenty-six near-sisters built over the next five years, shared a similar design, with a long, high forecastle deck with a turtleback and a breakwater designed to keep water out of the well deck in front of the pilothouse and bridge. The bridge was open and protected primarily by canvas stretched over the railing, but the officers were given the minimal comfort of a multi-pane windscreen to keep spray out of their eyes. (NHHC)

nationhood, there remained the need to establish the requirements for a navy and the definition of its purpose. Neither of these was necessarily obvious. The modern German state came into existence out of war with France and its continued existence was most obviously threatened by (and was threatening to) France in the west and Russia in the east. In a manner comparable only to Japan (and then later, for different reasons, Russia) the new German state, once it saw the need for a navy, had to build one from literally nothing. Like the French, the Reichsmarineamt (Imperial Navy Office – often referred to as simply the Marine Amt – the German equivalent to the British Admiralty), the new German Kaiserliche Marine had no extra funds to waste on idle or redundant craft. Every hull had to serve a purpose; even better if it served more than one. Having already experimented with the torpedo-carrying avisos *Zieten* and *Blitz* (described above), they went on to build a series of large torpedo boats designed to operate with their fleet against the only real naval threats the new German state faced – Great Britain, France, and, to a lesser extent, especially after the Russo-Japanese War of 1904–5, Russia. From the beginning, their torpedo boats were designed to operate in the open waters of the North Sea, some of the roughest on earth.

When, in the middle of the 1890s, the Royal Navy led the evolution of torpedo boats into destroyers, the Germans reacted quickly, sensing the threat these new ships presented to their torpedo boat flotillas. Rather than attempt to come up with a home-grown design that would counter what the Royal Navy was doing, the Germans took the simple expedient of ordering a destroyer from Thornycroft in 1896, somewhat more lightly armed, but otherwise similar to a British '30-knotter'. (From the designation this boat was given – SMS *D10* – it is obvious the Kaiserliche Marine did not think of this boat as a new type of craft but rather as the tenth in a series of larger torpedo boats intended to lead torpedo boat flotillas; more on this in the final section of this chapter below.) The Germans were impressed by this boat's mechanical features, but not by its overall design or its light construction or poor seakeeping ability, and when they ordered their first indigenous destroyers from the Schichau yard, the resulting ships looked quite different, with a high forecastle, followed by a well deck and then the bridge structure. This class had the following specifications:

SMS *S90*: 1901[49]

Displacement:	388 tons
Length (oa):	206ft 10in
Beam (wl):	22ft 11in
Draft (max):	9ft 4in
Powerplant:	3900ihp; two-shaft triple-expansion engines; three marine boilers
Speed:	26.5kt
Armament:	3 × 45cm trainable torpedo tubes (3 × 1); 5 × 45cm torpedoes; 3 × 5cm/40 SK L/40 (3 × 1)

The Germans followed *S90* with regular annual programmes of between six and twelve ships through to 1904, with most

By 1909, German destroyers had grown in size and seaworthiness. SMS *V183* from that year's program showed the changes incorporated in the design, including a longer, higher forecastle to improve seakeeping in rough weather. The well deck has virtually disappeared; the foremost torpedo tubes are now abaft the bridge, firing out below the bridge wings. The bridge itself is now fully enclosed, making for a greatly-improved command platform. The forward main-battery gun now is located on the forecastle. (NHHC)

of the orders going to Schichau, a few to Germaniawerft (Kiel), for a total of forty-eight ships in six subclasses over five annual programmes. Each subclass carried the same armament and shared a common silhouette, with a high forecastle, well deck, prominent bridge and two large funnels. They gradually increased in size up through the last subclass – *S126–S131* of the FY1904–5 programme – which displaced 447 tons. One of the ships of the previous programme, *S125*, was designated as an experimental design, being fitted with Parsons turbines. The Germans took their time making the switch over to turbine drive. One ship each of the 1905 and 1907 programmes – *G137* and *V161* (the letter 'V' indicating that the AG Vulcan yard at Stettin (Szczecin) had joined the programme) – were similarly singled out to test different turbine arrangements, before the Kaiserliche Marine committed to building turbine-powered destroyers with the FY1908–9 programme. That same year, the Germans added oil fuel as a supplement to coal, the oil being sprayed on the coal before it was burned to increase its energy content; they did not switch over permanently to oil fuel only until the FY1913 programme.

The size of German destroyers had increased gradually year-over-year through FY1910; the ships ordered that year displaced 660 tons. The ships since 1904 had steadily increased in size and armament, not always to the benefit of seakeeping. The S138 class ordered in FY1906 had such a high forecastle and long well deck that the bridge became unusable in heavy weather and the operation of forward guns and torpedoes was seriously hindered. Learning from this mistake, the following V150 class featured a longer forecastle and shorter well deck; the longer forecastle allowed one of the two big 8.8cm/35 guns to be mounted there. The Germans were well pleased with this design and built follow-on classes through FY1910; that programme comprised the six ships – a half-flotilla numbered *G192–G197* – ordered from Germaniawerft with the following specifications:

SMS *G192*: 1911[50]

Displacement:	660 tons
Length (oa):	241ft 6in
Beam (wl):	24ft 11in
Draft (max):	10ft 3in
Powerplant:	18,200shp; two-shaft Germania turbines; four Navy boilers
Speed:	32kt
Armament:	3 × 45cm trainable torpedo tubes (3 × 1); 5 × 45cm torpedoes; 2 × 8.8cm/35 SK L/35 (2 × 1)

This class represented a 'local maximum', the largest size that German destroyers attained for the next three years, because the Marine Amt, under impetus from RAdm Wilhelm Lans, had concluded that destroyers had grown too big for their intended role of carrying out and defending against night torpedo attacks. Against the wishes of many in the fleet, the next two programmes of destroyer construction, in FYs 1911 and 1912, were committed to the production of ships that were approximately 15 percent smaller in all dimensions, although they carried a similar weapons suite.[51] So different were these ships from their immediate predecessors that the Germans decided to restart the numbering sequence; the twenty-four ships of this class were *V1–V6*, *G7–G12*, and *S13–S24*. The result was predictable; the ships were far less seaworthy than their larger predecessors and far more cramped, because almost the only place from which space could be saved in these smaller ships was crew accommodation.

Not too surprisingly, the Germans reverted to a larger ship type as soon as they realised how badly Lans' experiment had failed. (The twenty-four ships of the class were retained despite their problems and some even served into and through the Second World War; however, they were always considered inferior to their larger half-sisters, being referred to as 'Lans cripples'.) In fact the FY1913 ships that followed were almost twenty-five percent larger than the G192 class, with the following specifications:

SMS *V25*: 1914[52]

Displacement:	812 tons
Length (oa):	257ft 7in
Beam (wl):	27ft 3in
Draft (max):	10ft 11in
Powerplant:	23,500shp; two-shaft AEG-Vulcan turbines; four Navy oil-fired boilers
Speed:	33.5kt
Armament:	6 × 50cm trainable torpedo tubes (2 × 1; 2 × 2); 8 × 50cm G/6 torpedoes; 3 × 8.8cm/45 Tbts LC/13 (3 × 1)

These were the first German ships to be completely oil fuelled. They were considered to be highly successful and remained the basis for destroyer design to 1915, increasing slowly in size until the G85 class, the last of this series, displaced 960 tons. Even after that, this design remained the pattern for the small number of 'war mobilisation' ships completed during the war.

Italy – Like Germany, Italy was a state that only came into existence in the second half of the nineteenth century. The *Risorgimento*, a multi-stage process that saw the multiple parts of Italy, many under foreign domination, coalesce under the leadership of King Vittorio Emmanuelle II of Sardinia-Piemonte and his brilliant prime minister Camillo Benso, Conte di Cavour, led to the creation of the Regia Marina (RM – Royal Navy) in November 1860. Not unlike Japan, when the Meiji Restoration brought together the existing fleets of the maritime *daimyos* into the nascent Imperial Japanese Navy, the Regia Marina originally comprised the fleets of Sardinia-Piemonte and Naples (Napoli), along with scattered units from other minor states.[53] The resulting fleet was a hotchpotch of mismatched units, both physically and administratively, that took a number of years to sort out. It was really only after the original fleet and its officers had been retired or replaced that the RM developed the character and functionality of a national navy, a process that took fifteen or more years.

This was not an easy process for a new nation with many demands on limited resources. Like the French, the Italians had to be intelligent in their expenditures on naval hardware, including having a willingness to invest in innovative designs.

The first primarily indigenous design of destroyers for the Regia Marina was the Nembo class ordered in 1900. This image shows the lead ship, RN *Nembo*, probably after 1905, when her main-battery armament was altered to include two 57mm/43 guns mounted side-by-side on a gun platform forward atop a pilothouse. (NHHC)

At the same time that the Germans were building SMS *Zeiten*, in 1875, the RM was laying down a ship with many similar ambitions, the designated torpedo cruiser RN *Pietro Micca*. Promoted by Navy Minister Simone Pacoret de Saint-Bon and designed by Felice Mattei, *Micca* had dimensions similar to many of the destroyers described above and was armed only with one torpedo tube and two machine guns, but was seriously underpowered and never exceeded thirteen knots, much less her designed speed of seventeen knots. For that reason, she never saw much active service. Nonetheless, the Italians were enamoured with the concept of the torpedo cruiser and built no fewer than fifteen before the end of 1895.[54]

At the same time, the Italians also followed a more normal path of torpedo craft development, opting to order their first torpedo boats from the British yards Thornycroft and Yarrow, starting in 1877, before placing their first order for an indigenously-designed boat in 1882 from the Fratelli Orlando (later Odero-Terni-Orlando (OTO), Livorno) yard. The first Italian destroyers – *cacciatorpediniere* – were likewise ordered overseas, from the Schichau yard, in 1899.[55] However, in this case, the RM, at the same time, ordered one ship of indigenous design from the Odero yard (Genoa). This ship, named RN *Fulmine*, was unsuccessful in service, never reaching her designed speed, and was quickly relegated to secondary tasks.[56] The first Italian-designed and built class of destroyers to enter service was the Nembo class (325 tons) of 1902–05.

The convoluted path by which Italy became a unified nation did not reach completion until 1870 with the occupation of the Papal States and Rome. This could only have happened because the French, who strongly supported the Pope's political independence, were distracted by the Franco-Prussian War. This also began a process of deterioration in Franco-Italian relations – the French had previously been strong supporters of Italian unification – that reached a nadir in 1881 with the French colonisation of Tunisia, which Italy had been eyeing. This made Italy receptive to the approaches of Germany's Bismarck, who offered alliance with the newly-unified Germany and Austro-Hungary as a protection against the French. The reaction in Italy was mixed, as the Germans were seen as a strong potential ally and a good counterbalance to France, but the Austrians were a perennial enemy; the two nations had actively fought as recently as 1866 and the Austrians still held large pieces of territory where Italian was the primary tongue (at a time when linguistic irredentism was still a potent political force). Nonetheless, the Italian anger at the French and distrust of the British, who had encouraged (or at least done nothing to impede) the French takeover of Tunisia, led them to agree to join what became the Triple Alliance in 1882.

This is more than just an interesting sidenote in European history, because it led directly to the Italians following a path towards building much larger destroyers, similar in size to those being built by the Germans and Americans, just before the outbreak of the First World War. This was because, as a member of the Triple Alliance, the Italians became privy to German planning for a possible war against France or Great Britain. The Germans informed the Italians in 1912 of their intent, should such a war appear likely, to deploy a small battle squadron to the Mediterranean, to be based at Messina.[57] The Italians agreed to provide escort vessels for this squadron when

Designed to support the German Mediterranean squadron, the three large *esploratori leggeri* of the Poerio class, including the lead ship RN *Alessandro Poerio*, seen here probably in 1920, were similar in size and function to flotilla leaders in other navies, but were built with their size and relatively heavy armament to support a high-speed, long-ranging unit. (NHHC)

it operated in the central or western Mediterranean. None of the Italian destroyers built to date or building would be capable of escorting this squadron at the speeds and ranges required, so in February 1913 the RM ordered three ships, half again larger than the Indomito class then being built, from the Ansaldo yard (Genoa). These ships had the following specifications:

RN *Alessandro Poerio*: 1915[58]

Displacement (normal):	1028 tons
Length (oa):	278ft 10in
Beam (wl):	26ft 3in
Draft (max):	9ft 3in
Powerplant:	20,000shp; two-shaft Belluzzo turbines; three Yarrow boilers
Speed:	32.3kt
Armament:	4 × 450mm trainable torpedo tubes (2 × 2; centreline); forty-two mines; 6 × 102mm/35 Mod 1914 (6 × 1)

The three Poerios were, actually, a larger version of a private venture 558-ton destroyer designed in 1912 by the Ansaldo yard of Genoa, laid down the next year for China; the ship was bought, in 1914, by the Portuguese navy who named her *Liz*. She was then sold in March 1915 to the Royal Navy, who commissioned her as HMS *Arno*. The Poerios were completed too late to serve their intended purpose of escorting the German Mediterranean squadron, which had, in fact, been active in the Mediterranean since the end of 1912. By the time these first Italian large destroyers were ready, the two ships of the German Mediterranean squadron had long since been chased into Constantinople, where they were turned over (at least nominally) to the Turks and formed the core of the Ottoman Navy for the remainder of the war.

The origin of the three even larger *esploratori leggeri* (light scouts) of the Mirabello class that followed the Poerios was every bit as convoluted. They had been planned in 1912 as 5000-ton light cruisers and laid down as such in 1914 before the beginning of the First World War. The British and French naval blockade, however, stopped the delivery of the high-tensile steel needed for their hulls. With an amount of steel in hand sufficient for only one of the three planned ships, the RM was faced with the choice of building two fewer ships or building three smaller ships; they chose the latter option. The design bureau at Ansaldo quickly recast the 5000-ton light cruiser design as a 1,570-ton light scout, and the three Mirabello class *esploratori leggeri* were all laid down between November 1914 and February 1915.[59]

The six large destroyers were designated *esploratori leggeri* by the Italians until the four survivors were redesignated destroyers after 1920. The two remaining Poerios were sold to the Spanish Nationalists in June 1938, in whose service they remained until after the end of the Second World War.

Japan – The navy of modern Japan had to be conjured into existence with the establishment of the Japanese Empire in 1868; in this way it bore considerable resemblance to the navies of Italy and Germany, but the Japanese had greater obstacles to overcome. Unlike those nations, the Japanese had hardly any pre-existing naval forces, such as the Prussian or Neapolitan fleets, to build upon, nor were there the industrial infrastructure or shipbuilding facilities, public or private, to support the indigenous development of a fleet. Of necessity, all warship production was initially done overseas and the training of the first cadre of Japanese naval officers and petty officers was done by foreign officers, primarily British or French, brought in for the task. While the nascent Imperial Japanese Navy (IJN) very much modelled itself on the British Royal Navy – a logical choice given that in the second half of the nineteenth century it was the largest and most respected navy in the world – it was nonetheless strongly influenced by the ideas of the French Jeune École and by the writings of the American Alfred Thayer Mahan.[60] The ideas borrowed from the French included attraction to the use of torpedoes and torpedo craft to help make up for a smaller navy's lack of larger ships. The French school of thought caught the attention of the Japanese after the naval successes of the small French squadron against the Chinese in several engagements during the Sino-French War of 1884, particularly after the Battle of Foochow (Fuzhou), 23 August 1884, at which two small French torpedo boats armed only with spar torpedoes sank the Chinese flagship and damaged another vessel. For some time after this, the Japanese switched their foreign orders for warship construction from British to French yards.[61] Mahan's writings convinced the Japanese that control of the sea was a necessary condition for national survival, let alone expansion. Anticipating early on that the Americans were a likely foe in any future conflict, the Japanese developed a strategic plan in which light forces armed with torpedoes would weaken any American fleet advancing across the Pacific to the point that the smaller Japanese fleet, waiting safely behind the Marianas, could meet it on equal or better terms.

When the first major naval expansion authorisation was funded by the Japanese Diet (Parliament) in 1882, it contained funds for twenty-two torpedo boats to be acquired over the following eight years, all of which would be ordered from foreign yards and shipped in sections to Japan, where they would be reassembled at the first Japanese shipyards being established at Tsukiji (Tokyo) and Yokohama. By 1885, the thinking had reached the point that the IJN called for a fleet force comprising mainly armoured cruisers and torpedo boats.[62] Twelve of the torpedo boats were supposed to be quite large and armoured, a considerable innovation for the time.[63] Funds were set aside for six of the these large boats to be laid down in 1885, and a delegation of two Japanese officers travelled to the Yarrow yard, where they found the British shipyard to be cautious about proceeding on such an ambitious project.[64] In part this was due to Yarrow's order book being full and in part due to concerns about the possibility that the Japanese might reject subsequent boats after testing the first. As a result, a contract was signed on 29 April 1885 for one boat to be built in sections and assembled at Yokosuka, with Yarrow guaranteeing that, if the Japanese opted to order five more, they would be built at the same price. The specifications of *Kotaka* were:

Kotaka: 1888[65]

Displacement (normal):	203 tons
Length (oa):	165ft
Beam (wl):	19ft
Draft (mean):	5ft 6in
Powerplant:	1217ihp; two-shaft horizontal compound engines; two locomotive boilers
Speed:	19kt
Armour (deck):	1in steel
Armour (belt):	1in steel (in way of engineering spaces)
Armament:	2 × 14in fixed torpedo tubes (2 × 1; bow); 4 × 14in trainable torpedo tubes (2 × 2; centreline); 4 × 1pdr (4 × 1)

Kotaka was indeed quite advanced at the time she was ordered from Yarrow in 1885. She was larger in terms of displacement than any contemporary torpedo boat and carried a heavier armament, but her speed was considered disappointing; marine engines at this point were not yet able to supply the power needed to give her the same speed as smaller, unarmoured boats. Nonetheless, *Kotaka* was considered to be a success, despite her lack of speed and manoeuvrability compared to smaller boats. These factors, and the negative opinion of the noted French naval engineer Louis E Bertin, who was then consulting for the IJN, led to none of the follow-ons being ordered.[66] While often pointed out as being a groundbreaking design and a foreshadowing of destroyers to come, *Kotaka* proved instead to be a dead end.

The Japanese, however, continued to be in the forefront of torpedo boat tactical employment and would show this to be the case during the Sino-Japanese War of 1894–5. Faced with Japanese advances on their main naval base at Port Arthur (Lushun), the Chinese moved their surviving fleet across the Bo Hai Straits to Wei-hai-wei (Weihai) on the Shandong peninsula. To keep a close watch on that fleet, the Japanese set up an advanced base at Yungching Bay (Rongcheng Wan) at the tip of that peninsula in January 1895.[67] As was typical of boats of this time period, particularly the case before the adoption of turbine powerplants,

Designed and built in sections to Japanese requirements at Yarrow's Poplar yard, but looking very French with her extended ram bow and pronounced tumblehome, the Imperial Japanese torpedo boat *Kotaka*, seen here in approximately 1890, participated in the Naval Battle of Wei-hai-wei in February 1895. (NHHC)

mechanical issues, including complete breakdowns, were common during sustained operations. The Japanese chose a radical solution to the problem, sending the repair ship *Genzan Maru* forward to Yungching Bay to perform maintenance on site, rather than rotating ships or entire flotillas back to Japan for upkeep. This was seen by many observers – the Sino-Japanese War was watched closely by Western naval powers – as a brilliant move by the Japanese, which improved the availability of their torpedo boats on station off Wei-hai-wei by a significant degree.[68] The Japanese captured the land side of the harbour with army forces, which trapped the Chinese naval forces – two armoured turret ships, two armoured cruisers, and numerous smaller and older craft – in the anchorage off Liukung (Liugong) Island in Wei-hai-wei Bay, the entrances to which were protected by floating booms. The island was far enough from the mainland that army artillery could not reach the Chinese anchorage, rugged enough that an opposed landing would be difficult, and well enough protected by forts and the guns of the Chinese ships that a bombardment by the Japanese fleet would be risky. The Japanese fleet commander, VAdm Ito Yukio, therefore ordered the three divisions of torpedo boats at Inshankou, a forward base set up in a shallow bay just east of Wei-hai-wei, to prepare to attack the Chinese fleet in sustained night attacks.[69]

The Naval Battle of Wei-hai-wei, 5–6 February 1895

The action known as the Naval Battle of Wei-hai-wei began on 30 January 1895 with operations intended to weaken the protective booms outside the Chinese anchorage.[70] The first attempts were unsuccessful, but on the night of 3 February, under fire from island forts, two torpedo boats managed to break off 100 yards of the southern end of the eastern boom, opening a sizeable gap in the protection. The following night, just after midnight on 5 February, the attacks began in earnest. Two divisions, a total of ten torpedo boats, set out from Inshankou, accompanied by two gunboats instructed to demonstrate outside of the booms to distract the Chinese. The Japanese waited until the moon had set at 0300 to attempt to enter the anchorage. The first four of the boats entered through the gap without problems; the remaining six boats missed the entrance and had a variety of problems with the booms. Two were able to 'jump' the booms by working up to speed and scraping over the top; a third attempted that manoeuvre, but failed to clear the boom and had to retire due to the damage she received. Two others strayed too close to the shore, encountered shoals and had to head back to Inshankou, one of them under tow by the remaining boat.

The six boats that did enter the anchorage were able to manoeuvre undetected for upwards of twenty minutes, searching for targets in the gloom, before the flashes of torpedoes being launched – at this time the most common method of launching torpedoes was with a gunpowder charge, which gave off a visible flash at night – gave away the locations of the attackers. After that, as might be expected in such an attack, confusion reigned.

The order in which boats attacked appeared to have little impact on their chance of success.[71] *TB-9*, a smaller, third-class torpedo boat, was one of the first to attack. She was one of the two boats that jumped the booms and, once inside the anchorage, had almost immediately encountered a small division of Chinese torpedo boats heading north and had attached itself to the end of the line. *TB-9* remained undetected, until its captain sighted a red light off his starboard bow; correctly identifying the source of the light as a major warship, he swung his small boat to bring his fixed bow tube to bear and fired a torpedo at approximately 200yds, still the maximum effective range of torpedoes at this time.[72] *TB-9* continued to swing to the right, firing its only other torpedo at the same large target from its trainable amidships mount at a range estimated to be 50yds, before heading south at high speed. Under fire from the now fully alerted Chinese fleet and pursued by the torpedo boats she had been trailing, *TB-9* was soon hit multiple times, putting her powerplant out of operation and leaving her dead in the water. At least one of the torpedoes fired by *TB-9* found a target; reports vary as to the number of torpedoes that hit the large armoured turret ship *Ting Yuen*, ranging from one to three, but regardless, one of those almost certainly was fired by *TB-9* and that one was sufficient to cause the Chinese flagship to sink gradually that day in shallow water.[73] There she was blown up by the Chinese to prevent her capture on 9 February. The derelict *TB-9* was sighted by *TB-19*, the other 'boom jumper'. This boat approached *TB-9* and attempted to set up a tow line, briefly towing *TB-9* towards the south; it became obvious that she would not stay afloat so *TB-19* took off *TB-9*'s crew and exited the anchorage.[74] *TB-5*, the fourth boat to enter through the breach in the booms, had attempted to attack along with *TB-9*, but fired her torpedoes from too great a distance and both missed; she too exited the anchorage successfully.

One of the newest Japanese boats, *TB-22*, leading the two remaining boats, turned north only after following the southern shore of the bay until level with the western end of Liukung Island, where the major Chinese units were anchored, at which point she turned north. She got off two or three torpedoes at an unidentified large target, probably the armoured cruiser *Lai Yuan*, at ranges described as 'quite close', but all appeared to have missed. This boat then escaped from the anchorage and on her way south towards the exit from the bay spotted a Chinese picket boat which she chased in towards the southern shore. There, *TB-22* grounded hard on rocks close under a fort that had been captured by the Japanese. By dawn, it was clear she was not coming loose and most of her crew were able to wade ashore as the tide was out; as the day went on, gunfire from the Chinese fort on Jih Island prevented any further rescue attempts for the ten men believed to remain on the boat. Early the following evening, 6 February, six men were rescued alive from *TB-22*; later that night, the Chinese were able to pull the boat off the rocks and tow her back to Liukung Island. She was apparently scuttled a few days later by the Chinese to prevent her recapture by the Japanese when Liukung Island was captured.

The two boats following *TB-22* in the attack on the Chinese anchorage had similar problems, but different luck. *TB-6* attempted to fire two torpedoes, but both failed to launch due to ice build-up in the tubes and the boat turned away. The last in the line, *TB-10*, also suffered icing problems, but was able to free one tube to fire; this torpedo may have hit *Ting Yuen*. *TB-10* then also turned away and headed towards the exit from the bay.

Most of the Japanese torpedo boats involved in this attack had received some degree of damage. *TB-9* and *TB-22* were lost; *TB-6* and *TB-10* collided while retiring, but both were able to reach Inshankou under their own power; almost all the other had scraped bottoms or gunfire damage or both. Against these losses, the Japanese could claim what they believed to be damage to the flagship of the Chinese Adm Ting Ju'chang (Ding Ruchang), which they had been unable to achieve at the Battle of the Yalu River in September the previous year. As best as can be ascertained, the Japanese attempted to fire eleven torpedoes, succeeded in firing eight, and achieved one or two hits, for the loss of two boats; given the state of the technology, that was not a bad result.

Not satisfied with these results, and willing to accept the risks resulting from what had to be assumed would be a heightened state of alert within the Chinese anchorage, VAdm Ito ordered a repeat attack the following night. Due to the damaged state of the eight surviving boats from the two divisions that had attacked the night before, this attack would be carried out by the five boats of TBDiv1, which included *Kotaka*. One of the five boats never entered the anchorage, suffering a steering failure and returning to Inshankou under manual control. The other four boats entered the anchorage and, despite the Chinese ships sweeping the area with searchlights and randomly shooting their smaller batteries into the darkness, three of them managed to line up and fire at least ten torpedoes at three targets. *TB-23*, another Schichau boat and the newest Japanese torpedo boat then in commission, fired all three torpedoes she carried; one of these hit a rock near the small mine vessel *Pao Fa*, which was so damaged by the resulting explosion that she took on water rapidly and had to be beached. *Kotaka*, following behind *TB-23*, fired four torpedoes, one or more of which hit the armoured cruiser *Lai Yuan*, which capsized and sank in short order. *TB-11* fired its two torpedoes, one of which hit the composite sloop *Wei Yuen*, being used as a training ship, which sank in place. One more Chinese ship was sunk, an unnamed twin-masted sailing ship, which exploded and sank in the anchorage. What can be certain is that she was not hit by torpedoes from the one other Japanese torpedo boat that entered the anchorage, *TB-13*, because that boat reported that she had searched for a target without success and had retired without firing any torpedoes. All four Japanese torpedo boats which had entered the anchorage retired without further incident. The net result of the second night's attacks had been nothing short of spectacular; four Chinese ships were sunk, including an armoured cruiser, for no loss and no damage to the attackers or injuries to their crews.[75] The reactions among observers were restrained. Any commentary on the action at Wei-hai-wei was limited to mentioning that torpedo boats had shown some potential.[76] The Japanese, however, seem to have appreciated more fully the implications of the attacks. Japanese torpedo boat forces, and the destroyer forces that succeeded them, adopted a slogan to describe the tactics they developed based on this experience – *nikuhaku-hitchu* (press closely, strike home).

Every torpedo boat with which Japan fought at Wei-hai-wei was foreign-built – although many were built in sections with final assembly in Japan, which gave the nascent Japanese yards some valuable experience – but the first boats completely built in Japan, albeit largely from imported parts, were *TB-24* and *TB-25*, the former to a Normand design and the latter a Schichau boat, both of which were completed in 1895, too late to participate in the war. News of the French *Forban* (see above) and the remarkable speed she attained on trials in October 1895 was as shocking to the Japanese as it was to the British, not surprising given the importance the Japanese placed on torpedo warfare. This led to the Japanese ordering four boats of the Cyclone class, a direct follow-on to *Forban*, from Normand, with a further six ordered three years later, all of these to be built in Japan.[77] Equally as noteworthy to the Japanese was the laying down of the first destroyers in England, starting with *Havock* in 1892. The Japanese watched these developments carefully and, as soon as it became clear that the first destroyers were indeed successful, they placed large orders for similar craft with specialist yards in Europe; six orders went to Thornycroft in 1897 and six over a span of two years (1897–8) to Yarrow, all of which were completed in England and delivered in 1899–1900. All were similar to Royal Navy 30-knotters. Four more similar ships were ordered from the two yards in 1900, after which production was shifted to the Kure and Yokosuka navy yards; all future destroyer production for the IJN would be in Japan.

None of those 'home-grown' ships would be ready in time to fight in the Russo-Japanese War of 1904–5. As had been the case with the Sino-Japanese War, this conflict was observed carefully by the world's navies for indications of the direction of progress in naval tactics and technology. There were two main destroyer actions in this war. One was a night attack on the Russian Pacific Fleet at anchor outside the harbour entrance at Port Arthur, which took place on 9 February 1904, several hours before Japan formally declared war.[78] Port Arthur has a large, natural harbour surrounded by hills with a narrow entrance; knowing that war was imminent, and concerned that a pre-emptive Japanese attack might trap his fleet in the harbour, the Russian VAdm Oskar Starck prudently anchored his major ships outside the harbour, protected by booms.[79]

The Japanese very quickly developed an indigenous design and shipbuilding capability that came to the aid of the Entente side in the First World War. Unable to find the labour or resources to build needed destroyers at home, the French turned to Japan, which built twelve ships of the Kaba class for France, completing all of them and delivering them before the end of October 1917, less than eleven months after they were ordered. This image shows MN *Annamite*, one of those twelve, in 1920. (NHHC)

Two Russian destroyers were patrolling in the shallow bay south-east of the anchorage; the remaining destroyers were inside the harbour.

The Japanese attack was initiated by ten destroyers in an orderly fashion against a force of unprepared, fully-illuminated targets, but poor communications between the attacking ships caused problems from the beginning. When the lead destroyers sighted the pair of Russian destroyers patrolling out in the bay, they swerved sharply north hoping to escape detection; this threw the following ships into utter confusion and caused two to collide. The damage was minor and did not prevent them from continuing with the attack, but any organisation among the attackers was lost. The patrolling Russian destroyers had in fact sighted the approaching Japanese, but were under strict orders not to fire first, so apparently they decided they had no option but to continue steaming their patrol; they failed to warn Starck of the enemy presence and played no further part in the action. In the disorganised attack that followed, as the Japanese destroyers approached in ones and twos, the first ones in had the most success because the Russian ships were slow to react, in part out of fear they might be firing on their own ships. After the first four Japanese ships had fired their torpedoes and turned away, the remaining attackers were met with increasing resistance; the first four launched eight torpedoes and scored three hits; the remaining destroyers fired between eight and twelve more torpedoes and, as far as can be ascertained, achieved no hits. The three torpedoes that did find Russian hulls damaged one protected cruiser (*Pallada*) and two battleships (*Retvisan* and *Tsesarevich*); all three were beached to prevent their sinking and, due to the lack of repair facilities at Port Arthur, would be out of action for months to come. Interestingly, many international observers saw this attack as pivotal in naval history. The renowned naval historian Julian S Corbett is quoted as stating that this was 'the first great torpedo attack', but in discussing the Port Arthur attack a few years before that attributed quote was written, Corbett stated that in assessing the odds of success of such an attack, 'the chances of success can scarcely ever be great enough to justify the risk'.[80]

The other well-known destroyer action of the Russo-Japanese War was the night action after the Battle of Tsushima, 27–28 May 1905. The remnants of the Russian Baltic fleet, already badly battered by the Japanese during the previous afternoon's engagement, hoped to escape in the darkness and deteriorating weather towards Vladivostok to the north, but were harassed by Japanese destroyers and torpedo boats. In the night attacks, which quickly became disorganised, the Japanese lost three torpedo boats, two to Russian gunfire and one as the result of a collision with a Japanese destroyer. The fifty-eight Japanese torpedo craft fired an undetermined number of torpedoes and laid some mines in the paths of several Russian ships, all of which accounted for two battleships and two armoured cruisers, either sunk directly or damaged to the extent that they were later scuttled to prevent capture. The night attack after Tsushima is generally considered to have been a meaningless postscript to the day's action.

The Japanese, however, saw the destroyer actions of the Russo-Japanese War as validating their belief that torpedo craft in general, and destroyers in particular, would be an important force equaliser in any future war against a stronger opponent. Being unable to out-produce in quantitative terms such potential enemies as Russia or the United States, the Japanese believed they would need to build ships that were individually as good as or better than any likely enemy counterparts. In the wake of the war, the Japanese worked on both developing better torpedoes – they introduced a 21in wet-heater torpedo in 1911 – and training continuously to develop better tactics; they continued to stress short range attacks even as torpedo ranges increased and experimented with various tactical organisations and formations, looking for the best combination of flexibility and control.[81]

The Russo-Japanese War found the Imperial Navy in desperate need of more destroyers; an Emergency War Programme, in addition to an already planned budget request, was funded by the Diet in 1904 that eventually authorised the construction of thirty-two destroyers, essentially up-gunned repeats of a 1901 Thornycroft design. At 381t normal displacement, the ships of the Asakaze class were somewhat small by contemporary standards, but, although all were completed too late to serve in the Russo-Japanese War, they were good seacraft and served as the backbone of the Japanese destroyer flotillas well into the 1920s. In 1907, the IJN once again obtained approval to order new destroyers; this time, due to the advances in size and power of ships being built abroad, these would be significantly larger. In a desire to make these ships as close to the 'state of the art' as possible, the start of construction was repeatedly delayed as the plans were updated. They were finally laid down in 1909 and completed with the following specifications:

***Umikaze*: 1911**[82]

Displacement (normal):	1150 tons
Length (oa):	323ft 6in
Beam (wl):	28ft 1in
Draft (mean):	9ft
Powerplant:	20,500shp; three-shaft Parsons turbines; eight Kampon boilers
Speed:	33kt
Armament:	4 × 18in trainable torpedo tubes (2 × 2; centreline); 2 × 4.7in/40 41st Year Type (2 × 1); 5 × 3in/40 41st Year Type (5 × 1)

The delivery of the two ships ordered in 1907 was further delayed by the decision to power them with Parsons turbines manufactured by Mitsubishi, which were not completed until March 1910. When finally delivered in late 1911, these ships proved to be as good as expected, and the IJN immediately attempted to order a second pair, but an economic downturn intervened and the navy was forced to look for a less expensive alternative. Thus, the next pair of Japanese destroyers were one-third smaller, at 605 tons normal displacement, mounting one fewer gun of each calibre and sacrificing three knots of top speed. These two ships, the Sakura class, proved to be good value for the money and served as a model for the next three classes ordered by the Japanese, in particular the Kaba class of the 1914 War Programme, which had the following specifications:

***Kaba*: 1915**[83]

Displacement (normal):	665 tons
Length (oa):	274ft
Beam (wl):	24ft
Draft (mean):	7ft 9in
Powerplant:	9500ihp; three-shaft four-cylinder vertical triple-expansion engines; four Kampon boilers
Speed:	30kt
Armament:	4 × 18in trainable torpedo tubes (2 × 2; centreline); 1 × 4.7in/40 41st Year Type; 4 × 3in/40 41st Year Type (4 × 1)

The Sakura and Kaba classes, of which twelve were built in total, served well into the 1930s, as did the twelve Kabas built for France (see above). The second follow-on class, Momo, of the 1915 Programme, introduced geared turbines in the place of reciprocating engines. After one more class of mid-sized destroyers, the Japanese reverted to large destroyers, a follow-on design to the well-liked Umikaze class, the four 1227-ton ships of the Isokaze class, which were also ordered in 1915.

Russia – The Russians, from the beginning, faced two unique and difficult problems building a modern navy. The first was geographic; Russia had three disconnected coastlines – four if one counted the Caspian Sea, five if one considered the northern European ports of Murmansk and Arkhangelsk as being effectively separated from the Pacific port of Vladivostok by a long Arctic coastline – which required separate fleets well out of mutual support distance from each other. This was similar to France, which had three separate coasts, but far worse, because the separations were substantially greater. Russia's second problem was economic. Like Japan, Russia began this period with only rudimentary industrial capacity, facing the need to order boats abroad while attempting to develop an indigenous shipbuilding infrastructure. The differences from Japan, however, were numerous; while Japan was a small, crowded nation with few natural resources, but a strong national ethic that emphasised education and self-improvement, Russia was vast, with seemingly unlimited natural resources, but with a social system that discouraged the advancement of the lower class, meaning many of Russia's 'technocrats' were from non-Russian minorities or of foreign birth. All this slowed the development of the Russian navy

Working to a German design, and using a powerplant supplied by the AG Vulcan yard, the Russians built a large, fast destroyer named *Novik*, seen here on her trials in 1913. The completed ship was more than 10 per cent heavier than designed, a fact that was concealed from the engine maker, causing her to fail her initial trials. Vulcan modified the powerplant to produce more steam, raising the ship's top speed and sustained high speed to levels that briefly made her the fastest warship in the world. (NHHC)

and its native shipbuilding capabilities, excepting only when strong personalities, such as Stepan O Makarov, came along with the influence to push a slow-moving bureaucracy.

Makarov and his protégés were influential enough to push the torpedo as a weapon, and the torpedo boat as its delivery system, to the point that more than 100 small Yarrow-designed torpedo boats were built for the Russian navy abroad and at home in 1877–8 alone, in response to the Russo-Turkish War.[84] One thing Russia did have that both France and Japan lacked was relatively unlimited funds, spent at the discretion of a largely unconstrained autocrat, which could greatly benefit the navy as long as the tzar was interested. The result of this system was growth in the Russian torpedo forces which came in irregular bursts of activity. For example, after ordering their first destroyer from Yarrow in 1894, the Russians then ordered a large, twenty-six-ship class of small destroyers – the Puilki class – with units being laid down in Baltic shipyards between 1896 and 1900. Twelve of that total were built in sections, shipped by rail to the Pacific coast and assembled at Port Arthur. There they formed the basis of the destroyer squadrons that faced off against the Japanese in 1904–5.

The losses to the Russian navy in the Russo-Japanese War in terms of capital ships and cruisers had been devastating; in terms of destroyers, the losses were less catastrophic. Eighteen destroyers were lost in that war, and three captured by the Japanese, but that was out of a total of fifty split between the Russian fleets; the Russians were in the midst of a large building programme which went a long way towards replacing those losses within a year or two. Because of the war, the Russians also placed large orders at foreign yards, primarily in France and Germany, so that they more than made up their war losses. With an eye on foreign developments, the size of Russian destroyers increased roughly in step with the increases happening in Royal Navy ships. The ships ordered starting in 1904 increased in average normal displacement from roughly 350 tons to 570 tons. The next big jump came in a ship ordered in 1909, but not laid down until July 1910. The design, in fact, originated in 1905 – this was the same timeframe in which the first Royal Navy Tribals were being designed – using monies left over after eighteen destroyers had been funded by public subscription.[85] The original proposal called for a turbine-driven, oil-fuelled 1400-ton ship capable of thirty-six knots, specifications quite advanced for the time. Design tenders were solicited from Russian yards and the award was given to the Putilov Works (St Petersburg), which had submitted a design from AG Vulcan, which supplied the powerplant and many other components for the ship. She was built to the following specifications:

***Novik*: 1913**[86]

Displacement (normal):	1280 tons
Length (oa):	336ft 3in
Beam (wl):	31ft 3in
Draft (mean):	9ft 10in
Powerplant:	40,000shp; three-shaft AEG turbines; six Vulcan watertube boilers
Speed:	36kt
Armament:	8 × 18in trainable torpedo tubes (4 × 2; centreline); 4 × 4in/60 Pattern 1911 (4 × 1); fifty mines

The design had originally called for only two 4in guns; for that reason and because the shipyard had frequently used heavier materials than the design had called for, the ship displaced approximately 140 tons more than designed, which increased the ship's trials draft by a foot. The shipyard attempted to disguise the problem by simply moving the draft marks and boot-topping up a foot. What they could not disguise was the drag caused by the additional mass and wetted area; the initial trials speed was well below the 36kt called for in the contract. AG Vulcan, apparently unaware of the deception pulled by the shipyard, took responsibility for the deficit and modified the boilers, lengthening the watertubes, to increase steam output. When tested again in August 1913, *Novik* achieved a maximum speed of 37.2 knots and a sustained speed of 36.86 knots, which made her the world's fastest warship. The Russians followed *Novik* with orders for fifty-five similar ships in five classes, many of which were not completed due to wartime shortages of materials.

United States – The US Navy had no apparent need for torpedo boats, much less torpedo boat destroyers, for most of the early period covered in this chapter. Since the War of 1812 (a Western Hemisphere dust-up distantly related to Britain's ongoing wars with Napoleon), the Americans had not faced a foreign enemy of consequence. until they belatedly joined the imperial 'scramble' towards the end of the nineteenth century. Only then did the need emerge for torpedo craft, particularly destroyers, not so much in a defensive, but in an offensive role. The first American destroyer, so described because of its size and armament, laid down in 1897, was still designated by the US Navy as a torpedo boat.[87]

The Spanish-American War of 1898 was something of a rude awakening for the Americans. While the US Navy easily defeated their Spanish opponents at both of the major naval engagements of that war, in one of those battles – the Battle of Santiago de Cuba, 3 July 1898 – the Spanish order of battle included two new destroyers built by the Clydebank yard.[88] The fact that the Spanish had been able to send destroyers across the Atlantic shocked the Americans; they had refuelled at the Canary Islands and two arrived safely to join Adm Pascual Cervera y Topete's squadron which was eventually bottled up at Santiago.[89] The fact that a US Navy squadron comprising battleships, armoured cruisers, and a pair of gunboats converted from private yachts managed to annihilate Cervera's squadron with virtually no losses or damage did not change the fact that the result could have been radically different had the Americans been just a bit slower to react or the Spanish a bit luckier. As an after-action statement by one officer – Lt Manuel Bustamente y Barrena, *Furor*'s executive officer – made clear, the Spanish destroyers, which were the last of Cervera's squadron to exit the narrow channel from the harbour that day, had torpedoes prepared to fire at the approaching enemy ships, but both were abandoned before the Americans came within range.[90] Even before the battle, in a report released in March 1898, the Assistant Secretary of the Navy, Theodore Roosevelt, Jr (TR), stated that: 'We especially need torpedo-boat destroyers, and every effort should be made to procure them.'[91]

Spurred on by the war, and possibly also by TR's exhortation, the US Congress passed an act on 4 May 1898 that authorised the construction of sixteen destroyers, so designated in the authorisation. Although the Navy's Bureau of Construction & Repair developed a detailed design for these ships, the individual yards were allowed to modify that design or substitute their own as long as they met the required trials

USS *Flusser* (DD 20), one of the Smith class 700-ton 'flivvers', is seen in a shiny, fresh coat of paint at the 1909 Hudson-Fulton naval review held at New York City. The long indent in the forecastle was to allow the 3in/50 guns mounted just aft of the forecastle and ahead of the first funnel to fire very nearly directly forward. (LoC)

Wearing a well-worn dazzle camouflage applied during her service in the European theatre in the First World War, USS *Balch* (DD 50), a Cassin class 'thousand-tonner', is seen sometime probably in 1918. Note the angled wind and spray deflector mounted on the face of the bridge in an attempt to protect command personnel from the worst of North Atlantic weather. The forward 4in/50 gun has a short shield similar to that seen on contemporary British destroyers, protecting the gun crew only from the waist up. (SLV – Green Collection)

speed.[92] The process was further complicated by a requirement written into the law that no yard could be awarded more than four ships to build. The result was that the Americans' initial batch of sixteen destroyers were split into six distinct subclasses built by seven different shipyards. They all were similar in size, speed, and armament – nine ships were built, more or less, to the BuC&R plans – so the specifications of the lead ship will suffice to represent them all:

USS *Bainbridge* (DD 1): 1902[93]

Displacement (standard):	420 tons
Length (oa):	249ft 9.8in
Beam (wl):	23ft 5in
Draft (mean):	6ft 6in
Powerplant:	8000ihp; two-shaft vertical triple-expansion engines; four Thornycroft boilers
Speed:	28.45kt
Armament:	2 × 18in trainable torpedo tubes (2 × 1; centreline); 4 × 18in Mk 2 torpedoes; 2 × 3in/50 Mk 2 (2 × 1); 5 × 6pdr (5 × 1)

Seven of this initial series of ships were completed with turtleback forecastles in the place of the high forecastle deck forward as in the BuC&R design. While these ships were well received in the fleet and served on through the First World War, they in fact proved to be too lightly built to handle the stress of escorting the fleet over long distances. When the US Congress was persuaded to authorise additional destroyers again in 1906, it was clear that a larger, more capable ship would be needed. This led to the ordering of the Smith class and two immediate follow-on classes, construction of which started in 1908. In total, twenty-six ships were built in these three classes, all of approximately 700 tons standard displacement. The gun armament was standardised, first as five 3in/50 guns and later as four 4in/50s. The torpedo armament was increased to three torpedo tubes in the Smith class and six in the two following classes by the simple expedient of adopting twin torpedo tubes. Perhaps the most important innovation in the 700-ton series of ships was the adoption of turbine drive, with the expected improvement in reliability, particularly at high speed.

In 1909, the General Board of the Navy, comprising senior admirals serving their last billet before retirement, which had in the past had an important role in recommending design characteristics for new ships, now took over primary responsibility for defining these specifications. For the first year that it held this full responsibility, FY1911, which required input in 1910, the board could only recommend minor improvements to the 700-ton design, leading to the third class in that series, the Monaghan class. But for the next year, the board called for a major increase in size, range, and fighting capabilities. The FY1912 ships, the seven ships of the Cassin class, had the following specifications:

USS *Cassin* (DD 43): 1914[94]

Displacement (standard):	1010 tons
Length (oa):	305ft 3in
Beam (wl):	30ft 4in
Draft (mean):	9ft 6in
Powerplant:	16,000shp; two-shaft Parsons turbines; four Normand boilers
Speed:	29kt
Armament:	8 × 18in trainable torpedo tubes (4 × 2; sided); 4 × 4in/50 Mk 9 (4 × 1)

These ships all attempted to solve the problem of the inefficiency of turbines at cruising speed by shipping one or two small reciprocating engines that could be clutched to the propeller shafts for cruising. While this was a workable solution, it was complex and heavy, and USS *Wadsworth* (DD 60) of the follow-on Tucker class, launched in 1915, had an experimental geared turbine powerplant that pointed the way to the obvious solution to the efficiency problem. *Cassin* and her descendants were dubbed the 'thousand tonners' for obvious reasons; the thirty-six ships in the 700-ton series became known as the 'flivvers', a somewhat derogatory term for a vehicle in bad condition. The name did not reflect opinion within the destroyer 'community' which found the flivvers to be useful; most served well into the 1930s.

The leadership problem

The earliest destroyers were essentially little more than hulls wrapped around powerplants. A look at the spacing of the funnels reveals that just the boilers alone often took up half or more of the length of the hull. (The hulls of these early destroyers were too narrow for the boilers of the time to be placed side-by-side; of necessity, boilers were placed in tandem; the number and spacing of funnels was a function of the number of boilers and their orientation. Between two and four funnels was common, with three being the most frequent arrangement.) What little space was left inside the hull was mainly given over to accommodation for the crew – more than two-thirds of which comprised the stokers and trimmers who tended the massive boilers, until the adoption of oil fuel eliminated the need for the labour-intensive handling of coal and clinkers – or to magazine space for the guns.[95] The problem of communication between ships in a squadron, or between the squadron and any higher command, was for the most part not addressed, as it was assumed that such communication would be by signal flags during the day or by lamps at night. Communication beyond visual distance, or when visibility was impaired by fog or smoke, was discounted. The invention of radio and, more importantly, its introduction to warships in 1895, brought a whole new potential for command and control, but also a whole new range of problems for the designers of destroyers, because early radios were large, fragile, and required a dedicated crew of technician/operators for whom space would have to be found.

Nevertheless, the increase in the size and capability of torpedo craft, and their evolution from coastal torpedo boats into destroyers large enough to accompany a fleet or even venture out to sea on their own, introduced the leadership problem. In their earliest days, when torpedo boats operated from tenders or harbours, tactical command of torpedo boat formations, rudimentary as it was, was exercised by the senior boat commander, who could attempt to signal his intent by whatever means were at hand to the other captains, who were undoubtedly preoccupied with their own difficulties. What little battle experience accumulated in those early years made it clear that more was needed.

Probably the first navy to address the leadership problem directly was the German Kaiserliche Marine. While the French had been developing their *torpilleurs de haute mer* and the British had countered with the first torpedo boat destroyers, the Germans had been dealing with some of the same issues as the French, attempting to build a credible naval force, one that would be capable of holding off a blockade by the British (or the French) along its short North Sea coastline. Like the French, the Germans saw at least part of the solution to their problems in building large torpedo boats, capable of accompanying their fleet into the North Sea. Unique among the world's navies in the mid-1880s, the Germans foresaw the problems that would be faced controlling those boats once they put to sea. Beginning in 1885, the Marine Amt began ordering a number of what they called *divisionsboote* (division boats) from the Schichau yard, which were torpedo boats more than double the size of the standard craft then being ordered – 295 tons vs 115 tons – but similarly armed and with similar speed. The specifications of the first of these boats is as follows:

SMS *D1*: 1887[96]

Displacement:	295 tons
Length (pp):	184ft
Beam (wl):	21ft 8in
Draft (max):	11ft 2in
Powerplant:	2020ihp; one-shaft vertical triple-expansion engine
Speed (trials):	20.5kt
Armament:	3 × 35cm trainable torpedo tubes (3 × 1); 4 × 35cm C/84a torpedoes; 3 × Hotchkiss 4.7cm/43.5 (3 × 1)

The extra displacement was given over entirely to accommodation and facilities for a division officer and his staff, approximately a dozen men and not, as noted above, to greater armament or speed, so that these boats, which the Germans intended should be provided one per division of ten boats, were most definitely not the ancestors of the first torpedo boat destroyers, which they anticipated by seven years. The first two *divisionsboote* were followed at yearly intervals by orders

The Imperial German Kaiserliche Marine might have been the first to build larger torpedo boats specifically to accommodate the command staff of a half-flotilla, rather than expecting to assign small cruisers this task. Some consider these *divisionsboote*, such as SMS *D1*, seen here in 1900, to be the precursors of the destroyer type, but the Germans did not, arming them and assigning them to flotillas just like ordinary torpedo boats. (NHHC)

for two more similar boats from the same yard until 1890, with slow increases in size until the last pair, *D7* and *D8*, displacing 404 tons, and in speed up to 22.5kt. After that, four years passed before a single additional *divisionsboot*, *D9*, was ordered again from Schichau, with again somewhat increased speed and size, but with armament similar to the torpedo boats she was intended to lead. The series came to an end with *D10*, mentioned above, which was, in fact, a destroyer ordered from Thornycroft which displaced almost 100 tons less than *D9*, despite being eleven feet longer, indicating how much more lightly built British torpedo craft were than their German counterparts.

The Royal Navy understood the problem that the German *divisionsboote* were built to address, but took a very different path to address it. The 'scouts' mentioned above in a 1905 Parliamentary debate were eight small cruisers ordered by the Royal Navy in 1903 to lead destroyer flotillas. They were ordered in pairs from four different yards with a good deal of leeway allowed to each as to how it met the requirements. The requirements each yard was expected to satisfy included a top speed of twenty-five knots, a 1.5in protective deck (or equivalent protection), shallow draft (so they could accompany destroyers close inshore), and armament of two torpedo tubes, ten 12pdrs, and eight 3pdrs. All of the scouts met these requirements at delivery in 1905, but they were already inadequate for the task for which they were intended, because destroyers being delivered at that time were a good five knots faster, and that deficit would only increase as destroyer designs progressed. The Royal Navy would have to look elsewhere to resolve the leadership problem.

One solution came along in the form of HMS *Swift* (see above, the second of that name) laid down in 1906. While somewhat of a disappointment as a fleet destroyer, *Swift* was seen as ideal for the destroyer leader role. In 1912 she became the flagship of Captain (D) of the 4th Destroyer Flotilla.[97] She had more than adequate space for the flotilla commander and his staff, but once war broke out she was found to be too lightly built and was retired from Grand Fleet duties to lead the 6th Destroyer Flotilla in the somewhat more tranquil Dover Patrol. Regardless of *Swift*'s relative lack of success, the idea of a large destroyer serving as the flagship of a destroyer flotilla, which the Germans had shown worked very well, had been implanted in the Royal Navy. In a naval staff conference in 1913 the idea was raised that a Tribal class destroyer be given long-range radio equipment and enlarged bridge structure allowing her to accommodate a Captain (D) and his staff.[98] The Controller instead proposed that new ships be built, not to exceed 1800 tons, that would accommodate a staff of eleven, but would otherwise have characteristics similar to the destroyers they commanded.[99] Two ships of this type were authorised in December 1913 to the following specifications:

HMS *Lightfoot*: 1915[100]

Displacement (normal):	1607 tons
Length (oa):	324ft 10in
Beam (wl):	31ft 9in
Draft (max):	12ft
Powerplant:	36,000shp; three-shaft Parsons turbines; four White-Forster boilers
Speed (trials):	34.5kt
Armament:	4 × 12in trainable torpedo tubes (2 × 2; centreline); 4 × 4in/40 QF Mk IV (4 × 1); 2 × 2pdr/39 Vickers QF Mk II (2 × 1)

These ships were so highly anticipated that two more were ordered in April 1914 and a further three in November. They proved to be highly successful as command ships and also as general-purpose destroyers, as their size gave them stability in a seaway far beyond that of standard destroyers of that time, which displaced 500 tons less. Henceforth, and for the duration of the First World War, the Royal Navy regularly included funds for leaders that were approximately 50 per cent larger than the standard destroyers being built, at a ratio of approximately one leader for every ten standard destroyers.

The Royal Navy took the building of flotilla leaders very seriously, when they began doing it just before the First World War. Realising only then that they simply had so few light cruisers to perform the myriad tasks already expected of the type – scouting, escort duty, blockade support, etc – that there were not enough left over to serve as flotilla leaders. The answer was to follow the German model and build a percentage of enlarged destroyers designed to accommodate a Captain (D), his staff, and their communications equipment. The result was a ship like HMS *Kempenfelt*, a Lightfoot class leader, seen here during a 1917 fleet exercise, with a flotilla of 'R' class destroyers. The leader is the fourth ship from the foreground in the middle group, most easily distinguished by having four funnels compared to the three funnels of the 'R' class ships. (NHHC)

Chapter 3

The test of the battle, Part 1: 1917

THE OUTBREAK OF THE First World War in August 1914 found most of the combatant navies prepared to fight what all expected to be a short, sharp clash, along the lines of the most recent major European conflict, the Franco-Prussian War of 1870. That had been a war of manoeuvre on land, little influenced by events at sea. The conventional wisdom was that technology, in the form of machine guns, rifled artillery, motorised transport, and railways, had made the movement of men and the concentration of firepower so efficient that sweeping offensives would lead to rapid victory or defeat. As is often the case in warfare, the conventional wisdom was wrong.

The First World War rapidly settled down to a long, bitter slog along lines of trenches running through Belgium and France in the west and from the Baltic to the Carpathians in the east. This shift from a war of rapid movement to one of stalemate had tremendous repercussions from a naval point of view. Now the war at sea became, potentially, the decisive theatre of operations. In the east, the Germans were occasionally able to achieve local naval superiority over the Russians – still hobbled by their disastrous losses a decade earlier – and move troops along the Baltic coast, sidestepping Russian defensive positions. In the west, they faced a much more difficult situation, where the Royal Navy held an insurmountable advantage in capital ship numbers that no amount of qualitative advantage on an individual ship basis could offset. Only by luring a subset of the Grand Fleet into a situation where it could be overwhelmed by the High Seas Fleet could the Germans hope to free themselves from the geographic trap where France and the British Isles effectively blocked German access to the open ocean and vital international trade.

While the massively costly battlefleets of Great Britain and Germany spent most days idling in their harbours at Scapa Flow, Inverness, Rosyth, and Wilhelmshaven – costly in every possible sense: money, metal, men, and materiel – the sea war continued on a daily basis at chokepoints where commercial shipping concentrated and along heavily trafficked convoy routes across contested waters, such the English Channel and the narrows between Shetland and Norway in the west, and the Otranto Strait in the Mediterranean. In these narrow waters, it was not capital ships that ruled the waves. Control of these waters was contested between much lighter forces – light cruisers and destroyers. It was in these sometimes daily (or often nightly) encounters, that it was found that larger destroyers, combining as they did some of the greater firepower and survivability of light cruisers with the high speed and manoeuvrability of destroyers, represented an ideal platform for combat.

Size does matter

As described in the preceding chapter, many nations had been trending towards the construction of larger destroyers, in particular to handle the task of leading flotillas. By the time the war began most naval powers were building standard destroyers, the kind that composed the working flotillas that escorted a nation's main battlefleet or blockaded an enemy's coastline, in excess of 1000 tons. It did not take much activity to show that these larger ships handled combat and even the rigours of day-to-day operations on the high seas better than smaller ships.

Interestingly, some of the largest destroyers accepted into the combatant navies in the first years of the war had been ordered by other nations, but were taken over by the building nation at the start of the war. One example of this was a class of eight large destroyers that had its origins in sets of turbines being built by Blohm & Voss for Russian ships of the Leitenant Ilin class.[1] At the outbreak of war in 1914, the nearly-complete turbine sets were seized and offered to the Kaiserliche Marine as the basis of a class of very large destroyers to be built by Schichau. At first the Marine Amt was reluctant, thinking the war would be over long before the ships could be ready and that their greater size would make them incompatible with the existing flotillas. Two yards responded to this hesitation on the part of the navy with revised, accelerated construction schedules, and the four large ships – two to be built by Blohm & Voss and two by Vulcan – were ordered with the enthusiastic support of the head of the Marine Amt, Adm Alfred Tirpitz. At over 1300 tons displacement, the B97 class would be the first ships designated destroyers (*zerstörers*) by the Germans. The specifications of the lead ship were:

SMS *B97*: 1915[2]

Displacement (design):	1374 tons
Length (oa):	321ft 6in
Beam (wl):	30ft 10in
Draft (max):	11ft 3in
Powerplant:	40,000shp; two-shaft Navy turbines; four Navy oil-fired boilers
Speed:	36.5kt
Armament:	6 × 50cm trainable torpedo tubes (2 × 1; 2 × 2); twenty-four mines; 4 × 8.8cm/45 Tbts LC/13 (4 × 1)

So influential was Tirpitz, and so powerful his support for this design, that four more ships were ordered from Blohm & Voss before the first two were completed. These eight ships, which operated together as the II TFl (2nd Torpedo Boat Flotilla) from the middle of 1915 onwards, acted as the HSF's 'firemen', moving from theatre to theatre as needed to face the enemy's best destroyers.

At the same time that the Argentines ordered the four large destroyers from the French, described briefly in the preceding chapter, they ordered like numbers from Great Britain and Germany. All eight of those ships were completed before the war, but only the German-built ships ended up in Argentine service; the four British-built ships were sold to Greece before they were completed. To replace the four ships sold to Greece, the Argentines, being well pleased with the four they received from Krupp and Schichau in 1912, ordered four more ships from Germaniawerft. They were nearing completion when war broke out in Europe and were taken over by the Kaiserliche Marine. While not quite as big as the B97s, the four ships of the G101 class, at 1098 tons, were also larger than their German-designed contemporaries. The Argentines had planned on these ships being fitted with cruising diesels to improve endurance, but this was not done by the Germans, in the rush to complete these ships and bring them into service. All four of the G101 class, and the seven surviving members of the B97 class, had their 8.8cm/45 guns replaced in early 1916 by 10.5cm/45 Tbts LC/16 single-mount guns.

The Italians also received a contract before the war to build a class of four large destroyer leaders for a foreign navy; in this case, it was the Pattison yard at Naples, which took in an order from Romania for four ships to be built to a design by Eng Luigi Scaglia and armed with three 4.7in guns and five torpedo tubes. At Italy's entry into the war, they were taken over in July 1915 and three were completed with the following characteristics:

RN *Aquila*: 1917[3]

Displacement (normal):	1594 tons
Length (oa):	310ft 6in
Beam (wl):	30ft 10in
Draft (max):	11ft 6in
Powerplant:	40,000shp (design); 38,925shp (actual); two-shaft Tosi turbines; five Thornycroft boilers
Speed:	34kt (design); 36.44kt (actual)
Armament:	4 × 450mm trainable torpedo tubes (2 × 2; sided); 3 × 6in/40 EOC (3 × 1); 4 × 76.2mm/40 Ansaldo (4 × 1); 2 × 6.5mm MG (2 × 1)

The first three ships of this class were fitted with old pattern 6in/40 main-battery guns with the intent that they would be able to defeat the Austro-Hungarian (hereafter referred to as simply Austrian) scouts of the Novara class, a trio of small, fast 3500-ton scout cruisers, which were armed with nine 10cm/50 single-mount guns. The choice of the old Elswick 6in guns for the large destroyers, which the Italians designated *esploratori leggeri*, proved unfortunate, as action would soon show.

Baptism of fire: Otranto Strait, 15 May 1917

War in the Mediterranean in 1917 was far different than it would be twenty-five years later, because even though two of the three main Central Powers had direct access to that sea, the Turks were far more interested in facing the Russians in the Black Sea than attempting to engage the Royal Navy in the Aegean, and the Austrians, despite good access to the Adriatic, were relatively easily bottled up at the Otranto Straits by an Allied patrol force maintained by the British, French, and Italians. The actual bottling was done by a permanent patrol of armed drifters – fishing trawlers generally under 100GRT – spread across the narrows between Capo Santa Maria di Leuca on the Italian side and Fano Is (Othonoi) on the Greek. On the night of 15 May 1917, the line comprised forty-seven drifters.[4] This 'barrage' had become a serious impediment to submarine operations out of the major Austrian bases at Pola (Pula, Croatia) and Cattaro (Kotor, Montenegro). As early as mid-March 1917, the Austrians had attempted to disrupt the barrage with a sweep by four destroyers, which had failed in its mission, and two other similar operations in April and early May had similarly failed. Undaunted, the Austrians planned a larger, two-pronged operation out of their southernmost base at Cattaro for mid-May. One force, comprising two destroyers – SMS *Csepel* and *Balaton* – would head south down the coast, hoping to intercept one of the small convoys run nightly by the Italians between Otranto and Valona (Vlorë, Albania).

In this they were successful, finding a convoy of three small merchantmen escorted by the old Italian destroyer RN *Borea*, not far off the Albanian coast. In these days before radar, on a calm, clear, moonless night, the battle often went to the side with more, or more alert, lookouts, and in this case, it was the Austrians who sighted *Borea* off their starboard bow and were moving past the Italian destroyer to close the trailing steamers, when they were sighted and challenged at 0324. They responded by illuminating *Borea* with searchlights and opening fire. The initial salvos by the 10cm main-battery guns of the Austrian flagship *Csepel* hit *Borea* hard, crippling her powerplant and starting a serious fire. The next ship in line, a 1600GRT freighter, was hit as well, setting off its cargo of munitions. Warned that a strong Allied force was approaching from the north, the Austrian destroyers turned and headed back up the coast, leaving *Borea* and the freighter sinking and a second freighter burning.

The Allied force – actually three separate forces: the first a fast squadron led by the Italian *esploratore leggero* RN *Carlo Mirabello* followed by four French destroyers; the second comprising two British light cruisers HMS *Dartmouth* and

Mirabello was forced to swing a wide circle around Valona to avoid an enemy submarine. Horthy's scouts began slow, ineffectual fire at *Mirabello* at 0710; the range was too great for the Italian leader to reply.

Meanwhile, Acton's cruisers, heading north-east from Brindisi, stumbled on the two Austrian destroyers, which had been ordered south to join Horthy at 0615. They caught sight of the Allied squadron at 0730 and turned back to the north. Again, it took some time for word to reach Acton and appropriate orders to be dispatched. *Aquila* was ordered ahead at 0750 to attempt to intercept the Austrian destroyers which were now heading towards Cattaro at best speed. The brand-new *Aquila* quickly pulled away from the division of destroyers she was leading, even the equally new RN *Giovanni Acerbi*. Armed with her big, slow-firing 152mm/40 Elswick Pattern gun forward, *Aquila* soon closed the range and opened fire at 0815 at a range of 12,500yds. *Aquila* was able to get off only a dozen rounds, none of which hit, except for some shell splinters, as the range dropped. When the range dropped to 10,500yds, *Csepel* was able to respond with her 10cm/50 guns; one round hit *Aquila* at 0832, detonating in her main fire room, killing seven men and destroying the steam lines leading to her turbines; drifting to a stop, *Aquila* could only watch helplessly as the two Austrian destroyers escaped into Durazzo.

As Acton's remaining forces continued on to the north-east at their best speed, the chase going on to the south was rapidly overtaking them from off their starboard quarter. At 0900, *Bristol*, bringing up the rear of the Allied second squadron, sighted smoke off to the south-east. It soon became apparent that the Austrians would pass Acton's squadron well to the east and, along the way, come within range of the disabled *Aquila*. Acton ordered a turn to the east, but it was

The three Austrian scouts that did the main damage to the drifters of the barrage itself were sister ships of the Novara class, such as SMS *Saïda* seen here running trials in August 1914. More than twice as big as the destroyers they faced, although armed only with destroyer-sized 10cm/50 main-battery guns, these scouts were considered by some to be one of the precursors of the super destroyers that emerged a decade later. (NHHC)

already too late for most of the Allied force to intercept the fast-moving scouts. Only one of the British cruisers, *Dartmouth*, was able to keep up with the Austrians, and she continued the action on her own, in the process forcing Horthy to pass by *Aquila* without further interference. At one point, Horthy slowed his ships, allowing the range to shorten so his scouts' 10cm guns could reach *Dartmouth*, but after trading shots briefly – during which period a 6in shell hit *Novara*'s bridge, killing the Austrian second-in-command, Cdr Szuboritz – Horthy became concerned that the *Mirabello* force might be overtaking them from astern, and ordered speed increased again. *Mirabello* managed to achieve one hit on *Helgoland* at 0950, but the 4in round failed to penetrate the Austrian ship's armour. A few minutes later, *Mirabello* discovered that water had leaked into some of her oil tanks, contaminating her boilers, obliging her to stop, and another of the French destroyers suffered a condenser failure, effectively eliminating the chase group.

With Horthy's scouts again at full speed, *Dartmouth* seemed to be losing touch with the enemy when she put a 6in/50 round into *Novara*'s after engine room, dramatically reducing her speed and killing eleven men. *Dartmouth* was hit by an Austrian 10cm/50 shell moments later. Concerned about a fire in the forecastle caused by this hit, *Dartmouth*'s captain ordered her forward magazines flooded. Harassed by a pair of Austrian aircraft flying from Cattaro, *Dartmouth* turned away to starboard, thus effectively ending the battle.

All of the Austrian scouts made it back to Cattaro in various states of damage. The Allied forces straggled back to Brindisi. The only losses in ships that day had been the fourteen drifters destroyed by the Austrian scouts, *Borea* and

Facing the Austrian scouts in the engagement following the attack on the Otranto Barrage on 15 May 1917, was an Entente force that included a brand new *esploratore leggere*, RN *Aquila*. This image shows Aquila's sister ship RN *Nibbio*, in the foreground, with *Aquila* behind her. As was the case with *Mirabello* when she was re-armed later in 1917, the 152mm/40 gun *Aquila* and *Nibbio* carried forward on their forecastles was powerful, but too slow-firing and heavy for ships of their size. (NHHC)

the one freighter lost earlier to the Austrian destroyers, and a French destroyer lost to a mine laid by an Austrian submarine. The battle had been a clear Austrian victory although it did not in any way improve the Austrians' strategic or tactical position in the naval war in the Mediterranean. The ease with which *Aquila* was disabled showed simply that large destroyers were vulnerable, like their smaller brethren, to damage to their propulsion systems that would deprive them of one of their greatest assets, which was speed. The bigger lesson, not heeded by most who studied this battle, was the need to avoid the temptation to arm large destroyers as if they were small cruisers; this was not the last time large destroyers would be armed with large, slow-firing main-battery weapons, to the later regret of the men who served in them.

There was considerable political fallout after this battle that impacted the region in the inter-war years. As noted above, Miklós Horthy went on to become de facto ruler of the Hungarian state that emerged from the disintegration of the Austro-Hungarian Empire. Anglo-Italian relations, never the best up to this point, were in no way helped by finger-pointing after the fact, with the British blaming Acton for the failure to catch and destroy Horthy's cruisers. Acton, who led the RM for two different stretches post-war, helped develop the Italian fleet into one designed to take on the Royal Navy as well as the MN in any future war.

Interestingly, the ships that were operated with the greatest success in this engagement were the three Austrian scouts. In the eyes of some, they were simply larger destroyers, displacing 3500 tons and mounting 10cm/50 main-battery guns, a weapon that would serve well as a destroyer's main battery. During the war, they were used often as flotilla leaders, a role in which the Royal Navy often used light cruisers. But the design lineage of the Austrians scouts makes it clear that these were not overgrown destroyers, but rather undersized light cruisers; they had armour and powerplants like a light cruiser. There was little about them except their size to recommend them to a designer of the next generation of destroyers.

Fighting fires: The Second Norwegian Convoy Action (aka Pellew Convoy), 12 December 1917

So versatile were the eight large ships of the II TFl to the Kaiserliche Marine that they were often assigned to operate as independent four-ship half-flotillas in co-ordinated operations. Such was the case in mid-December 1917, when the Germans initiated the third in a series of operations designed to disrupt the convoys running between Great Britain and Norway. The blockade by the Royal Navy, which was designed to cut Germany off from all international trade, had by 1917 seriously reduced imports even from nations as close as Norway, which had, before the war, been a major importer of foodstuffs and raw materials. The ever-tightening blockade meant that Norway, a neutral since the beginning of the war, could now export only to Britain or to neutral states, and did so at the risk of her ships being attacked by German submarines or other raiders. As part of their general response to the U-boat threat, the British had organised their seaborne trade with Norway into regular convoys running between Lerwick in Shetland and Bergen to the east. Most of the active patrolling by the German squadrons based in Flanders or at North Sea ports had been limited during the summer months by the short nights, but the return of autumn and longer nights had increased activity again. A highly successful sortie in October against a Norwegian convoy by two fast minelayers had been followed by another, less successful raid in November through Dogger Bank. These attacks had led to plans to reroute the convoys on a shorter route, but this had not yet been implemented by mid-December.

The Kaiserliche Marine decided to try a dual operation, returning to both areas, in mid-December, this time using the powerful II TFl to deliver the attack.[5] The eight large destroyers departed Wilhelmshaven at 0245 on 11 December 1917, negotiated the cleared channels up the Jutland coast to the Horn's Reef lightship and then turned west. Once they reached the north-western tip of Dogger Bank, the flotilla split into its constituent half-flotillas of four ships each; 4.HFl split off to the west, where they were to attack traffic off Tynemouth and north of there, while 3.HFl – SMS *G101* (flag, LCdr Hans Kolbe), *G103*, *G104*, and *V100* – stood on to the north, where radio traffic analysis indicated a pair of small convoys were being run between Lerwick and Bergen, and a coastal convoy was headed south from Lerwick. The 4.HFl missed the coastal convoy, but nonetheless had a successful sortie, sinking two stragglers and two out of a group of four small trawlers it encountered by chance.

The northern half had even more success. It steamed north until it sighted the Norwegian coast south of Stavanger shortly after dawn on the 12th and turned north-north-west and then north-west into rising seas. The half-flotilla was still heading north-west into the wind when, at approximately 1000, Kolbe ordered a turn to the south due to the deteriorating weather. At about this time, *G104* reported a condenser problem which reduced her sustainable speed to 25kt. Rather than send her back alone, Kolbe reduced the squadron's speed to match. At about 1130 smoke was sighted on the southern horizon, off its starboard bow. This was soon seen to be a destroyer on the

The four destroyers of the G101 class, one of which is seen here running trials in 1915, were among the most active and effective destroyers in the Kaiserliche Marine. They carried the same armament as the somewhat bigger B97 class; they had been ordered by the Argentinian navy, but were taken over by the Germans at the beginning of the war. (NHHC)

port bow of a small convoy of six merchantmen, shepherded by four armed trawlers and guarded by another destroyer on its starboard bow, which was not yet in sight. The further destroyer was HMS *Pellew* (G64) with LCdr J R C Cavendish in command of the convoy. The nearer destroyer, HMS *Partridge* (G30), was, like *Pellew*, a war-construction Admiralty-design destroyer of the Repeat 'M' class, giving her the following specifications:

HMS *Partridge*: 1916[6]

Displacement (legend):	1025 tons
Length (oa):	273ft 7in
Beam (wl):	26ft 10in
Draft (max):	9ft 5in
Powerplant:	25,000shp; three-shaft Brown-Curtis/Parsons turbines; three Yarrow watertube boilers
Speed:	34kt
Armament:	4 × 21in trainable torpedo tubes (2 × 2; centreline); 4 × Mark IV 21in torpedoes; 3 × 4in/40 QF Mark IV (3 × 1); 1 × 2pdr/39 QF Mark II

As can be seen from a simple comparison of these specifications with those of the attackers approaching from the north, on a ship-by-ship basis there was little to choose between the vessels coming into contact. The British ships had somewhat fewer, but bigger guns, while the Germans had more guns and more torpedo tubes. What was more important was that the Germans had twice as many destroyers in play.

The British did have heavier support nearby. Ever since the first, successful raid in October, the Grand Fleet had offered IICS, a squadron of armoured cruisers, as additional protection. Because there were often multiple convoys running simultaneously, as was the case this day, this covering force, this day made up of cruisers HMS *Shannon* (flag, Capt Vincent B Molteno) and *Minotaur*, and four destroyers, was in a supporting position heading south-south-east approximately 60nm south-south-west of the eastbound convoy. Molteno was well-regarded, having commanded HMS *Defence* intelligently at the 'Windy Corner' at the Battle of Jutland in June 1916.[7] Here, he was faced with a different problem, namely how to protect two convoys heading in opposite directions. His orders were first to make contact with the westbound convoy, which would be approaching Lerwick, and then attempt to catch up with the eastbound convoy before it approached the entrance to Bjørnafjorden, the southern approach to Bergen. None of the Royal Navy elements at sea this day had any specific information that German units were active in the North Sea.

Partridge had the approaching Germans in sight at about 1145 and attempted to flash a challenge, but her signal light malfunctioned and it took her ten minutes to ascertain these were enemy ships. The range had dropped to 8000yds before *Partridge* was able to signal a warning to *Pellew*, get off a report to Grand Fleet HQ at Scapa Flow, and turn away to close the convoy. Cavendish, at this point, had no good options. He ordered the convoy to turn away and scatter, while he turned *Pellew* across the convoy's bows towards the north-east, hoping to draw the Germans away. Neither Cavendish nor *Partridge*'s commander was aware of the presence of the trailing cruiser squadron; it is not clear whether, in any way, this knowledge would have altered their actions this day.

The three leading German destroyers began firing at the two British destroyers passing eastward in front of them; the fourth – *G104*, whose speed was reduced – was detailed to attack the convoy. Almost immediately, *Partridge* was dealt a mortal blow; a shell burst in the forward end of her engine room, severing her main steam lines and killing all in the compartment. Attempts to enter the space to render aid or inspect the damage were uniformly unsuccessful until after the ship had been ordered abandoned. Another shell hit her aft

This is an interesting image that demonstrates the growth progression of German destroyers after the experiment with smaller ships begun in 1911 – the so-called 'Lans cripples' – was cancelled. The three ships on the left in this image are German destroyers taken over by the US Navy as reparations after the end of the First World War. (The ship on the right is an American patrol boat.) From the left, the ex-German ships seen here are: SMS *V43*, a 1914 war mobilization plan destroyer of 852 tons; *G102*, one of the 1136-ton G101 class, as seen in the preceding image; and *S132*, a 1916 war mobilization plan destroyer of 919 tons, similar in most respects to *V43*, but somewhat longer and with larger main-battery guns. (LoC)

main-battery mount, putting it out of action. With no power to her turbines, she drifted to a stop. She was able to get off a torpedo at short range that survivors claimed hit *V100* but failed to explode; it is far more likely the torpedo missed. *V100* and *G103* fired torpedoes at *Partridge*, one of which hit her engine room and another reportedly hit further forward. Her CO, LCdr R H Ransome, saw that his ship was settling and ordered the crew off, at the same time instructing the surviving engineering officer to assure the ship sank rapidly; Eng Cdr Burt finally succeeded in entering the engine room and, groping in the darkness through rising water, he found the starboard condenser and opened its valves, releasing an inrush of water. That effort turned out to be unnecessary, as, not long after Burt followed the rest of the crew overboard and was pulled out of the freezing water by German destroyermen, *Partridge* was hit by another torpedo, which caused a massive explosion, after which *Partridge* disappeared beneath the waves. Three officers and twenty-one of her crew were rescued by the Germans; five officers and ninety-two men perished.

Pellew fought on against the odds, trying to protect the convoy. There was not much she could do. Not only was she hopelessly outgunned, but the wind blew a steady spray of icy water in the faces of her gunners as they attempted to fire on the approaching enemy, of which often only the masts and funnel tops could be seen between the wave crests. It appeared inevitable that *Pellew* would meet the same fate as *Partridge*; she was hit by an 8.8cm shell in her engine room that knocked out all but one turbine, but at what appeared to be the most desperate moment a dense rain squall enveloped the destroyer, hiding her from her pursuers. Rather than risk following her into the blinding rain, the three German destroyers turned back and joined *G104* in attacking the convoy. The result was the complete destruction of all six steamers and the four armed trawlers: HMT *Commander Fullerton* (FY3063), *Livingstone* (FY256), *Lord Alverstone* (FY3047), and *Tokio* (FY313). After dispatching the convoy and pulling seventy-five survivors from the water, the 3.HFl received word at 1240 that more enemy forces were approaching and sped off, first to the south-south-west and then, when west of Stavanger, they turned south-east so as to head into the Skagerrak.

HMS *Partridge*, a Repeat 'M' class destroyer, launched in March 1916, was part of the escort of the Pellew Convoy in December 1917. She had the misfortune to encounter the four ships of the German 3.HFl. This dramatic image shows her being hit by one of the two or three torpedoes that struck her during the attack on that convoy, most likely by the last of that set, which reportedly caused a massive explosion. (NHHC)

Molteno's armoured cruiser force had intercepted *Partridge*'s signal at 1200 and increased speed towards her location. The two old cruisers could not work up to much more than 20kt, so when *Shannon*'s radio room intercepted another, even more urgent message from *Partridge* at 1215, he sent his destroyers on ahead. The destroyers arrived on the scene at 1400 and found nothing more than flotsam and a few survivors in the lifeboats. *Pellew* had departed the scene under cover of the rain squall which hid her from the Germans and found safety in Selebjörnsfjorden, the southern entry into Bergen, where she met and was taken in tow by a Norwegian torpedo boat. There, *Pellew* hid out until she could be escorted back to Great Britain. The four destroyers of 3.HFl reached Kiel safely with no casualties.

Adm John R Jellicoe, First Sea Lord of the Admiralty, was collateral damage in this affair. It was believed at the highest level of Britain's government that Jellicoe had been privy to intelligence warning of the attacks on the Norwegian convoys in October and December and had failed to protect them adequately. This brought dissatisfaction with him over other issues, mainly the U boat campaign, to a head and led to his rather unseemly dismissal on Christmas Eve, 1917. Of more direct military consequence, the C-in-C of the Grand Fleet throughout this period, Adm David R Beatty, began assigning a division of battleships to escort the Norwegian convoys for the remainder of the war.

From the perspective of the continuing development of large destroyers, this engagement stands out as validating the notion that size did indeed make a real, tangible difference in the tactical capability of destroyers. The foul weather in which this battle was fought, which the larger German destroyers handled with little reported difficulty, certainly helped make that point; the somewhat smaller Repeat 'M' class destroyers facing them held up almost as well. The primary reason why the Pellew Convoy was overwhelmed was the simple fact that the defenders were outnumbered two-to-one and outgunned by a higher ratio. Because of better tactical positioning and more rigorous gunnery training, the Germans shot far better than the British in this engagement. As the Royal Navy's official historian put it: 'The Germans made admirable use of their advantage; and, as usual, their fire was extremely accurate and rapid.'[8]

Chapter 4

The immediate precursors: 1917–1918

WHILE MOST LIKELY SIMPLY validated, rather than influenced, by the actions that took place in 1917 described in the preceding chapter, the net effect of wartime development of destroyers was a steady increase in size as the need for greater offensive power, speed, endurance, and survivability all were served by a larger ship. As the war continued, the Royal Navy ordered a steady number of standard destroyers and leaders. Early war experience quickly showed the need for better handling in weather – Royal Navy destroyers and leaders, in particular, were found to have an overly rapid roll period, the sign of a ship with too high a metacentre – and for more firepower forward. The Parker class leaders, five of which were then under construction, were big (1660t) and had the room to accept several proposed design changes intended to address these concerns, but all those ordered as of mid-1915 had been laid down and a redesign at this point would seriously delay their entry into service. The decision to add a sixth Parker class leader in the December 1915 Seventh War Programme offered the opportunity to make major design changes.

The last ship in the previous programme, HMS *Saumarez* (G12), had been altered by raising her forecastle by one foot; it was hoped that this would help make the forward main-battery gun more serviceable in rough weather, but no one thought this was a sufficient solution. Therefore, DNC came up with a more radical solution for the sixth ship, HMS *Anzac* (G60).[1] By rearranging the command spaces, specifically

HMS *Anzac* (G60), was the last of the Parker class leaders to be laid down and, therefore, had all of the accumulated changes and improvements that had been considered for the class applied to her design. These included two main-battery guns forward with one sited superfiring on a deckhouse ahead of the heightened bridge, the reduction from the four funnels of the Kempenfelt class leaders to three with the first funnel larger because it was serving two boilers, the third main-battery gun repositioned between the second and third funnels, and the raked and flared bow. These images show *Anzac* after 1919, when she transferred to the Royal Australian Navy. (SLV – Green Collection)

moving the radio room inside the hull, shifting searchlights to the bridge wings, raising the bridge one deck, and putting the open bridge above the wheelhouse, it would be possible to move the entire forward superstructure aft 7½ft. This was, however, not considered enough of a move by fleet representatives, who complained that more had to be done to make ships' command spaces workable at high speed or in foul weather. To move the bridge further required rearranging the fire rooms. The standard arrangement of the four boilers required to generate the steam necessary for a leader-sized ship at this time was to arrange them in separate fire rooms, each boiler served by its own funnel. By merging the two forward fire rooms into one, flipping the orientation of the foremost boiler 180°, and trunking the funnels serving the two forward boilers together, the bridge could be moved aft 13ft.

This rearrangement displaced the No 2 gun of the main battery, which in the preceding Kempenfelt class had been sited between the two forward funnels and supported by the bulkhead between the first two fire rooms. It turned out there was an easy solution to the relocation of this gun that addressed a number of problems at the same time. By moving No 2 gun to a superfiring position ahead of the now-heightened and repositioned bridge, there would be more forward-firing guns and, at the same time, a new deckhouse was created that provided space for officers' accommodation and offices. To further improve weather handling, *Anzac* was given a raked bow with greater flare, a feature that had been tried out experimentally on a Repeat 'M' class ship earlier with great success. This design change was immediately approved and was ordered applied to the five ships of the Parker class already building. Depending on how far along they were, not all of the changes could be applied to all of the six ships; only *Anzac* received the full redesign.

In the 'mainstream' of Royal Navy destroyer development, the 'R' class followed the Repeat 'M' class in February 1915, differing mainly in having geared turbines. These ships, which would enter service in late 1916, boasted a top speed of 36kt, faster than the Parker class destroyer leaders then being ordered. This pushed the Admiralty, in early 1916, to ask for a design of a leader type specifically for flotillas of 'R' class destroyers. These ships, which became known as the 'V' class leaders, had geared turbines and three boilers, like the 'R' class destroyers, but had four main-battery guns superfiring both forward and aft. At 1188t legend displacement, they were less than three-quarters the size of the preceding leaders and, in service, proved unable to achieve the 36kt speed desired to keep up with 'R' class ships. Fortunately, the Admiralty had been hedging its bets. At the same time that the 'V' class leaders were being designed in early 1916, Thornycroft had proposed to the Royal Navy a design for a class of leaders that would incorporate all the design advances made in *Anzac*. These ships, named the Shakespeare class leaders, had the following specifications:

HMS *Shakespeare*: 1917[2]

Displacement (legend):	1554 tons
Length (oa):	329ft 1in
Beam (wl):	31ft 9in
Draft (max):	12ft 6in
Powerplant:	40,000shp; two-shaft Brown-Curtis geared turbines; four Yarrow watertube boilers
Speed:	36kt
Armament:	6 × 21in trainable torpedo tubes (2 × 3; centreline); 6 × Mark IV 21in torpedoes; 5 × 4.7in/45 BL Mark I (5 × 1); 1 × 3in/45 20cwt QF HA Mark I; 2 × 2pdr/39 QF Mark II (2 × 1)

Part of the impetus to build these bigger leaders and the larger standard destroyers of the 'V' class that followed were the

The Royal Navy's Shakespeare class leaders, proposed in early-1916 by Thornycroft, embodied all advances found in *Anzac* in a somewhat longer hull. The most notable difference in appearance is the reduction from three to two funnels. Less obvious, in this image of HMS *Spenser*, probably dating from 1923, but of more import militarily, was the change from 4in/45 to 4.7in/45 as the main-battery weapon, in an effort to give these ships more firepower. Note the splinter mats around the open bridge and bridge wings, intended to offer some minimal protection against shell splinters and small-arms fire. (SLV – Green Collection)

Royal Navy mainstream destroyer development incorporated many of the same features embodied in *Anzac* and the Shakespeare class leaders in a smaller hull and retaining the 4in/45 main-battery gun, at least through the 'W' class destroyers. This image shows HMAS *Voyager* (D31), seen here after her transfer to the Royal Australian Navy in 1933. (SLV – Green Collection)

persistent rumours reaching England of extremely large and heavily-armed destroyers under construction in Germany at this time.[3] These rumours reached such influential figures as Jellicoe and led to steady pressure being applied to push both the size and weight of armament of British destroyers. Hence the increase not only in the number of main-battery guns between the Parker class and the Shakespeares, from four to five, but also the increase in calibre from 4in to 4.7in. Based on reports that the Germans were arming their new destroyers with 5in guns, Jellicoe had requested a similar calibre for these new ships, but there was no gun of that size in the Royal Navy's inventory; the closest available weapon – the 4.7in/45 BL Mark I – was an army gun hastily converted with a naval mount that entered service in 1918.

Only two of this class were completed in time to see service in the First World War; three others were completed after the war's end and two more were cancelled incomplete. A second class of almost identical ships were ordered from other yards, the Scott class, to an Admiralty design based on Thornycroft's, but these yards lacked the skill of a builder that specialised in destroyers and none of the eight ships of the Scott class achieved the speed or the endurance of the Shakespeares.

The more interesting side-effect of the experience of designing the 'V' class leaders, followed immediately by the Shakespeare class, is that it became clear to the Admiralty that in order to counter the possibility that the Germans might indeed be building very large destroyers, it should build a new class of standard destroyers with the same size, speed and armament as the 'V' class leaders. Thus, in June 1916, it was decided that, in order to save time, the next programme of destroyer orders, a total of twenty-six ships from eleven yards, would be very nearly repeats of the 'V' class leader design. Almost the only differences would be the deletion of some of the officers' accommodation built into the leaders and the reversion to 4in guns for the main battery, but with the higher-rate-of-fire Mark V model. DNC wanted to fit these ships with the same triple torpedo tubes mounted on the Shakespeares, but they were not available in sufficient numbers and, rather than delay the start of construction, the initial ships were ordered built with double tubes, but with their decks strengthened so that, when triple tubes became available, they could simply be substituted. As built, the 'V' class destroyers had the following characteristics:

HMS *Vanoc*: 1917[4]

Displacement (normal):	1325 tons
Length (oa):	312ft
Beam (wl):	29ft 6in
Draft (max):	10ft 6in
Powerplant:	27,000shp; two-shaft Brown-Curtis single-reduction geared turbines; three Yarrow watertube boilers
Speed:	34kt
Armament:	4 × 21in trainable torpedo tubes (2 × 2; centreline); 4 × Mark IV 21in torpedoes; 4 × 4in/45 QF Mark V (4 × 1); 1 × 3in/45 20cwt QF HA Mark I

The basic design pattern established with the 'V' class – four single-mount main-battery guns superfiring at the forecastle and stern, two centreline torpedo tube mounts, and two (later one) funnels would, with few exceptions, become the standard layout of Royal Navy destroyers well into the Second World War.

The ships of the 'W' class and of the preceding 'V' class were the oldest destroyers retained by the Royal Navy long enough to serve in the Second World War. As in other navies, many of these old destroyers were modified for specialised tasks, as was the case with HMS *Vimy* (D33, later I33) seen here, which served as a minelayer, with mine rails on either side abaft her second funnel. 'A' and 'Y' guns of her main battery were removed to compensate for the weight of the mines.

Growth elsewhere, as well

Other nations involved in the First World War built destroyers and most saw the same general increases in size and capability as the British, for the same reasons. As noted in the preceding chapters, the Italians built three classes of *esploratori leggeri* initially to accompany the German Mediterranean squadron and later to counter the Austrian scout cruisers; all would later be redesignated as destroyers (*cacciatorpediniere*).

Japan was another nation that continued to build destroyers during the war. As described previously, they had by this point achieved so much excess shipbuilding capacity that they were able to build a flotilla for France while continuing to expand their own fleet. The Imperial Navy followed the four large ships of the Isokaze class, laid down in 1916, with two ships of the Tanikaze class, which introduced the 21in (53.3cm) Type 6 torpedo, but which otherwise were largely repeats of the Isokazes. A far more substantial change in design philosophy came at the war's end, when the IJN decided to follow the Tanikazes with a very different design specifically intended to handle long-endurance open ocean missions, which called for exceptional rough weather handling and the ability to maintain speed in all conditions. To achieve this goal, the Japanese looked to German pre-war designs, which had gained a good reputation for their seaworthiness. The features the Japanese designers found particularly worth adopting were the high forecastle with the forward main-battery gun sited there, a well deck immediately aft of this where the forward torpedo tubes were located, and the bridge sited aft of the well deck to provide the captain as much protection as possible from weather coming over the bow. The resulting design had the following specifications:

***Minekaze*: 1920**[5]

Displacement (normal):	1345 tons
Length (oa):	336ft 6in
Beam (wl):	29ft 8in
Draft (max):	9ft 6in
Powerplant:	38,500shp; two-shaft Parsons geared turbines; four Kampon boilers[6]
Speed:	39kt
Armament:	6 × 53.3cm trainable torpedo tubes (3 × 2; centreline); 6 × Type 6 53.3cm torpedoes; twenty mines; 4 × 12cm/45 3rd Year Type (4 × 1); 2 × 7.7mm MG (2 × 1)

The Imperial Japanese Navy built their own large destroyers, inspired by the example of the large destroyers being built by the Germans, which had earned a solid reputation for being able to handle the harsh weather of the North Sea. This image shows *Shimakaze* of the Minekaze class, apparently soon after commissioning in 1920. The 'German-like' features, some of which the Germans themselves had abandoned during the war, included a high forecastle and a well deck before a tall bridge.

By the time the Japanese adopted many of these design features, in particular the well deck and torpedo tubes forward of the bridge, the Germans had already moved on to a structurally superior design, carrying the forecastle uninterrupted all the way back to the bridge and the British had adopted a much better arrangement for a four-gun main battery for destroyers, with superfiring guns forward and aft. Nevertheless, the Japanese were well pleased with the Minekaze class, of which they built fifteen, all of which served into the Second World War, many converted to patrol boats or to other support functions. Indeed, they built two more classes of destroyers to this same basic design, abandoning it only when they decided to build the first of their 'super destroyers', first proposed in 1922, but not laid down until 1926.

As just mentioned, German destroyer design continued to evolve during the First World War. Very happy with the success of the large destroyers of the B97 and G101 classes, the Germans designed and laid down three classes of even larger destroyers, a total of twelve ships, all with design displacements over 2000t. All were intended to be armed with 15cm/45 guns, giving them the firepower to engage enemy cruisers, although their lack of armour would have made that an unwise choice in all but the most favourable of circumstances. As it was, these ships were only laid down in early 1918 and only a few of them were completed before the end of the war. One of those few, *S113*, had the following specifications:

SMS *S113*: 1918[7]

Displacement (design):	2060 tons
Length (oa):	347ft 9in
Beam (wl):	33ft 6in
Draft (max):	11ft 1in
Powerplant:	45,000shp; two-shaft Schichau turbines; four Navy boilers
Speed:	39kt
Armament:	4 × 60cm trainable torpedo tubes (2 × 2; centreline); 4 × H8 60cm torpedoes; 40 mines; 4 × 15cm/45 Tbts KL/45 (4 × 1)

Two ships of this type were surrendered at the end of the war and passed into the hands of Allied navies as reparations. *S113* went to France, where she was renamed MN *Amiral Sénès*. The French found what the Germans had during *S113*'s trial, that 15cm guns were too heavy and caused significant instability. In fact, the French were impressed with the handling of these guns, which featured a Krupp-designed horizontal sliding wedge breech block, which, even with separate ammunition, allowed these guns to sustain a rate of up to 8rpm.[8] When *S113* was transferred to the MN, no 60cm torpedoes were

Probably the truest precursors of the first super destroyers of the 1920s were the 2060-ton ships of the SMS S113 and V116 classes, of which only the lead ships of each class were completed before the end of the First World War and Germany's surrender, caused all destroyer construction to end for sixteen years. This image shows *S113* running trials in 1918. What stands out immediately is the size of the 15cm/45 main-battery guns compared to the large hull (for a destroyer) on which they were carried. This was, in effect, the main-battery armament of a small cruiser on the hull of a large destroyer; it is no wonder these ships had stability issues throughout their careers. This ship was ceded to France at the end of the war and, renamed MN *Amiral Sénès*, serving until 1936.

SMS *V116*, near-sister to *S113*, was ceded to the Regia Marina at the end of the First World War. She is seen here in Italian service entering the inner harbour at Taranto. Renamed RN *Premuda*, she served until 1939. (USMM via Enrico Cernuschi)

included and the French were unable to find the design of the German torpedo, so the 60cm torpedo tubes were replaced by standard German 50cm torpedo tubes for some weight savings. As such, *Amiral Sénès* served as leader of French Atlantic destroyer flotillas well past 1930 and was not written off until 1936.

V116 was also completed and was transferred to the RM in 1920. She was renamed RN *Premuda*. Like *Amiral Sénès*, she was refitted with 50cm torpedoes and retained her 15cm guns at least until 1932. Unlike her French counterpart, the Italians found her endurance and handling inadequate; she led a much less active post-war life, acting mainly as a cadet training ship until stricken in 1939. The Italians also took over SMS *B97*, which they renamed RN *Cesare Rossarol*. Unlike *Premuda*, *Rossarol* had seen continuous active war service in the Kaiserliche Marine and was definitely 'war-weary' by the time she was taken over by the Italians. In the RM she was used mainly as an experimental platform; in 1932, her after boiler was removed and replaced by a Sperry gyro-stabiliser. Like *Premuda*, she was stricken in 1939.

The importance of these ships was not their actual abilities or accomplishments, but what they represented as models for future development. The Italians, as described previously, had already been building large destroyers, which they designated *esploratori leggeri*, to counter Austrian scout cruisers. Experience with the German war prizes simply reinforced the

Ordered by the Russians, but taken over at the outbreak of the First World War, the SMS B97 class destroyers were smaller than the S113 class ships, but larger than the mainstream G101 type; available from 1915 on, they saw considerable action during the war, so that when *B97* was ceded to Italy after the Armistice, becoming RN *Cesare Rossarol*, seen here, she was already considered 'war-weary'. The grid-like device in front of the forward 10.5cm KL/45 main-battery gun is for bore-sighting the gun, aligning it with its sighting mechanism. (USMM via Enrico Cernuschi)

positive attitude towards large destroyers, with their ability to carry heavy weaponry and operate in any weather. The French had, on the other hand ceased all new construction at the outbreak of the First World War, but had kept a close eye on Italian, as well as German, developments and, at the end of that war, resumed planning for what they now determined would have to be larger and more powerful destroyers to maintain their position in the Mediterranean, as well as facing the Royal Navy. As early as March 1919, the French Naval General Staff (EMM) submitted a plan for the construction of destroyers of much greater size than previously considered.[9]

. . . And trailing the field . . .

The Americans had been late entering the First World War and in ramping up their economy for war production. As described above, the Americans had been late deciding to build 1000t class destroyers, not placing the first of the Cassins into commission until 1914, but, once energised, the Americans proved to be a massive engine of warship construction. After the pre-war Cassins, three similar early war classes added eighteen more ships, but with these the Americans were only just getting started. In mid-1917, the first of six Caldwell class ships were laid down; these were flush-decked ships with a variety of different engineering layouts tried out as the Americans sought the powerplant that would provide the best combination of speed and endurance. The flush deck increased hull strength and would, it was hoped, decrease rolling and pitching of the ship in heavy weather.[10] Forecastle height forward was maintained from previous classes, so freeboard amidships increased, which again should have had benefit, but in practice the ships had shallow draft and were excessively wet and slower than expected because their screws were often not completely submerged. The yards building these ships – five yards building the six ships – were given considerable latitude in interpreting the supplied plans and the ships differed widely in detail. The BuC&R design called for three shafts – the middle shaft driven by a high-pressure turbine with a geared cruising turbine clutched to it – and three funnels, but only two ships were built with this arrangement of turbines; the rest had two geared turbines of various makes and three had four funnels.

With its complex system in which the authorisation and funding of warship construction were separate acts often occurring in separate fiscal years, it was not surprising that the wartime surge in American destroyer construction was slow to gain momentum. While it is common to refer to US Navy war-production destroyers – generically known as 'flush-deckers' or 'four-pipers' – as if they were a uniform group of ships, in fact they comprised 267 ships in two distinct classes, with multiple subclasses. The first 111 were Wickes class ships, built to two basically similar designs, one generated by Bath Iron Works (Bath, Maine), the other by Bethlehem Steel (Quincy, Mass). Although all Wickes class ships were supposed to have two shafts with a geared turbine on the port shaft, a shortage of gear-cutting machinery led to some ships being delivered with two direct-drive turbines. A general specification for a Wickes class destroyer is as follows:

USS *Wickes* (DD 75): 1918[11]

Displacement (normal):	1090 tons
Length (oa):	314ft 4in
Beam (wl):	30ft 11.5in
Draft (max):	9ft 2in
Powerplant:	24,200shp; two-shaft Parsons turbines (with a geared turbine on the port shaft); four Normand boilers
Speed:	35kt
Armament:	12 × 21in trainable torpedo tubes (4 × 3; sided); 12 × Mk 8 21in torpedoes; 4 × 4in/50 Mk 9 (4 × 1); 2 × 1pdr AA mg

The first of the American mass-produced 'flush-deckers' (also known as 'four-pipers', although some preceding classes also had four funnels and the immediately preceding Caldwell class had flush decks), was USS *Wickes* (DD 75), name ship of her class, seen here in the early 1920s. (NHHC)

This stern view shows another Wickes class flushdecker, USS *Evans* (DD 78), immediately after the end of the First World War, probably at New York City, 27 December 1918, in what appears to be a victory celebration. The cage masts that appear to rise from her amidships belong to a battleship docked beyond her; there are three more battleships visible in the distance to the left. Two triple banks of torpedo tubes can be seen on the near side in this image. (NARA)

The destroyers of the Wickes class suffered from poor endurance, a problem acutely embarrassing to the US Navy, which saw its role as a force of power projection spanning great distances, particularly across the Pacific Ocean. To address this issue, the Wickes class was succeeded in shipyards by the Clemson class in mid-1918, with similar dimensions and appearance, but 35 per cent more fuel bunkerage. With these longer-legged escorts, a US Navy squadron steamed from the west coast of the United States to Australia in 1925 as a gesture of goodwill and, no doubt, to send a message to the Japanese. Eight battleships were accompanied by cruisers and destroyers, including USS *Stoddert* (DD 302) and *John Francis Burnes* (DD 299), seen in Sydney harbour sometime between 23 July and 6 August 1925. (SLV – Green Collection)

Although the US Navy began building new destroyers in 1933, a great many flushdeckers, particularly of the Clemson class, remained in service up to and through the Second World War. Most had been converted to perform duties other than general destroyer functions by the time America entered that war, but some, like USS *Noa* (DD 343) seen here at Annapolis, MD, on 1 May 1941, sporting an experimental camouflage scheme, had not. (NHHC)

Even among the flushdeckers that retained the destroyer designation during service in the Second World War, such as USS *Tarbell* (DD 142), seen here on 2 July 1943 at New York City, many were given an extensive conversion to serve in an escort role. This conversion included removing one boiler and its associated funnel to increase fuel stowage, replacing the main battery of four 4in/50 guns with six 3in/50 guns, and the landing of two banks of torpedo tubes to compensate for the addition of depth charge throwers and 20mm AA guns. (NARA)

As soon as the first of the Wickes class ships entered service it became obvious that there were great differences in performance, particularly in endurance, between the ships from different yards. Ships built to the Bethlehem design, which meant those built at that company's Squantum and San Francisco yards – a total of fifty out of 111 Wickes class ships – were particularly prone to poor endurance, although some Bath design ships also exhibited this problem as well. To address this issue, BuC&R produced a new design for ships starting with USS *Clemson* (DD 186), which applied the simple expedient of modifying the Wickes class design by increasing fuel tankage by 35 per cent. Additionally, both shafts were now driven by geared turbines. Nevertheless, many of the 156 Clemson class ships suffered from mechanical issues, including disappointing speed and endurance. This remained a particular problem with ships built at the two Bethlehem yards and at the US Navy's Mare Island Navy Yard.

Despite the involvement of seven private shipyards and three navy yards in the building of these ships, the pace of the programme had built up more slowly than the US Navy had hoped; only approximately seventy of the Wickes class ships had been commissioned by war's end. The end of the war did not cause any mass cancellation of construction orders (as would be the case at the end of the Second World War); all of the ships then building were completed and only six of those not yet started were cancelled.[12] Some of the Clemson class ships were not launched until well into 1921. The decision not to cancel the construction of more than 200 destroyers would have impacts felt by the US Navy for years to come.

The extra time and reduced pressure that came with war's end did not necessarily solve all of the quality issues that plagued these ships. A large number of ships of the two classes were completed with Yarrow boilers, which tended to wear out rapidly in service; in 1929, it was decided that the expense of reboilering these ships would be too great and that they would simply be disposed of.[13] In all sixty-six ships, all built by one of the three Bethlehem yards, had the Yarrow boilers, but six had been lost in September 1923 in the Honda Point Disaster, when a squadron of destroyers ran aground near Santa Barbara, California in heavy fog, leaving sixty to be written off. Six of these were Wickes class ships launched in 1918; the remaining fifty-four were Squantum- and San Francisco-built ships of the Clemson class, including some launched as late as April 1921.

One of many common conversions of flushdeckers was to fast minesweepers. This image shows four such converted ships, comprising MinDiv4, three of which can be identified – from the front, they are: USS *Perry* (DMS 17, ex-DD 340), *Trever* (DMS 16, ex-DD 339), and *Wasmuth* (DMS 15, ex-DD 338); the last ship in the row is almost certainly *Zane* (DMS 14, ex-DD 337), which was the fourth ship in MinDiv4 – at Pearl Harbor on 17 January 1941. Note the unusual roundel insignia at the bow of each of these ships. (NARA)

Perhaps the most immediately useful of the flushdecker conversions was that of USS *Manley* (APD 1, ex-DD 74), a Caldwell class destroyer converted into a fast transport, seen here in 1942. Two boilers were removed to make room for more fuel and for accommodation for the troops and their stores to be carried. Large davits were fitted to allow the carriage of four LCPLs (Landing Craft Personnel – Large, aka Higgins Boats), replacing the torpedo tubes. These ships proved so useful that a total of thirty-six flushdeckers received similar conversions. (NARA)

Chapter 5

The first of the true super destroyers – Washington swings and misses: 1922–1928

THE QUESTION OF WHICH ships were the first super destroyers and which nation built them is one for academics to debate. It depends entirely on how one defines the term 'super destroyer'; no simple, precise, universally agreed-upon definition exists. As mentioned above, the first British Tribal class of 1910 might be considered the first super destroyers because of the extent by which they exceeded previous ships in size, speed and firepower. Using that as a definition of the type, it could be argued that any of a number of other ships in the years after 1910 through the end of the First World War could equally be so designated. For the purposes of this book, we will adopt a working definition of the super destroyer designation that describes a ship of at least 2000 tons standard displacement, with a main battery of at least 4.7in (120mm) guns, and a speed of at least thirty-five knots. An upper limit on main-battery gun calibre of perhaps 5.9in (150mm) and 3000 tons displacement should be included just to ensure that ships such as the Austrian scouts of the Admiral Spaun and Helgoland classes are not considered in this discussion. Not every ship that is included in the following discussion of super destroyers will meet every one of these criteria; some latitude is given, particularly in terms of displacement, in order to include ships that clearly do belong here.

With this working definition to hand, it is possible to conclude that the first ships to meet these criteria were the wartime German S113 class destroyers and their sister classes, but the author is reluctant to grant them that honour for a number of reasons. First and foremost is the fact that they were a very unbalanced design, mounting 15cm main-battery guns on a displacement just barely topping 2000 tons, making for ships that were inherently unstable. Yet the two ships of this design that were completed did see long service in the navies to which they were assigned post-war, so they cannot be considered unsuccessful. Nonetheless, I prefer to reserve the distinction of being labelled the first super destroyers for the Italian Leone class leaders first ordered in 1917. The intent had been to lay down all five ships in this class at the Ansaldo yard at Genoa immediately; however, wartime shortages of steel caused that order to be cancelled, but not forgotten.[1]

The Leones, as was the case with the preceding Poerio, Mirabello, and Aquila classes, were conceived in response to the Austrian scouts as part of the 'Adriatic arms race' between Austria and Italy. They, however, represented a significant growth in size and displacement over *Aquila*, although, since the Aquila class had been generally disappointing, the Leones' design was based on the Mirabellos'. Wartime experience with the 6in main-battery guns in *Aquila* and her first two sisters showed that this weapon and its ammunition were too heavy and that firing it put too much stress on a lightly built destroyer hull. The last ship of the Aquila class – RN *Falco* – which was not completed until January 1920, was fitted with a main battery of five newly designed 120mm/45 Schneider/Canet/Armstrong Model 1918 guns in one single mount and two twin mounts. This gun proved to be very successful, but not necessarily the mountings, which proved troublesome. After two of the Aquila class were turned over to Romania in July 1920 to compensate for the confiscation of two ships in 1915, the remaining ship – the name ship of the class, RN *Aquila* – was taken in hand and her three 6in guns were replaced by a like number of the new 120mm guns and one was taken from *Falco*, leaving both remaining Aquilas with four of the lighter gun.

The end of the war allowed the plans for the Ansaldo ships to be revisited. With the pressure of war past and the immediate threat from Austria now eliminated, the need for five new large *esploratori leggeri* was called into question for economic reasons. It was not, however, as if the horizon was cloudless for the Italians. Even before the end of the war, the British had signalled their intent to re-establish total dominance in the Mediterranean. Their rivals, on a global scale, were first of all the Americans, whom the British felt had taken advantage of their late entry into the war to overtake Great Britain's dominant position in world trade. The British reaction took many forms. For example, referring to American production of merchant tonnage outstripping Great Britain's, the British Comptroller-General of Merchant Shipping, William J Pirrie, was quoted on 7 September 1918 as stating:

> If owing to the excess of our devotion to the common cause we cannot make up the wastage of our tonnage and are left when the war ends with a merchant fleet so reduced by war losses as to be incapable of meeting our needs as a great maritime country, the position will be disastrous.[2]

This feeling of impending doom led to talk of introducing a system of Imperial Preference that would lower tariffs on trade within the British Empire, to the disadvantage of nations on the outside, which raised hackles in Washington.[3]

How, the reader might ask, might this impending post-war dust-up between the Americans and the British over mercantile trade in any way involve the Italians and lead, even indirectly, to the construction of the first super destroyers? The answer starts in 1915 with the inducements that led to Italy's entry into the war in the first place. Italy had been nominally allied to the Central Powers – Germany and Austria – before the First World War began, but had managed to avoid taking sides at the outbreak of hostilities.[4] (Italy was able to stay neutral because her alliance with Germany and Austria required her to join them at war if either were attacked, which was clearly not the case in 1914. The Germans seemed willing to accept this graciously; the Austrians made it clear they were very displeased. Since the creation of the Triple Alliance in 1882, the Italians and Austrians had been at odds on multiple occasions; their interests frequently clashed in the Adriatic – over the Trentino region – and in the wider Mediterranean region – for example, in the degree to which Italy supported Austrian attempts to colonise the Sinai Peninsula in 1904 and 1906.) A solid majority of public opinion in Italy favoured remaining neutral, especially after the war on the western front rapidly devolved into a stalemate.

However, Italy, as a neutral, was in a precarious position. Should the Central Powers prevail in the war, Italy would be isolated with an angry Austria at its north-eastern border. Should, on the other hand, the Entente win, they would lose out on the opportunity to profit from the defeat of the Austrian and Ottoman Empires. With this in mind, Italy approached London in March 1915 with a list of territories it would want to annex in return for joining the Allied side. Additionally, the Italians asked for a British loan of £50M and assurances that her position in the post-war Mediterranean 'balance of power' would be recognised. It took over two month of haggling to reach an agreement, complicated because the British were simultaneously attempting to lure Bulgaria into joining the war against Austria, offering them some Serbian territory, in recompense for which, in turn, they offered the Serbs part of the Dalmatian coast, which was on Italy's list of desired accretions. In the end, the Italians got almost all they wanted in the secret Treaty of London signed on 26 April 1915; besides the loan and the 'guarantee' of a post-war Mediterranean power balance, the Italians were to get the Trentino and Alto Adige (Süd Tirol) in the north, Gorizia, Trieste, and Istria in the north-east (minus only Fiume (Rijeka)), and the northern half of the Croatian province of Dalmatia, plus most of the Dalmatian islands.

All this set the stage for what happened at war's end to the consternation of the Italians. As the war was winding down and it was becoming clear that promises made years earlier would need to be fulfilled, the British, for multiple reasons, found themselves unwilling or unable to follow through. In the case of Italy, the British found themselves in a 'double bind'. With the Americans now a partner with a strong, possibly even equal, say in shaping the post-war world, the positions taken by their president, Woodrow Wilson, could not be ignored, and the first of Wilson's 'Fourteen Points' he proposed for the peace settlement ending the war, called for:

Initially ordered in 1917, but not started then due to wartime shortages, three of the five *esploratori leggeri* of the Leone class were re-ordered by the Regia Marina in October 1920. As can be seen in this image of RN *Leone* under construction at the Ansaldo yard at Genoa, one way that the Italians determined to increase the firepower of their large destroyers was to borrow a practice commonly employed in the design of capital ships and large cruisers, namely the use of twin mounts for the main battery, a novel idea for destroyer-sized ships at this time. (Fulvio Venturi)

'Open covenants of peace, openly arrived at, after which there shall be no private international understandings of any kind, but diplomacy shall proceed always frankly and in the public view.'[5] While this did not specifically reject pre-existing secret pacts, such as the Treaty of London, it was quite clear that Wilson had no intent to honour them without question. That was especially true in this case, in that the ninth of Wilson's points stated: 'A readjustment of the frontiers of Italy should be effected along clearly recognisable lines of nationality.'[6] Even allowing for the vagueness of this declaration, much of the territory promised Italy by the Treaty of London would be difficult to recognise as 'clearly' Italian, particularly the Istrian peninsula and Dalmatia, where, in the last pre-war

RN *Pantera*, a super destroyer of the Leone class, is seen soon after her completion in 1924. Another novel feature introduced by the Italians in this class, again borrowed from common usage in much larger ships, was the installation of a large (and heavy) fire control director atop *Pantera*'s bridge. (USMM)

census, the Italian-speaking populations were 50 per cent and 10 per cent respectively.

But this was only one of many complicated diplomatic corners into which the British had painted themselves with promises made during the war, and was probably nowhere near the most potentially embarrassing. Before the war Italian trade with the outside world had been dominated by imports of British raw materials, particularly coal, and finished goods, carried on British bottoms, but the war disrupted this state of affairs. The Americans stepped in and took advantage of this situation, much to the annoyance of Great Britain. Intent on re-establishing the status quo ante by any means necessary, on 3 September 1918 the British began detaining Italian- and American-flagged shipping bound for Italy at Gibraltar. The condition for release of the detained ships was the ceding of partial interest in multiple state-owned enterprises in Italy, most particularly railways. So dependent was Italy on those imports that she was able to hold out only a week before agreeing to terms.[7]

All this historical background is by way of explaining why, despite having been on the winning side of the recent Great War and with nothing to fear now from her former primary naval rival, the Italians still felt the need to proceed, on 30 October 1920, to re-order three of the five ships of the Leone class originally ordered from Ansaldo (Genoa) in 1917, with the following, revised specifications:

RN *Leone*: 1924

Displacement (normal):	2195 tons[8]
Length (oa):	372ft 1in
Beam (wl):	33ft 10in
Draft (max):	11ft 10in
Powerplant:	42,000shp (design); 45,667shp (actual); two-shaft Parsons turbines; four Yarrow boilers
Speed:	34kt (design); 33.7kt (actual)
Armament:	6 × 450mm trainable torpedo tubes (2 × 3; centreline); sixty mines; 8 × 120mm/45 Model 1918/19 (4 × 2); 2 × 76.2mm/40 Ansaldo (2 × 1); 2 × 6.5mm MG (2 × 1)

Compared to the two Aquilas remaining in Italian service, the Leones were one-third larger, somewhat slower in actual service, but significantly more heavily armed due to the adoption of triple torpedo tubes and twin mounts for the successful Schneider/Canet/Armstrong 120mm gun. (This gun mount, like many Italian gun mounts that followed during the inter-war period, was characterised by the relatively narrow spacing between the two gun barrels; they were only 22.8in apart. This led to persistent dispersion problems caused by muzzle blast interference during salvo fire. It was ameliorated only after considerable effort by introducing a delay loop in the firing circuit, to prevent the two barrels from

Alongside the Leones, the Regia Marina produced three classes of smaller, 1000-ton *cacciatorpediniere*, intended to restock the flotillas now filled by generally smaller pre-war and war-construction destroyers. The third of those classes was the 1070-ton Turbine class, of which the name ship is seen here, probably on 5 May 1935 at Naples (Napoli). These smaller ships carried fewer and smaller guns than the *esploratori leggeri*, but otherwise had similar characteristics, including the large fire control director. (NHHC)

firing at exactly the same time, and reducing the size of the powder charge.)

The contracts for three of the five Leones were confirmed in 1921, mainly because machinery and weapons had already been manufactured and the governments of those years wanted to alleviate social unrest by granting some works to the yards. The RM was satisfied with the Leones, but felt no immediate need to build more super destroyers of similar size. Instead, the Italians followed them with a much smaller ship, which would be less expensive to build in larger numbers and be better suited to the fleet defence and flotilla attack roles. This was the Sella class laid down starting in 1922, four ships displacing just under 1000 tons and carrying half the main battery of the Leones. As such, they fitted the needs of the RM quite well and the next twelve destroyers built by the Italians, laid down in 1924 and 1925, were only slightly enlarged repeats of the Sellas. It was not until 1926, when, in response to the French *contre-torpilleurs* of the Jaguar class (described below), the Italians responded with their own new super destroyers, the Navigatori class ordered in 1926, which are covered in a later chapter.

In service, the Leones proved to be popular for their size and stability. They served on into the Second World War; by 1936, they were considered second-line units and were designated for colonial service, to be home-ported at Massawa, Eritrea. After Italy's entry into the war in 1940, the three Leones, which formed DesRon5 (*5a* Squadriglia Cacciatorpediniere), were cut off from support from home, leaving them without fuel and maintenance supplies. When it appeared Massawa was about to be captured by British forces, the Italians were left with little choice but to conjure a suicide plan, a pair of missions that would likely lead to the loss of the ships. If after all that the ships were still afloat, they were to be scuttled. The first mission, an attempted raid on Suez (As Suways), led to the loss of *Leone*, which ran aground on an uncharted rock on 1 April 1941, causing that mission to be cancelled. The second raid, two days later, aimed at Port Sudan (Bur Sudan) was intercepted by British aircraft that sank two smaller destroyers, but left the two remaining Leones afloat. Rather than return to Massawa and probable capture, RN *Pantera* and *Tigre* headed across the Red Sea towards the Saudi shore, where they were scuttled and their crews interned.

More supers?

The Leones would be far more important in the impact they had on other navies than they had on the RM. The navy upon which they had the most immediate impact was the MN. Even more than had been the case with the Italians, the French shipyards had been effectively paralyzed by wartime restrictions, being limited almost entirely to repairs and maintenance. They had been unable to lay down new destroyers for the duration of the First World War. Not that the enforced layoff did not give the naval architects at the STCN, the design bureau of the État-Major de la Marine (EMM – Naval General Staff) a chance to look at destroyer development in the rest of the world's navies and sharpen their pencils.

Designs dating from 1914 languished on drawing boards as the war dragged on.[9] These had called for ships in the 1500-

The French delayed building large *contre-torpilleurs* due to the approaching Washington Naval Disarmament Conference, but did proceed with their plans, delayed by the war, to build new *torpilleurs d'escadre*. The lead ship of the first class of these new squadron destroyers was MN *Bourrasque*, seen here in the Mediterranean, probably in 1930. These ships were large, at 1319 tons, but were rather conventionally armed with single-mount 130mm/40 guns with a high trunnion (in order to achieve 36° elevation) which led to a slow rate of fire. (NHHC)

ton range, mounting two of a newly designed 14cm/25 main-battery gun and eight 45cm torpedo tubes. This design was revisited in April 1917, after the French had first-hand experience operating alongside Italian *esploratori leggeri* in the Adriatic and Royal Navy destroyer leaders in the North Sea. They found much to admire in both ship types, in particular their generally good seakeeping and large command spaces, but thought they could have been better designed to handle the often rough North Atlantic weather. When the MN began the process of designing new destroyers again in earnest after the end of the war, they benefitted from additional experience with German war prizes, particularly the former SMS *S113*, despite its issues with inherent instability due to the size and weight of the large main-battery guns.

By early 1920, the French economy had recovered from the war to the point that France was once again ready to consider seriously proposals for new destroyers. A proposal was put forward to the French Parlement in April for the construction of two large *contre-torpilleurs* of 1750 tons, armed with five 100mm/45 single-mount guns of pre-war design.[10] These were intended to be experimental prototypes, testing out new technologies such as geared turbines, but the legislators would have none of it and turned the request down on the grounds that they were underarmed for their size. All of this led STCN to increase the normal displacement for the proposed new ships to greater than 2320 tons and to offer six different armament schemes ranging from four 138.6mm

MN *Ouragan*, another ship of the Bourrasque class, is seen from the starboard quarter, probably in 1934. Her 75mm anti-aircraft gun, which had been sited on the amidships deckhouse, has been removed and replaced by a pair of single 37mm/50 mounts, one on each side, alongside the after bank of torpedo tubes. Note the rails aft to allow depth charges to be rolled off over her sloping stern. (NHHC)

main-battery guns to eight 130mm guns.[11] The *contre-torpilleur* design approved by the EMM on 25 May 1921 would mount six or seven of a new 130mm/40 gun – single-mounts forward, aft, and amidships with superfiring positions forward and aft, one or both of which would be twin-mounts. At the same time, EMM approved a design for a smaller 1430-ton *torpilleur d'escadre* (squadron torpedo boat) to make up the main component of the destroyer flotillas. These would become the Bourrasque class, whose construction was approved in the same 1922 programme as the larger *contre-torpilleurs*. The plans for the large *contre-torpilleurs* were put on hold, however, by the calling of a conference by the Americans to prevent a new naval arms competition, such as had preceded, and, in the eyes of many, was at least partially to blame for, the outbreak of the First World War. On the other hand, the twelve Bourrasques, which were laid down in 1923–4, were followed by fourteen nearly identical follow-on *torpilleurs d'escadre*, funded in the budgets for 1924–6.

What appears to be a typical day's activity at the Marine Nationale base at Toulon, sometime between 1931 and 1934. At the left, one of the later *contre-torpilleurs*, MN *Albatros*, is on her way out of the harbour. The aircraft carrier MN *Béarn* is alongside a pier in the centre. On the right is one of the L'Adroit class *torpilleurs d'escadre*. These were improved Bourrasques, with a newer model 130mm main-battery gun. (NHHC)

Smoke on the Potomac, but no flames

The naval arms limitation conference convened in Washington, DC on 12 November 1921 was not caused by concern at the growth in size or numbers of destroyers.[12] The conference had been called primarily to limit the numbers of capital ships then being built or planned, and the Americans naïvely assumed that the ratios of tonnage applied to capital ships would simply carry over to 'Auxiliary Surface Combatant Craft'.[13] The specific opening proposal of the American Secretary of State, Charles Evans Hughes, delivered on 12 November 1921, which had gone into excruciating detail regarding which capital ships of each of the three original participants' navies would be retained and which scrapped, became quite vague when it came to other types of ship. All existing unarmoured warcraft under 3000 tons were to be exempted from the terms of the treaty, which effectively covered virtually all destroyers built to date.[14] No future auxiliary combatant craft would fall under the proposed treaty's restrictions unless they displaced more than 3000 tons, were faster than fifteen knots top speed, *and* carried more than four 5in guns. For those existing auxiliary combatant craft that were not exempted by the limits just stated, Hughes proposed that the same tonnage ratios planned for capital ship tonnage would apply.

In fact, it was the insistence of the British that the French accept a significantly lower tonnage ratio than Japan's, that led to the conference failing to adopt an overall tonnage limit on auxiliary combat craft, or a specific upper tonnage or gun calibre limit for destroyers as they finally did for cruisers. The inclusion of the Italians in the same 'bucket' as the French came about because, by this point, the Italians were solely interested in ensuring that the French gained no advantage over them.[15] The Italian position was simply that they insisted on the right to maintain parity with any other naval power in the Mediterranean, which was understood to mean France

(because the Italians had neither the financial or industrial power, nor the desire, to maintain parity with the British, who were the other Mediterranean naval power).

The Americans presented their basic proposal, that the United States and Great Britain be allowed 450,000 aggregate tons of cruisers, flotilla leaders, and destroyers, and Japan 300,000 tons, with France and Italy to receive substantially lower allocations, at the first Plenary Session of the conference on 12 November 1921.[16] Replacements of destroyers and leaders were to be allowed after twelve years, with keels laid on those replacements a year before the replacement date. While this did not become an immediate or critical issue in the discussions, it would become so as agreements were being reached on limiting the numbers, gun size, and tonnage of capital ships. By the time the scaled-down Committee on Limitation of Armament met for the fourth time on the morning of 22 December, most details on limiting capital ships and aircraft carriers had been agreed upon by all parties, except the French, who were unwilling to accept what amounted to third-rank status among world powers.

Frustrated at what he saw to be French intransigence, Hughes sent a note on 16 December to Aristide Briand, who was at that time both Prime Minister and Foreign Minister of France.[17] In it, Hughes complained that the French position amounted to a deal-breaker for the entire proposed arms reduction agreement:

> In these circumstances, I feel that the suggestion that has been made that France should build 10 new capital ships in replacement with a tonnage of 300,000 tons or more suggests a program of such magnitude as to raise the greatest difficulties. In fact I regret to say that after canvassing the matter thoroughly and taking the best information I can obtain, I am compelled to conclude that it would not be possible on this basis to carry through the agreement.[18]

Hughes knew Briand to be a much more reasonable man than any of the French delegates at the conference, particularly the leading French naval representative, VAdm Ferdinand-Jean-Jacques de Bon, Chief of the EMM, who took a hard line in the negotiations. As Hughes hope, Briand proved more conciliatory, accepting the capital ship limits that Hughes had proposed, certainly aware that the construction of ten new capital ships in the near future was well beyond France's present capabilities, but there was a point beyond which he would not be pushed. His reply to Hughes on the 18th was quite clear on this point:

> As regards naval armament, it is not the offensive, but solely the defensive, point of view with which France is preoccupied.
>
> With regard to the tonnage of capital ships—that is to say, ships of offense, which are the most costly—I have instructed our Delegates in the sense you desire. I am certain that I shall be sustained by my Parliament in this course.
>
> But so far as defensive ships are concerned (light cruisers, torpedo boats and submarines), it would be impossible for the French Government, without running counter to the vote of the Chambers, to accept reductions corresponding to those which we accept for

> capital ships subject to this formal reservation which you will certainly understand.[19]
>
> The dominating idea of the Washington Conference is the restriction of offensive and costly naval armaments. But I do not believe it to be any part of its program to restrict a nation, which, like France, has a large extent of coasts and numerous distant colonies, in means essential to its communications and security.[20]

Hughes read these cables to the delegates at the 22 December meeting and, cutting through the diplomatic niceties, he summarised his understanding of the situation as:

> The Chairman stated that he had had no desire to detract in any manner from the French reservation or to build anything on a phrase, but he had not thought it necessary to wait for a decision regarding lesser craft before reaching a provisional agreement on capital ships'[21]

Which was simply a fancy way of stating that Hughes was elated at Briand's acceptance of the 1.75 ratio, which effectively ensured Italian acceptance as well, and wanted to nail down the progress he believed had been made regarding capital ships, letting future discussions resolve the hanging question of how, or even if, to limit auxiliary warships.

Having thus effectively separated the issue of capital ship tonnage from that of other warship types, a major concession on Hughes' part, he tossed the ball back to the French. Hughes stated:

> . . . his understanding was that France was ready to accept the limit of 175,000 tons for capital ships, but distinctly reserved her decision with regard to auxiliary vessels and submarines and was not willing to have her acceptance taken in any way as implying an agreement to a corresponding figure for auxiliary vessels.
>
> The Chairman then pointed out that the Committee was uninformed with regard to what tonnage of lighter crafts, destroyers, submarines, etc., the French Government desired.[22]

The French now became much less forthcoming. Hughes was asking for specific numbers and le Bon stated he was in no position to give an answer; it was obvious that Hughes' tactic of going around the back of the French delegation and communicating directly with Briand, something that, at the time, was just not done in 'polite' diplomatic circles, was causing irritation in the French delegation.[23] This left le Bon nonplussed and stalling for time, stating:

> . . . correspondence was in progress with the French Government, and a telegram was expected which would make known the results of the decisions reached. The question was a very serious one, and before definite figures could be presented they should receive the sanction of the Government.
>
> If concessions should be demanded in the matter of light craft and submarines, the great concession that had been made in regard to capital ships should be borne in mind and the security of France and the limitations of her normal naval existence should not be lost sight of.[24]

With that bit of handwaving, the subject of a precise tonnage limit for auxiliary warships was dropped, at least for the moment. As the almost daily meetings continued over the Christmas holidays, the submarine debate absorbed all the energy of the delegates, leaving little time left over for any other subject. It was not until the ninth meeting of the naval arms limitation subcommittee, on 28 December, that the issue of auxiliary warship tonnage arose again. At that meeting, the first to speak was Albert-Pierre Sarraut, head of the French delegation, who stated:

> At a meeting of the Cabinet and of the Supreme Council of National Defense [CDN], the situation was examined and discussed with the most earnest desire to do whatever seemed possible to further the aim of the Conference and assist in reaching results . . .
>
> . . . After examining . . . the composition of the forces needed by France in auxiliary craft and submarines, which are especially intended for the protection of her territory and its communications, the Cabinet and the Supreme Council of National Defense have reached the conclusion that it is impossible to accept a limitation below that of 330,000 tons for auxiliary craft and 90,000 tons for submarines, without imperiling the vital interests of the country and of its colonies and the safety of their naval life. The French Delegation has been instructed to consent to no concessions in regard to the above figures.[24]

Not very surprisingly, the reaction to Sarraut's statement was hardly favourable, least of all by the British, who were fighting hard to retain their position as, if not *the* leading maritime power, then one of the leading powers, and they made it clear they would not, having fought a war to defeat one rival naval power across the North Sea, quietly submit to the emergence of another, even if in the form of an erstwhile ally. Arthur James Balfour, President of the Privy Council and head of the British delegation, reacted with undiluted sarcasm:

> He did not feel himself that the sacrifice on the part of France was in itself of an overwhelming character, even in regard to capital ships, for he thought that if the naval strength of a nation was to be estimated in relation to the naval strength of other nations, it would

> be found that the relative strength of France under the arrangement already accepted in regard to capital ships would be increased. He did not begrudge her that increase; he rejoiced in it.
>
> But when he turned from the matter of capital ships to the matter of other craft he confessed that a very different picture met the eye. . . . It had further to be noted that their French colleagues accompanied their view of the necessity of submarines with the announcement that they intended greatly to increase the tonnage of their auxiliary craft. It must be acknowledged that this constituted a somewhat singular contribution to the labours of a Conference called for the diminution of armament.[26]

This from the representative of a nation that was being asked to sacrifice little beyond a few projected, future battleships and a number of obsolete and obsolescent vessels that would likely be deleted in the normal course of events. The British, who had come into the conference hoping to convince the other participants to ban the construction and use of submarines altogether – based on their recent war experience during which their maritime lifeline had come under such dire threat from German U-boats – now saw their hopes not only dashed but seemingly turning into a redoubled danger.

> [Balfour] must, however, dwell shortly on the effect which the French declaration of naval policy must inevitably produce upon British opinion. It was perfectly clear that if at Britain's very gates a fleet of 90,000 tons of submarines (60,000 tons of which were to be of the newest type) was to be constructed, no limitation of any kind on auxiliary vessels capable of dealing with submarines could be admitted by the Government which he represented. Public notice had now been given in the most formal manner that this great fleet was to be built on the shores nearest to Britain, and it would necessarily be a very great menace to her. He had no doubt, if the occasion ever arose, that Britain would be equal to it, but that it was on condition that she reserved the full right to build any auxiliary craft which she considered necessary to deal with the situation.[27]

Hughes was confronted with what he feared would be an escalating problem with the participating nations demanding increased total tonnage of auxiliary warships and exceptions for special needs. He was not prepared to address the issue as framed by Sarraut and Balfour, as one centred on tonnages and numbers of submarines and the ships needed to counter them (ie, destroyers). His response expressed his frustration at a conference that seemed to be spinning out of control. It was necessary to obtain, at a minimum, agreement as to what distinguished a capital ship from an auxiliary ship:

> [Hughes] assumed that Mr. Balfour, in referring to the entire liberty of action of Great Britain in this respect, did not intend to include capital ships, nor did he understand that it was intended to build, under the guise of auxiliary ships, vessels which might possibly come within the category of capital ships.
>
> He desired to present, for the consideration of the Committee, the suggestion that, if it were not possible to reach a satisfactory agreement for the limitation of the total tonnage of auxiliary craft, some arrangement might perhaps be made defining the tonnage limit of individual ships. He therefore desired to propose the adoption of the following Resolution:
>
> 'No ship of war other than a capital ship or aircraft carrier hereafter built shall exceed a total tonnage displacement of 10,000 tons and no guns shall be carried by any such ship with a caliber in excess of 8 inches.'[28]

These numbers may have seemed to have been plucked out of thin air by Hughes on the spur of the moment, and contemporary accounts can find no specific source for them other than to note that they were 'in the air'.[29] These specifications had, for example, been raised in discussions in late 1919 and 1920 by the US Navy's General Board when considering future cruisers to follow after the Omaha class, in response to the Royal Navy's Hawkins class. What it did not address in any direct manner was the issue of tonnage or armament of destroyers individually or collectively, nor would any other discussion at the Washington Conference. When the resulting treaty was signed on 6 February 1922, it contained no limitation on the number or size of destroyers allowed any of the signatories.

Green light in Paris

With the signing of the Washington Treaty, the French felt free to reactivate the *contre-torpilleur* project approved by the EMM on 25 May 1921 that had been put on hold in November of that year. If anything, the position so publicly and forcefully taken by Sarraut in the name of France at Washington could only redouble the determination of the MN to pursue the construction of auxiliary warships, in particular the Jaguar class *contre-torpilleurs*, which would become the first French super destroyers.

Even before the Washington Treaty was signed, the Conseil de la Défense Nationale (CDN) had not only established the 330,000-ton target for auxiliary warships, but had suggested that 500M FF – approximately $37.3M at that time – be spent annually towards the construction of warships. The plan was that six large destroyers would be included in the first two-year tranche, with the goal of having thirty-six by 1940.[30] What has changed dramatically is how much the allocated money would buy. Twenty-six million FF was budgeted for each of these new destroyers, or approximately $1.925M

This port quarter view of MN *Tigre* gives a good look at the original main-battery gun shields, identifiable by their rounded front and flat sides. Note also the guide rails aft of the sliding doors at the aft ends of the depth charge belts. (NHHC)

Running her high speed trials on 25 October 1925, the Jaguar class *contre-torpilleur*, MN *Tigre*, was able to exceed 36kt for a hour of sustained steaming. Note the original, short half-shield on her 130mm main-battery mounts, which would be replaced when they were found to offer poor protection against weather for the gun crews. (NHHC)

each.[31] In today's dollars, based on when this manuscript was written in 2022 and accounting for inflation, that would be approximately $32M per ship. By way of comparison, the US Navy is now ordering Burke class destroyers at approximately $1.8B each, a fifty-six-fold increase in cost per ship. To be fair, what goes into a ship has changed dramatically in the intervening century. A twenty-first century destroyer is wrapped around sophisticated sensor and weapons suites unthinkable a hundred years ago. The powerplant exists mainly to generate electricity for the sensors and weapons. Crew spaces are almost an afterthought and in the not too distant future may become unnecessary.

The first two ships to this new design were ordered on 18 April 1922 from the Arsenal de Lorient, a government-operated shipyard, and the remaining four on 26 February 1923 from three different private yards. Unlike previous classes of French destroyer, they were to be built to a specification sheet provided by STCN and final designs generated at Lorient, in order to ensure a uniform squadron of six ships of similar characteristics. In clear and direct response to the Italian Leone class, these French destroyers had similar – in many cases identical – names of large, predatory cats.[32] The specifications to which the Jaguar class were built are as follows:

MN *Jaguar*: 1926[33]

Displacement (standard):	2126 tons
Length (oa):	415ft 11in
Beam (wl):	37ft 2in
Draft (max):	13ft 5in
Powerplant:	50,000shp (design); two-shaft Rateau-Bretagne (4) or Breguet-Laval (2) single-reduction geared turbines; five du Temple boilers
Speed:	35.5kt
Armament:	6 × 550mm Mle 1920T torpedo tubes (2 × 3; centreline); 2 × depth charge chutes; 6 × 200kg depth charges (plus 4 reloads) per chute; 4 × Thornycroft Mle 1918 (4 × 1 depth charge throwers; sided); 6 or 7 × 130mm/40 Mle 1919 (1 × 2; 4 × 1 or 2 × 2; 3 × 1); 2 × 75mm/50 Mle 1924/1922 (2 × 1); 2 × 8mm Hotchkiss Mle 1919D MG (2 × 1)

As the experimental ships proposed in 1920 had been cancelled, these were among the first ships to enter the MN with geared turbines and, as such, some teething problems were to be expected. Four of the ships had Rateau-Bretagne impulse turbines and, after minor initial issues were resolved, proved to be reliable; the two built at AC Loire – *Léopard* and *Lynx* – had Breguet-Laval reaction turbines which suffered from a long litany of problems ranging from failed turbine blades to saltwater leakage in the condensers, which caused those ships to be the last of the six to be delivered by almost a year. However, once these initial issues were resolved, all six ships in this class proved to be reliable steamers, able to maintain thirty knots even late in their careers.

The main-battery gun selected for these ships – the 130mm/40 Mle 1919 – was intended to be available in both single and twin mountings. The initial plan had been for *Léopard* and *Lynx* to be armed with two twin mounts and the other four with one, but initial tests with a prototype of the twin mount showed that it was poorly designed, with constant interference between the men serving the two barrels, leading to a rate of fire slower than could be obtained from two single mounts. In August 1923, before any of the six had been launched, it was decided to discontinue development of the twin mount and equip the ships with five single mounts. Nor was this the end of the issues with the Jaguars' main batteries. Because the gun had a relatively short (40cal) barrel, chosen to save weight and for ease of handling, it had a relatively short range. It also had a Welin interrupted screw breech, which made for slower loading despite employing a cased powder

The Jaguar class *contre-torpilleur* MN *Léopard* is seen in this image from 1939. Most obvious are the larger (and heavier) main-battery gun shields substituted for the narrower and shorter model originally fitted to these ships. Less obvious is the longer-base 5m OPL rangefinder at the back of her bridge, which replaced the 3m model originally fitted. (Fulvio Venturi)

Three Jaguar class *contre-torpilleurs* are seen during a public event at an Atlantic coast port sometime in 1938–39. They are (from the left): MN *Jaguar*, *Chacal*, and *Léopard*. Notice the concentration dials visible above 'Y' gun mount on the two nearer ships. A small number '2' is mounted to the port side of *Léopard*'s 'Y' mount in such a way that it would be visible from astern. The purpose of this device is not clear, but it appears to be some sort of tactical marking, nor is it clear whether other ships carried a similar marking at this time. Curiously, only *Jaguar* appears to have visitors on board. (NHHC)

charge.[34] And, as a further disadvantage, the 130mm gun had a high trunnion – five feet above the deck – intended to allow the guns to elevate to 36°, unusual for a gun of this size at this time, but which made loading the gun difficult at lower elevations due to the weight – 70.5 lb SAP; 76.83 lb HE – of the man-handled rounds. By September 1923, this gun had been officially declared obsolescent and its replacement by the 138.6mm/40 Mle 1923 then under development was authorised, but delays in the design of the new gun led to the Jaguars being completed with the original guns in place.

The 130mm single-mount gun was designed with a close-fitting half-shield that covered the front of the mount and the top and side only just past the trunnion. Early trials showed this to be inadequate in protecting the gun crew in any weather and the Lorient dockyard designed a replacement that was both broader and deeper; trials on *Panthère* in 1927 showed this did a far better job of weather protection. The only problem with these new gun shields, which were installed in *Léopard* and *Lynx* before their completion and replaced the original gun shields in the other three ships when available, was that, in aggregate, they added almost 2.5 tons in topweight, aggravating an already serious problem.

The MN liked almost everything about their new *contre-torpilleurs*. They were fast, well-armed, and their high forecastle and flared, clipper bow gave them good weather handling, but from their earliest trials, it was obvious that they were inherently top-heavy and prone to instability. They heeled excessively in turns and even in a crosswind, despite long, deep bilge keels on either side, and suffered from a rapid roll period (up to 25° in 8–10sec) in a crossing sea, sufficient to impact habitability and their effectiveness as gun platforms. The cause of the problem was determined to be the size and weight of the bridge structure, made large enough to accommodate the command staff for a destroyer flotilla and the communications suite required by a fleet scout, and the two tall tripod masts. The immediate impact of the discovery of this problem was the cancellation of the plans to replace the 130mm guns with the larger and heavier 138.6mm model, and the planned installation of the main-battery fire control director intended for the top of the forward superstructure. Instead, the 3m coincidence rangefinder Mle B.1926 temporarily installed on the open flying bridge became a permanent fixture.

Besides these weights *not* added above the waterline, there was little the MN could do to resolve the stability issues discovered in the Jaguars. Plans were made to replace the 75mm/50 anti-aircraft guns – large, heavy, slow-firing, long-ranging guns better suited to fending off high-altitude attacks by Zeppelins than more modern aircraft – with lighter, faster-firing 37mm/60 Mle 1925 guns and the tripod masts with shorter pole masts, but neither change took place due to lack of funds. The only significant changes that were implemented before the outbreak of the Second World War were the fitting of loading trays for the main-battery guns, which alleviated, to an extent, the problem of loading the guns at low elevations, the removal of the depth charge throwers, and the replacement of both the 75mm AA guns and the 8mm MGs by eight 13.2mm/76 Hotchkiss Mle 1929 in CAD Mle 1931 twin mounts. All these changes saved over seventeen tons of topweight, all of which aided stability.[35] The Hotchkiss machine gun was reliable and well liked, but by 1939 was too lightweight and short ranging to be an effective anti-aircraft weapon. Starting in late 1939, in five of the six ships – all except *Panthère*, for which time ran out – the Jaguars were taken in hand and fitted with sonar, and a pair of depth charge throwers were installed in better locations, one each side just forward of the after bank of torpedo tubes, where one of the twin 13.2mm MG mounts had been located. The two displaced machine gun mounts were moved to atop the amidships deckhouse, replacing No 3 main-battery mount. This mount had always been served only manually from the after magazine; once any ready-service rounds were expended, fire from that gun was far slower than the other mounts, depending on a human relay to pass rounds forward.

Three years passed after the 1922 building programme before the MN turned again to building *contre-torpilleurs.* In the meantime, they turned to filling out their flotillas with somewhat smaller *torpilleurs d'escadre.* A dozen Bourrasque class destroyers were ordered in 1922 at the same time as the Jaguars. They carried one fewer main-battery gun and had a designed top speed two knots slower than their larger counterparts, but otherwise were quite similar in their characteristics. Along with fourteen of the slightly larger L'Adroit class ordered in three tranches between 1924 and 1926, these *torpilleurs d'escadre* provide the bulk of the flotilla craft with which the MN went to war in 1939. Larger *contre-torpilleurs* were ordered again in 1925, when the first of three six-ship subclasses of somewhat larger super destroyers were ordered; these eighteen ships would collectively be known in the English-speaking world as '2400-tonners' because of their dis-

One of the last class of 'four-pipers' (*quatre-tuyaux*), the Vauquelin class, is seen here, sometime between 1934 and 1936. MN *Kersaint* was among the fastest and most powerful of the '2400-tonne' *contre-torpilleurs*, but the type was always limited by the lack of directed fire control. (NHHC)

placement – some 300 tonnes (295 tons) greater than the Jaguars – but would be known to the French as 'four-pipers', because they had four slender funnels serving their two widely-spaced fire rooms, a unique silhouette among modern French warships.[36] Because they displaced more than the Jaguars, with more length and beam, they suffered less from the stability issues of that class, despite mounting the heavier 138.6mm main-battery gun.

All six of the Jaguar class *contre-torpilleurs* served in the MN at the outbreak of the Second World War, when the weakness of their anti-aircraft armament became glaringly apparent. Despite the fact that, upon the declaration of war in September 1939, France was at war only with Germany, they eyed Italy with the greatest of suspicion and, from April 1940, kept the bulk of their naval strength in the Mediterranean. Only three of the six Jaguars – *Léopard*, *Jaguar*, and *Chacal* – were based at ports facing the North Sea. *Jaguar* and *Chacal* were sunk within hours of each other, one by torpedoes from German S-boats, the other by Luftwaffe bombers. *Léopard* was seized by the British at Portsmouth at the time of the French surrender and eventually turned over to the Free French. She went through a lengthy refit during which one boiler was removed and the space given over to oil fuel stowage, hoping to address the poor endurance suffered by all the French *contre-torpilleurs.* She was sent to the Indian Ocean to assist in the repatriation of the French colony at La Réunion in November 1942. After that she transferred to the Mediterranean, operating out of Alexandria, but her career there was brief. On her first escort mission with a convoy to and from Malta, she ran aground near Benghazi on 27 May 1943, and eventually broke apart there due to wave action.

The Imperial Japanese Navy continued to build destroyers at a steady rate. This view shows *Yunagi* of the Kamikaze class on 5 September 1936. The ships of this class differed from those of the Minekaze class shown above solely in having the bridge level enclosed in steel plating; to compensate for this added weight carried high above the waterline, four inches were added to the ship's breadth. The number carried in white at the bow indicated the destroyer division to which this ship belonged. (Fulvio Venturi)

The three other Jaguars were, after the surrender of France, placed in 'care and maintenance' status (*en gardiennage*) at Toulon, where on 27 November 1942 they were ordered scuttled to keep them from falling into German hands. Because they had only skeleton crews, only one of the three, *Lynx*, was effectively scuttled. The other two were seized by the Germans in more-or-less intact condition. *Panthère* and *Tigre* were captured intact, turned over to the Italians and moved to La Spezia in March 1943. *Panthère* was scuttled there at the time of the Italian armistice on 9 September 1943. *Tigre* was returned to French control on 28 October 1943 at Bizerte and was used as a transport and later as a gunnery training ship before being decommissioned in 1948, but was not stricken until 1954 and not broken up until the following year. The Italians considered these ships roughly equivalent to the Leone class *cacciatorpediniere*, but less stable.

A rising sun in the east

Like the Americans, the Japanese fought on the Allied side in the First World War and, like the Americans, they suffered none of the direct impact of war fought on or in the skies over their homeland. They were sufficiently pleased with the basic destroyer types they were building at the end of the First World War that, as mentioned in the preceding chapter, they continued building successor classes after war's end. This comprised two more, progressively improved classes authorised between 1921 and 1923, a total of twenty-one ships. All still shared a number of by now anachronistic features, such as a well deck ahead of the bridge and single-mounted 12cm (4.7in) main-battery guns at a time when

The last of the 'well deck' destroyers built by the Japanese were those of the Mutsuki class, three of which are seen here nested together in the Whangpoo (Huangpu) River at Shanghai in February 1932. The destroyer in the foreground is *Fumizuki*, nested together with two mates from DesDiv22. The ship in the left background, part of DesDiv30, is most likely *Mutsuki*, the lead ship of the class. Note the 'swan's neck' bow profile, as opposed to the simple curved profile of the earlier classes, intended to lengthen the forecastle and improve bad weather handling. (NHHC)

almost all navies, except Great Britain's Royal Navy, had adopted a larger main-battery gun for their larger destroyers.

The Washington Conference, and the lingering unhappiness in Japan at the treatment that nation received at those negotiations, proved to be the impetus necessary to push the Imperial Navy to move to a significantly larger and more powerful destroyer design. Starting in 1922, the IJN began planning for a war in which the US Navy, more than the Royal Navy, would be the primary enemy, and in which the Japanese would start with a significant (60 percent) disadvantage in capital ships, as mandated by the Washington Treaty. To give the Japanese a chance for a Tsushima-like dramatic victory when the opposing battle fleets met west of the Marianas, it would be necessary to reduce the expected American advantage in ships and firepower by attrition. The means of achieving this was to be the use of all the new and best technologies the new century had to offer, primarily torpedoes, carried by aircraft and ships. The plan was to subject the enemy fleet to repeated torpedo attacks as it advanced from the American west coast towards Japan. To accomplish this, the IJN concluded they needed multiple redundant torpedo delivery platforms capable of attacking the advancing fleet by day (primarily long-range aircraft flying from bases in Japanese-held island bases such as Saipan and Truk (Chuuk)) and from fast torpedo-armed ships at night. In order to make those attacks as effective as possible, the plan called for the development of both platforms carrying more torpedoes, so that each attack would have a greater chance of success, and faster and more powerful torpedoes, so that each hit would have a greater chance of causing serious damage.[37]

The ships that would have to carry out the 'interception-attrition operations' (*yugeki zengen sakusen*) envisaged in the plans drawn up by the Naval General Staff would have to be new types of cruiser and destroyer endowed with unusual seakeeping and armament. In considering the fleet required to conduct these operations, the naval staff, in a policy statement submitted to the prime minister on 28 February 1923, reported that the IJN required the construction of 400,000 tons of cruisers – this would be forty cruisers at the Washington Treaty maximum of 10,000 tons each – and 201,600 tons of destroyers – which represented the construction of 144 ships of 1400 tons each. Each of these cruisers and destroyers would be armed with the biggest appropriate main-

After the Mutsukis, the Imperial Japanese Navy took time to digest the impact of the Washington Treaty and responded with a totally new design that was one-third larger, one knot faster, and carried significantly heavier armament. To many, the Fubuki class, epitomised by the name ship seen here in an undated image, were the first true super destroyers. The hull dispensed with the well deck that had been a feature of Japanese destroyer design since the end of the First World War. In its place was a hull with a more raked swan's neck prow and a pronounced flare carried all the way aft. The new model 12.7cm/50 main-battery guns were carried in twin mounts in fully enclosed, gas-tight, splinter-proof gunhouses, a novel feature. (Fulvio Venturi)

A side-on view of *Usugumo*, one of the first group of the Fubuki class, seen in August 1928. This group could be distinguished by the tall, angular main-battery gunhouses, which were capable of low-angle fire only, and by the large horn-type aft-facing fire room ventilator hoods.

battery guns – 20.3cm (8in) guns for the cruisers, as mandated by the Washington Treaty, and 12cm guns for the destroyers, unless potential enemies were arming their destroyers with larger main-battery guns, in which case, 12.7cm (5in) guns or even larger should be mounted. The torpedo would be a 61cm (24in) model, then the largest and most powerful torpedo introduced by any naval power. The torpedo initiative eventually led to the development of the 61cm Type 93 Model 1 series of oxygen torpedoes that became known in the West as the 'Long Lance'.[38] This torpedo, which entered service in 1935, replaced the earlier Type 8 No 2 and Type 90 torpedoes of similar diameter which, using compressed air as their oxidiser, had significantly less speed and range.

The last class of destroyers designed before the implications of the Washington Treaty had been digested was the Mutsuki class authorised in 1923. These twelve ships were the last of the well deck ships, armed with two triple banks of 61cm torpedoes – they were the first to carry this larger torpedo – and four single-mount 12cm main-battery guns, and displaced 1315 tons. The task given to the Design Section of the Naval Technical Department, then headed by Constructor Capt Hiraga Yuzuru, was to design new cruisers and destroyers with exceptional speed, range, and armament on the smallest possible displacement. In the case of the cruisers, the task was made difficult because of the arbitrary limitation of 10,000 tons standard displacement imposed by the Washington Treaty, and by the constant pressure from the Naval General Staff to maximise the number of torpedo tubes and torpedo reloads carried relatively high up in the ships' structures, where they would adversely impact stability and represent a serious danger of cataclysmic explosion, because the torpedoes would be sited above the 'armoured box'. So intense did the arguments become between Hiraga and his superiors that he was, in 1922, 'kicked upstairs', meaning he was promoted to vice admiral and sent on an extended tour of foreign dockyards ostensibly to report back on the latest naval technologies, but mainly to get him out of the way so that his more pliable deputy, Cdr Fujimoto Kikuo, could take command of the IJN's Design Section.

Thus, it was Cdr Fujimoto who was primarily in charge of designing the successors to the Mutsukis. The instructions from the Naval General Staff, which led to Fujimoto's design proposal designated F42 and submitted for consideration on 18 April 1924, called for a ship of similar characteristics to

Uranami, the tenth unit in the Fubuki class, was a transitional unit between the first and second groups of the class. She retained the tall, boxy gunhouses of the first group, but introduced the lower 'mushroom-type' ventilator hoods around her funnels, which characterised all later Special Type destroyers. (USN)

the Mutsukis except they were to be significantly faster – 39kts vs 37.25kts – carry 50 per cent greater torpedo armament, and mount four of the new 12.7cm (5in)/50 3rd Year Type main-battery guns then under development. This design was predicted to have a standard displacement of 1900 tons. The reaction of the Naval General Staff to this design, after protracted discussion lasting months, was that the main battery needed to be increased by 50 per cent and that the displacement needed to be reduced significantly, both to make each ship less costly in terms of money and use of precious resources, and to make it a smaller, stealthier vessel better suited for night combat. The only concessions the staff were willing to make were to allow main-battery magazine capacity to be reduced from 200 rounds per gun (for four guns) to 120 rounds per gun (for six guns) for a total reduction of eighty rounds, and to allow the maximum speed to drop from thirty-nine knots to thirty-eight knots.

These requests were, in every imaginable way, a naval architect's nightmare. The weight savings permitted by the slightly smaller magazine size and somewhat lower speed requirement were almost entirely negated by adding a third twin main-battery turret with its enclosed gunhouse, which weighed 25.4 tons. The only way to save weight in any significant manner was to lighten the ships' structure dramatically; Fujimoto set his designers to saving weight wherever they could. The strategies they employed included utilising arc welding in the place of riveting at the extremities of the hull, decreasing the thickness of the shell plating, butt-joining rather than lap-joining the shell plating where possible (which not only saved weight, but also decreased the resistance of the hull passing through the water), and using a high percentage of light alloys such as duralumin in the upperworks, particularly the bridge. This way, Fujimoto was able to deliver a design on 24 February 1925, labelled F42–2, that had a designed displacement of 1650 tons. Still not quite satisfied, the Naval General Staff asked for greater range, accepting that, in order to achieve that, speed would have to be sacrificed. Fujimoto responded with design F43, similar to the preceding, but, at 1680 tons and with a powerplant producing 50,000shp, which would allow thirty-five knots maximum speed. This design was accepted and ordered into production. Five ships had been authorised in March 1923 and these were now to be laid down starting in 1926; a further programme was authorised in 1925, but chronic budgetary issues caused that to be cancelled by the Diet. Naval influence in the government was strong, however, and the Treasury gave special authorisation for four additional

One of the later units of the second group of the Fubuki class, *Sagiri* seen here soon after completion in 1930, shows some of the initial measures taken in response to the instability issues that were discovered even before the major incidents of 1934–35. Although she was completed with an even larger and taller bridge structure, which only added to the topweight she carried high above the waterline, she was given Type C main-battery gunhouses, which had thinner steel plating, a lower overhead on the gunlayer's position, and lower maximum elevation, all in an effort to save weight in the three main-battery mountings. In fact, the steel plating used in the gunhouses was so thin that it was found to deform in bad weather and two external horizontal steel braces were added to the forward mount to support the side plating. (Fulvio Venturi)

Ayanami, seen here from the starboard quarter, most likely in 1931, was the second unit of the second group of the Fubuki class. These could be distinguished by the 'mushroom-type' ventilators around the funnels and by the Type B main-battery gun mounts. These were lower and broader than the Type A mounts of the first group, and had a noticeably raised gunlayer's position on the left side of the gunhouse. This was necessary because this mount allowed the guns to be raised independently as high as 75°, making them a true dual-purpose mount. Note the ship's name in katakana characters on the side and in hiragana cursive script across the stern. (Fulvio Venturi)

hulls to be laid down in 1926.[39] Far from satisfied with this sop, the IJN came back the following August with a request for funding for sixteen more of these new, large destroyers – now called Special Type (*tokugata*) to distinguish them from the preceding designs – to be spread over three years, but the Diet was still feeling great fiscal pressure, and the multi-year programme was pared down to nine ships. Fate would intervene at this point. The ruling Emperor of Japan, Yoshihito (Taisho), had suffered from poor health all his life and had exhibited declining mental capability since at least 1918, which allowed the Diet to exert greater autonomy than had been possible during the reign of the founding Meiji Emperor.[40] Yoshihito's elder son, Hirohito (Showa), was appointed regent in November 1921, but had little power to guide events directly. On 25 December 1926, Yoshihito died and Hirohito immediately set about pushing policies he supported, including greatly strengthening the navy. A new session of the Diet was convened and the IJN's original request for sixteen Special Type destroyers was resubmitted. This time it was approved with a reduction of only one ship; fifteen would be laid down between 1927 and 1930. Thus, by the early 1930s, the IJN would have twenty-four Special Type destroyers, enough to equip six four-ship destroyer divisions.

The initial specifications to which the first subclass of Special Type destroyers were built – the Fubuki class – were as follows:

***Fubuki*: 1928**[41]

Displacement (standard):	1680 tons (design); 1750 tons (actual)[42]
Length (oa):	388ft 9in
Beam (wl):	34ft
Draft (max):	10ft 6in
Powerplant:	50,000shp; two-shaft Kampon impulse geared turbines; four Kampon Type RO boilers
Speed:	35kt
Armament:	9 × 61cm trainable torpedo tubes (3 × 3; centreline); 9 × 61cm Type 90 torpedo reloads; 6 × 12.7cm/50 3rd Year Type (3 × 2); 2 × 13mm/76 Type 93 MG (2 × 1)

The first subgroup of ten Special Type destroyers – the nine authorised in the 1923 and 1926 programs and the first of the fifteen ships in the 1927 programme – were fitted with the Type A twin 12.7cm/50 turret which allowed elevation from -7° to 40° and weighed 25.4 tons per turret. This turret mounted both guns in a single sleeve, which led to a relatively tall and narrow gunhouse, but the Naval General Staff wanted these large destroyers to be able to defend themselves against air attack and to contribute to the defence of capital ships, so a dual-purpose version of the turret was developed, the Type B, with separate sleeves for the barrels and the ability to elevate to 75°. Although these guns could elevate at 27°/sec, a rate considered adequate for use in an anti-aircraft role, they could be loaded at a maximum of 10° elevation and could train at a maximum of 6°/sec, both of which made their use against modern, high-speed aircraft problematic. To make matters worse, the Type B turret was significantly larger and heavier than the Type A, weighing 32.5 tons, which added to the instability issues that were noted with the Special Type destroyers from their entry into service in 1928. The Type B turrets were fitted in all the remaining Special Type destroyers, which, in part, distinguished the second and third subgroups from the first. (The tenth Special Type destroyer completed, *Uranami*, was a transitional unit; she mounted the original Type A turrets but replaced the large aft-facing horn-type fire room ventilator hoods alongside each funnel with the downward-facing mushroom-shaped hoods around the funnels' bases used in all later ships.) The last four Special Type destroyers comprised a third, separate subgroup visually distinguished by a narrower forward funnel. These ships had three larger boilers in the place of four smaller ones; the forward funnel was smaller because it served only one boiler.

The problems with instability came into sharp focus after the so-called Tomozuru Incident of 12 March 1934, when a newly built, lightly constructed torpedo boat capsized during a storm. Investigation revealed that this had been the result of inherent instability due to reduced metacentric height caused by weight added high in the ship's structure during construction. A ship's metacentric height (GM) is the distance between a ship's centre of gravity (G – the centre of mass) and its metacentre (M), which is a point on a line perpendicular to the waterline drawn through G where it is crossed by a line drawn vertically from the ship's centre of buoyancy (B – the centre of mass of the water displaced by the hull). G remains fixed as a ship rolls with waves, while B moves in the direction of the roll, and thus M represents the action of buoyancy on a ship's mass as it rolls. The distance between G and M is a ship's metacentric height and it describes to a great extent how a ship will behave in weather, particularly in cross seas. In general, the higher a ship's GM, the more stable it will be in weather, with a shorter, more rapid roll period. While an excessively high GM can cause a ship to be 'stiff', to roll too rapidly and thus be a poor gun platform and uncomfortable to inhabit, it is rarely dangerous. The opposite, an excessively low GM, however, can cause a ship to be 'tender', to roll too slowly, to lag behind rather than rolling with the waves, which can lead to dangerous wallowing in heavy weather. This, the investigation concluded, led to the capsizing of *Tomozuru* and, further, threatened a large number of the newest ships built for the IJN, which suffered from a low GM caused by excessive topweight, including all of the Special Type destroyers.

Before a remedy could be put in hand, a further grave problem was revealed by the so-called Fourth Fleet Incident in September 1935. In that case, a major part of the Japanese fleet passed through a typhoon with wind speeds reaching 78mph and wave heights 50ft, causing some ships to roll up to 70°. But it was neither the wind force nor the wave heights that proved to be most dangerous. Special Type destroyers suffered particularly as the distance between wave crests well exceeded expected maximums and stressed their hulls far beyond expected limits. Two units, one from the first subgroup and one from the second, lost their bow as far aft as their bridge; eight others sustained damage to their hull in varying degrees including separated or buckled shell plates and failed welds.

It was decided to tackle both the stability and hull strength problems at the same time. Between 1935 and 1938, all surviving Special Type destroyers were taken in hand for extensive modification. They were given a full-length ballast keel and forty tons of ballast was added to improve stability; funnels and forward superstructure were lowered in height, the nine torpedo reloads were reduced to three for the centre set of torpedo tubes only, voice tubes to turrets and torpedo tubes were removed, and on several of the ships, the Type B main-battery turrets were replaced by Type C turrets that restricted elevation to 55°, saving some weight.[43] Other second and third subgroup units had their Type B turrets modified by having the prominent rounded hoods over the sighting telescopes lowered and flattened. The hulls were strengthened by the extension of the rivetted and lap-jointed parts of the shell plating further towards the bow and stern and strengthening longitudinal members.

So modified, the Special Type destroyers now had a standard displacement of 2090 tons, which lowered their nominal top speed by a knot, but resolved the stability and structural issues to the extent that they served into the Second World War, with added topweight in the form of antennas and anti-aircraft weapons often mounted high in the ship. Workhorses of the fleet, the Special Types suffered severe attrition, particularly in the South Pacific in 1942 and 1943; only two – *Ushio* and *Hibiki* – survived the war and only *Hibiki* was still in operational condition.

The third group of the Special Type destroyers was given a separate class name, the Akatsuki class. Technological improvements allowed the number of boilers to be reduced from four to three, which allowed the forward funnel to be made thinner. The bridge was reduced in size in a further effort to save weight. Otherwise, the four ships of this subclass, such as *Hibiki*, seen here off the east coast of China in the late 1930s, were similar to the second group of the Fubuki class. *Hibiki* did have the distinction of being the first all-welded ship in the Imperial Japanese Navy. (NHHC)

Chapter 6

Trying to hold back the flood: 1927–1930

Most who walked away from the Washington Conference felt that, while good work had been done there restricting the size and numbers of capital ships and aircraft carriers, much was left undone, in particular regarding smaller warships. The League of Nations, brought into existence by the Treaty of Versailles, became the prime mover towards further naval arms restrictions in the immediate aftermath of the Washington Conference. The League, with full support from the United States, although the Americans never ratified the Versailles Treaty or formally joined the League, announced the intent to convene a General Disarmament Conference and began the first of many meetings of a Preparatory Commission in May 1926. By this time, the Americans, driven by internal politics – the Republicans, then as now, had a hard right wing that pushed more reasonable actors towards extreme positions – urged the president, Calvin Coolidge, to act without waiting for the seemingly endless League debates to call a conference with the specific goal of restricting 'auxiliary warships'. His message was straightforward; the following announcement was made in the US Congress on 10 February 1927:

> . . . the American Ambassadors at London, Paris, Rome and Tokyo will today present . . . a memorandum suggesting that they empower their delegates at the forthcoming meeting of the Preliminary Commission for the Disarmament Conference at Geneva to negotiate and conclude at an early date an agreement further limiting naval armament, supplementing the Washington treaty on that subject, and covering the classes of vessels not covered by that treaty. . . .
>
> The desirability of such an agreement has been apparent, since it was only to be expected that the spirit of competition, stifled as regards capital ships and aircraft carriers by the Washington Treaty, would sooner or later show itself with regard to other vessels not limited under the treaty. Actually, I do not believe that competitive building of these classes of ships has begun. Nevertheless, far-reaching building programs have been laid down by certain powers, and there has appeared in our own country, as well as abroad, a sentiment urging naval construction on the ground that such construction is taking place elsewhere. In such sentiments lies the germ of renewed naval competition.[1]

Coolidge precisely identified the problem, but made no clear statement in his address to Congress as to how to achieve a solution. What he did offer was to fall back on the only process that had worked in the past:

> The problem of the limitation of naval armament, while not regional in character or susceptible of regional treatment, has been successfully treated, in part, by an agreement among the five leading naval powers, and, in my opinion, can be definitely dealt with by further agreements among these powers.[2]

In the actual memorandum sent to the four other 'powers', Coolidge gave only the vaguest hint of the position the Americans would take at the conference he was inviting those nations to join:

> Although hesitating at this time to put forward rigid proposals as regard the ratios of naval strength to be maintained by the different powers, the American Government, for its part, is disposed to accept in regard to those classes of vessels not covered by the Washington Treaty, an extension of the 5-5-3 ratio as regards the United States, Great Britain and Japan, and to leave to discussion at Geneva the ratios of France and Italy, taking into full account their special conditions and requirements in regard to the types of vessels in question.[3]

It is not known whether Coolidge consulted the US Secretary of the Navy or the Navy's General Board before making even that vague proposal in his memorandum – it is known that he immediately turned to the General Board for advice when the reaction to his memorandum was quite different from what he expected – but it is clear that he was either unaware of, or chose to ignore, the fact that a very similar proposal by the Americans to extend the capital ship displacement ratios to cover 'auxiliary' warships at the Washington Conference had led directly to those ships not being restricted in total displacement in the Washington Treaty.[4] Coolidge had been correct in 'reading' the general atmosphere of international frustration with the slow pace of the League of Nations process and with the rising cost of naval armaments. Despite the 'battleship holiday' implemented by the Washington Treaty, both labour and raw material costs, held in check by the war, were now rising rapidly and, with the added cost of new technology, were making each new ship more expensive than the last. Most of the nations to which he had addressed his call to a new naval disarmament conference welcomed the idea, but his mention of the old capital ship ratios as a starting point for the new discussions tempered that reception in some capitals.

The problem, simply stated, was that, of the five 'powers' that had agreed to the Washington Treaty, with its displacement limit ratios for capital ships, only the United States and Great Britain – the two with the largest ratios – were truly pleased with the result. The other three nations, to a greater or lesser extent, were unhappy to be limited to a lower status and tonnage allowance, and each looked forward to a future conference to gain some redress for its sense of grievance. Thus, announcing that the United States took as its opening position the intent to extend the existing ratios to smaller warships was hardly likely to be greeted positively.

Each nation's reaction to Coolidge's invitation was, not surprisingly, reflective of that nation's attitude towards the Washington Treaty ratios. The president's invitations were sent out on a Thursday and, already, by Monday, 14 February, it was clear that international reaction was mixed and, in some cases, decidedly negative.[5] Already, speculation was being raised that Coolidge might have to proceed with a reduced-size conference, including only Great Britain and Japan, as the French and Italians were giving every indication that they would decline the invitation. By the next day, French rejection was official; Briand personally wrote the note that politely, but firmly, stated that France was committed to the League of Nations disarmament process, which included air and land forces as well as naval armaments, and involved the participation of not just the five major naval powers, but all fifty-five nations which were members of the League. It was this point, that so many more nations had a voice at the League than would participate in Coolidge's proposed conference, that formed Briand's most powerful argument:

> . . . the categories of ships to which the new proposed limitation would be applied are in fact those which for a majority of the powers are of the greatest importance. An accord limiting certain navies might be possible for capital ships, as in practice these powers are the only ones that possess this class of ships. But the situation is different when the question of lighter vessels is envisaged; all navies would have an interest in being associated in the consideration of this important problem.[6]

Briand's argument here was simple and logical and difficult to refute, unless the goal was, as was the case for the Americans and British, to assure the maintenance of the status quo while reducing arms expenditures. Undoubtedly, the French arguments were more than a little self-serving. Without question, the French were still stung by having been relegated by the Washington ratios to a status not only lower than the British, their traditional rivals, but also only half that of the Japanese, a nation all Western 'powers' reflexively thought of as inherently inferior. Also, there was no question that the French enjoyed the position of relative influence they held at the League in Geneva, where a number of smaller nations looked to the French to speak for them and to defend their positions, while there was no question that at a 'five-power' conference, they would be cast in the role of a minor player, which seriously offended their *amour propre*.[7]

Thus, it could come as no surprise that the French response was to decline Coolidge's invitation. The Italians, who were by now already under Mussolini's leadership and rapidly becoming a Fascist state, were in no way allied to the French or led by France when it came to foreign or defence policies, but, nonetheless, their actions at this time often mirrored France's. At the time of France's rejection of the invitation, the Italians had not responded, but it was assumed, correctly, that they too would decline to attend, so plans were already being made for the conference proceeding with just the United States, Great Britain, and Japan in attendance.[8] In the event, Washington and London were able to pressure Paris and Rome to the extent that those nations sent 'observers' to sit in on the conference proceedings

It must be remembered that almost all of the talk between the participating (and observing) nations prior to the Geneva Conference convening on 20 June 1927 primarily concerned cruisers and, to a lesser extent, submarines; it was universally believed that if agreement could be reached on the numbers and sizes of these vessels to be allowed, then agreement concerning all other categories of ships, especially destroyers, would be easily achieved. However, the problem with reaching an agreement on cruiser limits rapidly proved intractable and the conference made little progress before sputtering to a disappointing finish on 4 August.[9] But the fact that cruisers dominated the attention of conference participants, and ultimately doomed the endeavour, did not mean that the issue of limiting destroyer displacements and weaponry was not discussed.

What became immediately obvious from the first day of the conference was the utterly different approaches to arms limitation taken by all three participating powers, but most particularly by the British and Americans. The American proposal, as presented by Ambassador Hugh S Gibson, who was allowed to speak first as the conference had been called at American instigation, was highly detailed, breaking down the remaining naval ship types not already addressed by the Washington Treaty into four groups: cruisers, destroyers, submarines, and all others.[10] Under Gibson's proposal, the destroyer type was defined as surface ships that displaced between 3000 and 600 tons with a top speed greater than seventeen knots. For ships of this type, the Americans proposed that they and the British be allowed between 200,000 and 250,000 tons, with Japan allowed between

120,000 and 150,000 tons. The age at which the replacement of destroyer-type ships would be allowed was to be set at fifteen to seventeen years. No specific gun calibre limit was proposed. All told, Gibson's proposal was not unexpected, fitting well within the parameters of Coolidge's remarks when invitations were sent to the participants.

The British head of delegation, First Lord of the Admiralty William C Bridgeman, spoke next. Every effort had been made by the British delegation to keep the content of his proposal secret before he rose to speak at Geneva, and, by all appearances, that effort succeeded.[11] Remembering very well the impact of Charles Evans Hughes' opening address at the Washington Conference and how he had set the agenda for the ensuing discussions by laying out detailed plans including tables of tonnages and gun calibres by ship type and lists of ships to be retained and those to be scrapped, Bridgeman wanted to do much the same. The effect of Hughes' remarks at Washington had been literally stunning and had gone a long way towards ensuring what success the conference achieved. The problem with attempting to reproduce a past success, however, is that the preconditions that enabled that success are rarely applicable when the second attempt is made, and such was the fate of Bridgeman's attempted *coup de théâtre*. The reaction to Bridgeman's presentation was described later by participants as causing a chill to pass through the hall, hardly the reaction for which the British were hoping.[12]

In terms of content, Bridgeman's presentation could hardly have been more different from Gibson's. Where the Americans had broadly proposed an individual ship tonnage limit and an aggregate tonnage limit for the destroyer category, but no gun calibre limit, the British proposal, which included eleven very specific provisions, two of which (Nos 9 and 10) addressed destroyers, called for the destroyer category to be divided into two subcategories – destroyer leaders and destroyers – with the former limited to 1750 tons displacement per ship and the latter to 1400 tons, with both limited to a maximum main-battery gun calibre of 5in.[13] The replacement age for all destroyer-type vessels would be twenty years. What the British proposal lacked was any aggregate tonnage limit for any of the 'auxiliary warship' types. This was a deliberate omission, as the

While the Leones had been a recasting of an earlier, wartime design, the Navigatori class *esploratori leggeri*, ordered in 1927, were an entirely new design, intended to carry very nearly the same firepower as the Leones on a smaller hull at a higher speed. The *Navigatori* carried two fewer main-battery guns, a move that was justified by the fact that these were a new model 120mm/50 gun designed in 1926 that offered 25 per cent greater range firing a heavier shell. This image shows RN *Leone Pancaldo* before most of the alterations intended to reduce topweight in these ships had been performed; for example, the tripod upper foremast has been replaced by a pole mast, but the funnels have not yet been shortened. (Fulvio Venturi)

British knew that their desired total cruiser tonnage far exceeded that desired by the Americans and that this would become a critical sticking point in the forthcoming negotiations.

The Japanese proposal was the least specific, proposing only that the displacement limit in three categories – surface auxiliaries over 3000 tons, those under 3000 tons, and submarines – be set for each nation as the aggregate of their current tonnage plus any then under construction. No additional construction would be allowed until an existing ship reached its replacement age, which for the smaller surface auxiliary category would be twelve years.[14] Unsurprisingly, neither the British or Americans took the Japanese plan seriously.

These proposals led to no actual result because, as stated, the conference immediately bogged down in stubborn disagreement between the British and Americans over questions of cruiser size and tonnage limits. Nonetheless, they are worth noting because they would, in fact, not be forgotten; when these same nations met again in London three years later, these proposals for destroyers, particularly the British proposals, served as the basis for the discussions that came to a more fruitful result, as will be seen below.

Business as usual, particularly in Rome

The failure of the Geneva Conference to reach any result was notable for its lack of impact on any nation's naval construction

plans. That was particularly the case with the Italians. More than perhaps any other navy, the RM was driven by internal economic and political forces, and a very simple view of force requirements that did not look far beyond the shores of the Mediterranean. Whatever animosity there was between the British and the Italians – there was much lingering resentment in Rome about the way the First World War had ended and the way most Italians believed they were treated in the Versailles Treaty – the RM saw Great Britain as a more distant threat to its immediate goals than was France, with its major naval bases at Toulon and Bizerte and with the island of Corsica and its colonies in the Middle East seemingly hemming in any dreams of expansion of Italian influence in the region.

Thus it was the MN that the RM watched most closely and to which it reacted in 1926 to news of the ordering of the first of French Guépard class destroyers, the larger follow-ons to the Jaguar class.[15] The design generated in that year by the Comitato Progetti Navi, the branch of the naval staff tasked with developing new designs for Italian warships, was a new type of *esploratori leggeri*, with better speed and armament packed into a smaller hull due to improvements in engineering plant and gun design. Twelve ships to this design were ordered to be laid down starting in May 1927; given the names of famous Italian navigators from history, they were built to the following specifications:

Navigatori class: 1929[16]

Displacement (standard):	1900 tons
Length (oa):	352ft
Beam (wl):	33ft 5in
Draft (max):	11ft 2in
Powerplant:	54,241shp (design); two-shaft Belluzzo/Parsons/Tosi geared turbines; four Odero/Yarrow boilers with superheaters[17]
Speed:	38kt (design); 39.5kt (actual)
Armament:	6 × 533mm trainable torpedo tubes (2 × 3; centreline); 6 × 120mm/50 Model 1926 (3 × 2); 2 × 40mm/39 Vickers/Terni 1915/1917 (2 × 1); 4 × 13.2mm Breda (2 × 2)

Shorter than the Leones by twenty feet and displacing almost 300 tons less, the *Navigatori*, like so many other early super destroyers, suffered from inherent instability due to excessive topweight. Within a year of their entering service, the units were docked and the height of their forward superstructure was lowered by one deck and the tripod upper foremast replaced by a pole mast.[18] At the same time, the height of their funnels was lowered, but this led to some issues with smoke interference with bridge operations; starting in 1932, a funnel cap was fitted to the forward funnel of each unit. In part because of stability issues and in part because of a chronic shortage of torpedoes, the triple torpedo tubes were replaced by double tubes in 1932–3.[19] They were designed to escort the fast, thin-skinned Condottieri, small *esploratori* (light cruisers) intended by the RM to perform high-speed sweeps of the western Mediterranean.

Due for modernisation and in need of strengthened anti-aircraft armament, the ships of this class were taken in hand starting in 1939 for a major reconstruction. Eventually ten of

The Navigatori class destroyer RN *Antonio Pigafetta* is seen here leaving the inner harbour at Taranto in 1934. Her funnels have been shortened, as has been the height of her forward superstructure. This led to smoke interference from the forward funnel, which, in turn, led to a funnel cap being fitted. (NHHC)

the twelve were rebuilt over the next year, with the bow given a noticeable rake and considerable flare added to the forecastle, adding over six feet to their overall length, while the hull was widened by 3ft 3in, the extra beam being used to increase fuel bunkerage. By the time these modifications were completed, which included restoring the triple torpedo tube mounts and replacing the 40mm and 13.2mm anti-aircraft guns with at least seven 20mm/65 Breda Model 1939 or 1940 single-mounts, any benefit of the increased beam had been largely negated.

The *Navigatori* were well liked in the fleet and hard-used, despite the fact that they never entirely overcame their basic instability problems; they were always tender in a cross-sea, leading to excessive rolling, and were wet forward before their 1939 reconstruction. Their newer model 120mm main-battery mounts suffered from the same design flaw as the older model: the two barrels, sharing a common sleeve, were mounted too close together and therefore suffered from salvo dispersion until the powder charge was reduced to lower muzzle velocity and a delay was built into the firing circuit so that the two barrels did not fire exactly simultaneously. Eleven of the twelve were sunk during the Second World War – seven before the Italian armistice (two by British submarines, one by 'friendly fire' from an Italian submarine, two in gun battles with British surface forces, one by British air attack, and one by mining) and four after (two were lost on 9 September 1943 attempting to oppose the transfer of German troops between Sardinia and Corsica, one was scuttled to prevent her capture by the Germans, and one actually entered German service, but was destroyed by Allied bombers). Only one – *Nicoloso da Recco* – survived the war and remained in Italian service in the post-war Marina Militare until stricken in 1954.

Fooling around: The Anglo-French Compromise of 1928

The failure at Geneva had stung, both the pride of the Americans, who had hopes for another diplomatic success like the Washington Conference, and the more general hopes for those around the world who genuinely believed in the process of negotiating the limitation of naval arms. The League of Nations Preparatory Commission, which had adjourned *sine die* when the Geneva Conference had begun in June 1927, did not meet again until 15 March 1928, and then got off to an inauspicious start due to a gross misunderstanding on the part of the British. This had its roots, apparently innocently enough, on 9 March 1928, when Austen Chamberlain, then Great Britain's Foreign Secretary, met with Aristide Briand, who was once again Foreign Minister in the ongoing game of French Cabinet roulette, in an attempt to form a united front at the upcoming League Commission meetings.[20]

> . . . Sir Austen Chamberlain discussed yesterday with M. Briand the disarmament position with reference to the forthcoming meeting of the Preparatory Commission on the 15th March.

Scion of a well-established political family, and older half-brother to an even better-known (and more notorious) politician, Austen Chamberlain, seen here in 1931, was British Foreign Secretary in 1928 during the Conservative government led by Stanley Baldwin, and helped craft the unfortunate Anglo-French Naval Compromise.

> . . . Chamberlain began by regretting that on the two crucial military and naval questions the French and British were diametrically opposed. British public opinion, with its traditions, believed that voluntary armies were defensive and that conscript armies implied offensive war, whereas he was fully aware that to the French compulsory military service seemed a guarantee of a pacific policy and a voluntary army appeared in the dangerous light of a Praetorian Guard. He went on to say that we had been reconsidering what we thought essential from the naval point of view, . . . British public opinion realised that concessions were necessary from all parties for a general settlement to be reached, and if he could point to a concession by the French in naval matters it would probably acquiesce in his giving in on the military side. . . . He did not want to be driven in the matter of the limitation of military reserves to vote against France with Germany, and still less with Russia.
>
> M. Briand said that this must on no account

Chamberlain's French counterpart in 1928 was Aristide Briand, one of a generation of French politicians who seemed to exchange roles in government as frequently as they changed shirts.

happen. Experience at Geneva showed that, if Great Britain and France stood together, what they agreed upon would be accepted by others. . . .

Turning to the navy, M. Briand asked for a personal copy of the revised Admiralty proposals of which Sir Austen Chamberlain had informed him. He said that he would put them to the French naval authorities and use all his influence to obtain their acceptance.

. . . He had often told the French Admiralty that a naval war with the British was out of the question, and he was quite prepared to meet the criticism he had encountered for agreeing at Washington to a very small battleship allotment; the French would, in fact, never vote money for large numbers of battleships. The real naval crux for France was Italy. At the Three-Power Conference the British and the Americans had disagreed over the basis of estimating cruiser requirements; France had *vis-à-vis* Italy much the same case that we had as regards the United States. In each case our oversea responsibilities were much greater and our sea communications much longer. It would be to our mutual advantage, when the proper time came, jointly to press that those factors should form the basis of assessment of the cruiser strength which States could justifiably claim.[21]

The 'revised Admiralty proposal' that Chamberlain wanted Briand to press his government to accept was not particularly new; it resembled fairly closely previous British proposals at Geneva, offering multiple categories of ships and lacking specifics on aggregate tonnages. It called for six categories of warships: 1) capital ships; 2) aircraft carriers; 3) warships between 10,000 and 7000 tonnes; 4) warships below 7000 tonnes; 5) submarines; and 6) smaller warships exempt from regulation. This was not much different from previous British proposals and was intentionally vague about where the line would be drawn between categories 4 and 6. Two 'stipulations' were included in an attempt to address known French sensitivities. One would have allowed for the transfer of tonnage downward across categories with the exception of categories 1 and 2. This very much aligned with French wishes. The other would have allowed any nation with a total aggregate naval tonnage in all categories less than 80,000 tonnes to be free of all restrictions. This was a sop to the French who saw their role in the League discussions, in part, as acting as a defender of the rights of smaller nations.[22]

Both sides treated this initial burst of enthusiasm with a good deal of caution, as both governments saw a good deal of risk in any such agreement. Were the British to accede to the French demand for continued army conscription, it would inevitably alienate the Germans, whom the British wanted to engage in the disarmament process, because they were restricted to a tiny, all-volunteer army by the Versailles Treaty. However agreeable the French idea of basing the allocation of a nation's smaller warships on the scope of that nation's 'overseas responsibilities' (ie, the size of its empire) and the length of its 'sea communications' might have sounded to the British, there is no question that it was understood in London that taking this stand would certainly antagonise the Italians and the Americans.[23]

This led to delay on the part of the British, but the French were encouraged enough by the prospect of an disarmament deal with Britain that, together with the British, they caused the newly reconvened Preliminary Commission session in Geneva to be adjourned again at the end of March before any serious discussions had taken place. At the beginning of June 1928, the French reopened the discussions with the British by ignoring the earlier British six-category proposal and offering a vague one of their own based solely on the length of lines of communications, stating that this was in accord with previously stated American and British proposals.[24] Chamberlain seemed happy with this idea and wired back to London on 3 June: 'I am to discuss question fully with M. Boncour and shall try to get an agreement on these lines . . .'[25]

The reaction in London could hardly have been more negative. The First Sea Lord, Adm Charles Madden, responded, pulling no punches:

> The Admiralty suggest that we are on unsound ground if we agree with the French that the length of the lines of communication should form the basis for estimating . . . strengths. The statement that such a basis is at the root of the arguments of ourselves and United States is not correct. We have always maintained that the length of the lines is only one, and not the most important, of the factors. . . .
>
> It is not clear . . . what the French mean by lines of communication. It is important to be certain of this. Do they mean the lines of communication along which the whole trade of the country passes, such lines going to foreign as well as colonial ports or do they mean the lines of communication only to overseas possessions? If the latter it is likely to raise serious resistance both from America and Italy and if we accepted it we should find ourselves aligned with France against these two countries and basing ourselves on an argument most difficult to justify.
>
> The former definition is considered the true one but is probably not that intended by France as thereby she would obtain no advantage over Italy in relative . . . strengths.[26]

The same day, 5 June 1928, on which Madden was shutting one door, soundly rejecting the formula that Chamberlain found so hopeful, it seemed that, miraculously, another one opened. It came out of a conversation in Geneva between the British naval delegate to the Preliminary Commission, VAdm William A H Kelly, and his French counterpart, Capt Deleuze. It was of such interest to Kelly that he reported it back to London before the day was out:

> I had a conversation this morning with the French naval representative on the subject discussed between Sir Austen Chamberlain and Monsieur Briand on Saturday last, 2nd June.
>
> I asked him what was his definition of 'lines of Communication' as a basis of naval strength, as I found it very difficult to understand. He was not very helpful . . . and pointed out the difficulty of comparing a line of communication to Tahiti, and the lines across the Mediterranean to Algeria. He also said that their long lines of communication to the Far East would have to be abandoned in case of war . . . He considered that it was useless striving to obtain an agreement which it was quite certain Italy would never accept with reference to France, and the United States with reference to Great Britain. He thought it would only cause bad blood without any compensating advantage to be got from it.
>
> He said the 3rd Bureau had worked out some sort of scale providing for cruiser strength on a lines of communication basis, but he personally thought it was worth nothing.[27] Of course, in this conversation he was not representing the views of his department, but his own personal views.
>
> What the French desire to find is a formula of any sort which will give them a superiority over Italy, but which at the same time will not hurt Italian pride, and this can never be obtained by categories.
>
> He agreed that in coming down to actual facts, the real cause of the breakdown of all naval negotiations is not Great Britain, but the impossibility of reconciling the French and Italian points of view, viz. the Italians will never put their signature to any document which does not admit their right to parity with France, and France, with her double frontiers in the Channel and Mediterranean and her distant colonies, cannot sign away her right to a considerably larger navy than Italy, on account of greater commitments.
>
> Every effort must be made to find some formula for limitation which will save the face of the United States when comparing with Great Britain, and Italy when comparing with France.[28]

So far, Kelly was simply reporting that the discussions between the British and French had failed to yield a usable formula for limiting the construction of 'auxiliary warships'. The British plan for six categories of warships and the French proposal for a scheme based on the length of lines of communication were both correctly seen as being more likely to raise antagonism than provide a basis for agreement among the Washington 'powers'. However, Deleuze had more to offer:

> He personally had had a new idea lately, which he thought might possibly meet the case, but feared it would be unacceptable to us. It was as follows:
>
> Owing to the possibility of arming merchantmen for which full provision has been made by Great Britain and Italy, it is obviously impossible to limit ships with 6" guns and below, and if this is the case why go on trying to do so.
>
> The Washington Treaty already limited battleships and aircraft carriers: why not go a step further and limit all ships below 10,000 tons mounting a gun of over 6", leaving surface vessels of 6" guns and below limited only by the necessities of the respective States.
>
> In ships below 10,000 tons armed with guns of over 6" calibre, Great Britain could accept parity with America and France with Italy, and in the smaller class all States would retain their liberty of action.
>
> Submarines would be limited in a separate category.
>
> I asked him whether the Ministry of Marine would accept this arrangement, and asked him to put it before them to obtain their views.[29]

It is unlikely that Capt Deleuze could have come up with a

proposal more likely to please the British, as it would restrict the big cruisers, which they greatly desired to cap at current levels, but leave them free to build as many smaller warships as they believed they needed. However, it could have not been more clearly stated by Kelly that this was in no way an official proposal from the French government. Deleuze stated he 'personally' had come up with these ideas; a point which Kelly clearly understood, given that he asked Deleuze whether the Ministry would endorse them. Nonetheless, the British hastened to accept the proposed plan, and, without any further word from Kelly on the matter, Chamberlain, with full authority from the Cabinet, on 26 June instructed the British Ambassador in Paris to inform Briand of his nation's agreement with the French proposal.[30]

The French reaction was uncharacteristically rapid and negative:

> Admiral Kelly learns from his French colleague that the French Government would not accept solution suggested by him in regard to which Cabinet decided to approach French government.
>
> It appears that opposition comes not so much from Ministry of Marine as from politicians who consider the public could not be brought to understand system of limitation which exempted so large a class of surface vessels.[31]

In fact, there were serious repercussions for Capt Deleuze, who was threatened with loss of his position before tempers cooled on the French side.[32] Indeed, tempers cooled quickly and the French proved willing to discuss the Deleuze plan with VAdm Kelly, talking with VAdm Henri Salaün and VAdm Louis-Hippolyte Violette, the former and current Chiefs of Staff of the MN. By 11 July, Kelly was able to report that the French were considering two alternative limitation schemes – one being essentially the Deleuze proposal calling for four regulated categories of warships and the other plan was a reworking of the old American plan from the Geneva Conference, which had been at least nominally accepted by the French and Japanese as well, with limited tonnage transfer added in to satisfy French desires.[33] When Kelly reported on his discussions, he stated that the French Council of National Defence, which would meet two days hence, would likely only consider the Deleuze plan, but even if it were approved by the council, that the French could not be the ones to present it to the rest of the 'powers', because it differed too greatly from their prior publicly-stated positions.

The deliberations of the French Council of National Defence resulted in a general acceptance of the Deleuze plan with requests for a few small adjustments that stretched the willingness of the British to compromise.[34] These included proposing that whatever tonnage limitation was agreed for large cruisers and large submarines would be equal for all of the 'powers', and that smaller submarines would be, as would be the case with smaller warships, exempt from limitation. Thus the long-sought Anglo-French Naval Compromise, in its final form, was submitted by Chamberlain on 27 July 1928 as follows:

> Limitations which the Disarmament Conference will have to determine will deal with four classes of warships:
> (1.) Capital ships, i.e., ships of over 10,000 tons or with guns of more than 8 inch calibre.
> (2.) Aircraft carriers of over 10,000 tons.
> (3.) Surface vessels of or below 10,000 tons armed with guns of more than 6 inch and up to 8 inch calibre.
> (4.) Ocean-going submarines, i.e., over 600 tons.
>
> The Washington Treaty regulates limitations in classes (1) and (2), and the Disarmament Conference will only have to consider the method of extending these limitations to Powers non-signatory to this treaty.
>
> As regards classes (3) and (4), the final Disarmament Conference will fix a maximum tonnage applicable to all Powers which no Power will be allowed to exceed for the total of vessels in each of these respective categories during the period covered by the convention. Within this maximum limit each Power will at the final conference indicate for each of these categories the tonnage they propose to reach and which they undertake not to exceed during the period covered by the convention.[35]

The British Foreign Office sent this precise text in messages to the United States, Japan, and Italy on 30 July, with appropriate verbiage encouraging a sympathetic reading and a rapid reply before the next meeting of the general assembly of the League of Nations scheduled for 3 September.[36] It appeared the British feared that a session of the Preliminary Commission might convene before then and Chamberlain wished to have all of the Washington powers on board should that occur.[37]

The reaction from the three nations that received Chamberlain's message varied considerably. In Italy, the initial reaction can only be described as peevishness, even jealousy. The British Ambassador, Ronald W Graham, reported on 3 August that 'in conversation Signor Grandi [Italian Under-Secretary of State for Foreign Affairs] said that Signor Mussolini was painfully impressed by Anglo-French negotiations, Italy being left out in the cold'.[38] The reaction reported from Japan was favourable, but realistic:

> Naval attaché reports that Ministry of Marine still have matter under discussion, but that naval attitude is sympathetic and ready to support proposals, which they consider offer new hope of agreement. At the same time, they foresee many difficulties, particularly as regards United States.[39]

The Japanese prediction was remarkably accurate; the reaction of US Secretary of State Frank B Kellogg is best described as incredulity. There is no question that, on top of any unhappiness the Americans may have felt at being presented what looked for all intents to be a fait accompli – a deal between the British and French neatly tied up with a bow, presented as 'offer[ing] the best, if not the only prospect of making an advance from the present position' – for good measure, it included a deadline for a reply, making it sound like an ultimatum.[40] This was hardly the way to win favours from Kellogg, a man who was fiercely protective of America's newly won position of international strength. He literally could not believe what he was reading and asked for clarification on nearly all of the major points in a note to Chilton dated 2 August:

> In the note above referred to there are certain points which I do not fully understand. These points are as follows:
>
> One of the categories of war vessels is stated to be as follows: "(3) Surface vessels of or below ten thousand tons armed with guns of more than 6 inch and up to 8 inch calibre."
>
> I should be grateful if you would inform me whether this means that there is to be no limitation on any surface vessel armed with guns of 6 inch calibre or less. If this is the case, it appears that all destroyers might be built in unlimited numbers and that the same would be true of cruisers armed with 6 inch guns or less.[41]

There were several other points on which Kellogg requested clarification, but the first point was the critical one; the Americans were simply stunned that the British would come back, barely a year after the Geneva Conference had foundered on the issue of the British desire for large numbers of smaller, 6in-armed cruisers, with a proposal that not only repeated, but extended, the same basic idea. It was, to use modern jargon, a non-starter, as far as the Americans were concerned. That same day, Kellogg received direct instructions from Coolidge to refrain from committing the United States to any arms limitation scheme during his upcoming European sojourn to sign the Kellogg-Briand Pact.[42] His reply the next day assured the president that he had no intention of agreeing to the Anglo-French Compromise.[43]

Kellogg took his time replying to the Chamberlain, finally sending a long note on 25 September 1928, the crux of which is as follows:

> The position of the Government of the United States has been and now is that any limitation of naval armament to be effective should apply to all classes of combatant vessels. The Franco-British agreement provides no limitation whatsoever on six inch gun cruisers, or destroyers, or submarines of 600 tons or less. It could not be claimed that the types of vessels thus left without limitation are not highly efficient fighting ships. No one would deny that modern cruisers armed with six inch guns, or destroyers similarly armed, have a very high offensive value, especially to any nation possessing well distributed bases in various parts of the world. In fact, such cruisers constitute the largest number of fighting ships now existing in the world. The limitation of only such surface vessels as are restricted in Class 3 of the draft agreement, that is, cruisers of or below 10,000 tons, armed with guns of more than six inch and up to eight inch calibre, would be the imposition of restrictions only on types peculiarly suited to the needs of the United States.[44]

This effectively brought any chance of success for the Anglo-French Compromise to a close. The Japanese went on to signal their general acceptance of the terms of the compromise on 29 September, while the Italians announced their rejection of the plan on 6 October, neither of which, in the end, made much difference in the outcome. It would take the replacement of the American president in the election of 1928 and of the British Prime Minister in mid-1929 to clear the air enough for another international disarmament conference to become possible.

Trying again: London, 1930

The accession of Herbert Hoover in the United States and Ramsay MacDonald in Great Britain in 1929, and of their respective heads of foreign affairs – Secretary of State Henry L Stimson and Foreign Minister Arthur Henderson – brought a new spirit of co-operation and compromise into Anglo-American relations. Hoover was a Republican, like his two predecessors in the American presidency, but unlike them, he had much broader life experience to inform his actions; as a young man, he had worked as a mining engineer in Australia and Asia and, then, as a consulting engineer around the world. At the outbreak of the First World War, he had been in England, where he became involved in rescuing Americans trapped on the continent and, later, led the efforts to supply aid to the civilian population in Belgium. So successful was he at that endeavour that, after the United States joined the war, President Wilson recruited him to manage food distribution in America.

As president, Hoover had the misfortune to be in office when the world's economies collapsed beginning in October 1929, an event for which he was poorly equipped to respond, as a strict economic conservative, but that was still in the future when the Preliminary Commission of the League's General Disarmament Conference reconvened in April 1929. Once again, the United States, although not a member of the League of Nations, was represented at the meeting by Hugh

The American ambassador to London at this time, and a walking definition of the word 'polymath', Charles G Dawes is seen here in his uniform as a Brigadier General in the US Army's Corps of Engineers, a rank to which he rose during and immediately after the First World War. A banker by training, he won a Nobel Peace Prize for crafting the Dawes Plan in 1924, which eased Germany's reparations payments and helped stabilise the European economy. He was elected vice president of the United States in that same year. Years earlier, he had written a popular melody that, many years later, would top the music charts in both the United States and Great Britain.

Gibson, who took an active role. Gibson made it clear in his opening remarks at that session that the United States still insisted on extending the idea of limitation by categories to smaller warships.

> My Government believes that there can be no complete and effective limitation of armament unless all classes of war vessels, including cruisers, destroyers, and submarines, are limited. It could not agree to any method which would result in leaving any class of combatant vessels unrestricted.[45]

But, he added a new willingness on the part of the Americans to be flexible in their approach. In particular, he said they would consider schemes such as that proposed at an earlier session by the French, in which the aggregate tonnage for smaller warships would be agreed by treaty and the characteristics of those warships defined by category, but the allocation between categories within the aggregate total would be left to each nation.

> My country's defense is primarily a naval problem. The American Government has found no reason for modifying its view that the simplest, fairest, and most practical method is that of limitation by tonnage by categories, a method which has been given practical and satisfactory application in the Washington treaty . . .
>
> The American delegation has urged this view throughout the first reading, but, in view of the inacceptability to some other delegations of our unmodified thesis, my Government has sought in the various methods presented some solution which might offer the possibility of compromise and general acceptance. During the third session of the Preparatory Commission, the French delegation brought forward a method which was an attempt to combine its original total tonnage proposals with the method of tonnage by categories. Under this method, a total tonnage was assigned to each nation, and this total divided among categories of ships by specified tonnages. If I am not mistaken, certain modifications were suggested in informal discussions, so as to provide that the tonnage allocated to any given category might be increased by a certain percentage to be agreed upon, such increase to be transferred from any other category or categories not already fixed by existing treaty.
>
> In the hope of facilitating general agreement as to naval armaments, my Government is disposed to accept the French proposal as a basis of discussion. It is, of course, the understanding of my Government that this involves an agreement upon the method alone and not upon any quantitative tonnages or the actual percentages to be transferred from one category to another.[46]

Even further, Gibson mentioned a willingness to consider the application of a formula – a 'yardstick' – that would allow the tonnage of older or more lightly armed ships to be reduced in terms of how much they counted against a nation's aggregate tonnage.

> My Government is disposed to give full and friendly consideration to any supplementary methods of limitation which may be calculated to make our proposals, the French thesis, or any other acceptable to other powers, and if such a course appears desirable, my Government will be prepared to give consideration to a method of estimating equivalent naval values

While pushing for global naval arms reduction and struggling with a difficult post-war economy, the United Kingdom continued to attempt to maintain a navy as powerful as it had before the First World War. The Royal Navy understood this would not be an easy task, given the large number of pre-war and wartime construction destroyers in the fleet that were 'war-weary'. Not surprisingly, the Royal Navy looked to the two most experienced yards to produce prototypes of a post-war design that would be both affordable and take advantage of any technological advances that had occurred in the meantime. Thornycroft's prototype, laid down in January 1925 was HMS *Amazon* (D39), seen here at completion in 1927. In general, she did not look much different from the late-war ships of the 'W' and Modified 'W' classes, differing mainly in having a larger bridge, modelled after the tower bridge being used in the new cruisers being built at this time. *Amazon* was considered a success and served as a model for the succeeding 'A' and 'B' classes.

> which takes account of other factors than displacement tonnage alone. In order to arrive at a basis of comparison in the case of categories in which there are marked variations as to unit characteristics, it might be desirable in arriving at a formula for estimating equivalent tonnage to consider certain factors which produce these variations, such as age, unit displacement, and caliber of guns.[47]

There is no question this latter idea would appeal to the British, whose Royal Navy retained a large number of First World War-vintage light cruisers; a proper yardstick would allow those older ships to be heavily discounted, making room under any tonnage limit for more of the new, smaller 6in-armed cruisers they wanted to build.

There was something in Gibson's remarks to appeal to every listener, opening the doors to further discussions, which is exactly what Hoover and the Americans wanted. Although no one in Geneva that day foresaw the economic disaster looming on the horizon six months away, it is safe to say that every nation welcomed the prospect of reducing expenditures on armaments, especially warships, which were the most expensive items in each nation's military budgets.

When Ramsay MacDonald assumed the role of prime minister again in June, he made it abundantly clear that among his highest priorities was improving relations with the United States and he found Hoover eager to co-operate. But, still, the haggling over cruisers continued, the British continuing to insist on their need for as many as forty-five smaller, 6in-armed trade protection cruisers, while rejecting the American desire for more of the 'treaty maximum' 8in-armed cruisers, which better suited the US Navy's needs, on the seemingly-sensible grounds that if the Americans built more large cruisers, the Japanese would demand the right to do so as well.[48] As a sop that they thought would somehow satisfy the Americans, the British offered the possibility that they would agree to the US Navy building ten 10,000 ton cruisers armed with 6in guns, a type of ship the Americans had never wanted.[49]

Fortunately for the negotiating process, the Americans had an exceptional man as their ambassador in London, Charles G Dawes. A true polymath, a lawyer and banker by training, he joined the US Army in the First World War and rose to the rank of brigadier general based on his abilities as a logistics officer. (Thereafter, he would generally be addressed as 'General Dawes' in formal situations, even though he resigned from the army in 1919.) After the war, he was chosen by Warren G Harding to head the newly created Budget Bureau; in that role, he crafted the Dawes Plan in 1924, which rescheduled Germany's reparations debt and prevented an economic crisis, earning him a share of the Nobel Peace Prize of 1925.[50] From 1925 to 1929, Dawes served as vice president under Coolidge. When Hoover was elected, Dawes was offered and accepted the post in London, where he proved highly effective. Along the way, Dawes had, in 1912, composed a popular melody which remained in common circulation until 1951, when lyrics were written for it; seven years later, a recording of the song reached number one on both American and British music charts. He thus became the only person ever with the unique distinction of having written a chart-topping melody, served in American national elective office, and won a Nobel Prize.[51]

As seemed to be inevitably the case at this time, the main issue was the number and size of cruisers the Americans and British would allow each other. The subject of destroyers appears to have come up only in passing; in an 18 July 1929 message from Dawes to Stimson, which quoted a letter received from MacDonald as part of the ongoing run-up to what was hoped would be another five-power naval disarmament conference, the following was included:

Yarrow built a prototype at the same time, HMS *Ambuscade* (D38), which was not as successful as *Amazon* in the eyes of the Royal Navy. Other nations, however, found the design attractive, and ordered near-repeats from Yarrow. In the case of the Netherlands, four ships were built at Dutch yards to the Yarrow design, starting with ZRMS *De Ruyter* (later renamed *Van Galen*), seen here between 1928 and 1934 before the name change. Only the lower main-battery guns had narrow shields, covering their fronts only; the superfiring guns, covered here in canvas, lacked shields. (SLV – Green Collection)

> We [Great Britain] agree to parity in destroyers and in submarines, parity in this case being equal gross tonnage in each of the two categories.[52]

Stimson's reply to Dawes on 21 July was straightforward:

> On the assumption that parity in other classes is reached, we accept parity in destroyers. As the Prime Minister states, such parity should consist of equal gross tonnage and it is our suggestion that this be achieved by the scrapping of destroyer tonnage on the part of the United States until it is equal to the present tonnage of the British. This determination may encounter public opposition in this country, inasmuch as the United States now has a clear preponderance over Great Britain in the destroyer class such as it does not have in any other. However, when fleet parity is actually in sight, this is a real contribution toward armament reduction which this Government is able and prepared to make.[53]

From the moment he took office in June, MacDonald had been eager to arrange a visit to the United States to meet with Hoover. One of Dawes' tasks was to cool the prime minister's eagerness, knowing that an agreement in principle on the major points of a naval arms limitation accord needed to be worked out before the two heads of state met, or the world press would consider the meetings to have failed, something neither side wanted. The agreement by the American side to scale back their fleet of rapidly-aging 'flush-deckers' built at the end of the First World War was a major concession, but addressed only part of the problem. The other was determining at what aggregate tonnage limit the agreed-upon parity should be fixed.

The Americans at this time had 284 destroyers in commission or in reserve, whose completions dated from 1913 to 1921, but none of which could be considered particularly modern.[54] A very rough approximation would put their aggregate displacement at 315,000 tons. The US Navy had no destroyers under construction and little prospect that funding for new destroyers would be forthcoming from the parsimonious Republican-dominated Congress. The Royal Navy had 152 destroyers, which count included twenty then under construction. The oldest dated from 1916. A similarly rough estimation put their aggregate displacement at 204,000 tons. If that tonnage were taken as the agreed-upon limit, the Americans would have to scrap approximately 100 of their oldest, smallest, and most worn-out destroyers, something that would not be seen, in the short term, as a great sacrifice, given that at least that number were in reserve at this time due to funding and manpower shortfalls.

The twenty new destroyers being built by the Royal Navy were the result of a process that began with a meeting called by the navy's Controller (Third Sea Lord, responsible for procurement) in November 1923 to begin planning the designs of new ships.[55] When the discussion turned to destroyers, he made it clear that the intent was to start from the design of the Modified 'W' class of 1918, which had benefitted from the accumulated experience of four years of war and with which the navy was well pleased, and add any improvements garnered from technological advances in the intervening years. The list of new features or improvements desired by the meeting's participants, who included representatives of the technical departments as well as the fleet, grew quite lengthy. It eventually included a larger, 'tower' bridge, like that planned for the new cruisers of the Kent class, with sufficient volume to contain the radio facilities of a flotilla leader, the sound equipment for a sonar set, and the transmitting station associated with a fire control director to be mounted atop the bridge. Also on the list were increased ammunition for the main battery (190 vs 150 rpg), a more compact powerplant using preheating and/or superheating and geared turbines, and, perhaps, most challengingly, almost 60 per cent greater radius of action (an endurance of 5000nm at twelve knots, compared to 3120nm for the Modified 'W' class).

Two prototypes had been ordered and built to somewhat different designs starting in 1925. One, HMS *Amazon* (D39, later I39), built by Thornycroft, was notably more successful and served as the design model for eighteen 'A' and 'B' class destroyers ordered in 1928 and 1929. Visually, these ships were hard to distinguish from their First World War predecessors except for the larger bridge, the somewhat larger gun shields, and quadruple rather than triple torpedo tube mounts. Those eighteen ships comprised two annual programs of eight ships, enough to equip a flotilla, and two more of the 1929 'B' class built for Canada.

The Royal Navy also ordered a leader for each of the eight-ship flotillas under construction. During the hectic days of wartime emergency construction, the acquisition of standard destroyers and flotilla leaders became disconnected; on several occasions, destroyers were ordered in batches of more than thirty at a time, and there was no correlation between those orders and the flotillas to which the ships would be assigned, so the ordering of flotilla leaders was similarly disconnected from the ordering of 'mainstream' destroyers. In this new, more parsimonious, post-war world, the Royal Navy was grateful to get funding for the eight destroyers sufficient to constitute a homogeneous flotilla and adopted a policy of attempting to build a leader for each new flotilla. For the 'A' class of 1927, the leader was HMS *Codrington* (D65, later H65, D82 & I82); for the 'B' class of 1928, the leader was HMS *Keith* (D06, later I06). Two very different philosophies went into the design of these two ships. *Codrington* was 20ft longer and almost 220 tons heavier than her classmates, and carried one more main-battery mount along with a larger bridge and an extended after deckhouse to accommodate the

An interesting image of an incident from the Second World War illustrates the difference between the Royal Navy's late-First World War-construction destroyers and those that followed in the late-1920s. HMS *Anthony* (H40), foreground, and *Wishart* (I67) are seen in the process of sinking the German submarine *U761* (Geider), whose position had been marked by the smoke floats in the foreground dropped by an American PBY Catalina of VP-63 flying out of Gibraltar, 24 February 1944. *Anthony*, an 'A' class ship, looks much larger than Wishart, a Modified 'W' class Thornycroft Special, in this image, but most of that is due to foreshortening caused by distance, because *Anthony* was, in fact, only 11ft longer than *Wishart*. The difference in the size and shape of their bridge structures and the quadruple torpedo tubes of *Anthony*, compared to *Wishart*'s triples, are more obvious distinctions.

flotilla staff. She had bigger engines, giving her a somewhat faster top speed, but her greater length caused her to have a significantly greater turning radius than the standard 'A' class ships, leading to difficulties during tactical manoeuvres. However, these problems were not yet known in mid-1929, when the design of *Keith* was finalised. It was other reasons, primarily financial, that led to the decision to build *Keith* on the same hull as the standard 'B' class destroyer with the same main-battery armament and powerplant, but with the enlarged bridge and after deckhouse as on *Codrington*. The result was a ship that had very similar speed and handling characteristics to the other ships in her flotilla.

Hoover, in no way a 'Navy man', was eager to see an even greater reduction in the numbers of smaller warships, both to save money and, as a Quaker by upbringing and a true advocate for international peace, a means to lessen the likelihood of future wars. In a letter he addressed to MacDonald, dated 17 September, and sent through Dawes, he wrote:

> There are some other phases of the problem which seem, to me also of the utmost importance and could quite well be taken up on Mr. MacDonald's arrival here with view to making an announcement after his visit of an accord much more powerful from a world point of view. At various times in these discussions we have referred to the maximum destroyer strength of somewhere about 150,000 tons for each country. If we could agree on this figure, it would in itself mark a great tonnage reduction on both sides, although we would each require some construction for replacement.[56]

The British Prime Minister responded on 23 September in a letter to Dawes. He expressed extreme pleasure at the prospect of meeting Hoover, but, in reference to the desired five-power naval disarmament conference, which by this point had gotten at least tacit approval by all parties and which would be held in London as soon as possible, MacDonald felt compelled to point out, rather melodramatically:

> If the appearance of parity is to be obtained, neither of us can get away from the fact that the standard must be fixed by British needs. The tides of events swelling upwards and downwards, backwards and forwards, change our defence problems every year and with that the figures change.[57]

He added a further note of caution on the matter of destroyers, with an appropriately theatrical flourish at the end:

> I believe I should have no difficulty in closing at once with figures in the region of the President's proposals. But the tonnage in destroyers depends largely on the tonnage put by other powers into submarines. I am warned that certain other powers will not agree to a limitation in submarines. I might be willing to support something like the President's figures, but what can I do if the Five-Power Conference were to reject them?[58]

MacDonald sailed for the United States on 28 September 1929. From then, until the London Conference convened on 21 January 1930, all of the discussion before the conference began and through its first days was centred on cruisers and, to a lesser extent, submarines, with little time or energy spent thinking about destroyers. Although Hoover and MacDonald had readily agreed on cruiser limits that seemed satisfactory, neither nation's naval establishment was pleased with the deal they reached and the London Conference began with more tension within the British and American delegations than between them, and with the French and Japanese as adamant as ever that they needed higher allocations of these lesser warships than they had received in the capital ship ratios set in the Washington Treaty.[59]

It soon became obvious that the French and the Italians would not be coerced into signing on to any agreement for limiting smaller warships acceptable to the other three 'powers'. The result was a bifurcated treaty, the first two parts signed by all five nations and the last only by the United States, Great Britain, and Japan. The first two parts dealt with

The need for flotilla leaders still existed, and the Royal Navy ordered HMS *Codrington* (D65), seen here in 1932, to lead the 'A' class. She was 20ft longer than a standard 'A' class destroyer and carried an additional main-battery mount, located between the funnels. Other than those differences, she looked much the same as the ships of the flotilla she was intended to lead. (via Ken MacPherson)

Codrington's greater length caused her to handle differently from her flotilla-mates, a fact that was not yet known when the time came to order the leader for the 'B' class destroyers, which were identical in size and other characteristics to those of the 'A' class. HMS *Keith* (D66), seen here in 1931–32, was ordered on the same size hull and with the same armament as her classmates, differing only in having a somewhat larger bridge and a larger after superstructure to accommodate Captain (D) and his staff.

extending the 'battleship holiday' an additional five years (until 1936) and revising the definition of what constituted an aircraft carrier. Certain limits were agreed to regarding the size of aircraft carriers and submarines. Any mention of destroyers was notably lacking from the first two parts of the treaty signed on 22 April.

The third part of the treaty, agreed only by the British, Americans, and Japanese, attempted the much harder task of limiting the size and aggregate tonnage of smaller warships. The real substantive heart of the treaty can be found in Articles 15 and 16. In the first of those articles, destroyers were defined as '[s]urface vessels of war the standard displacement of which does not exceed 1,850 tons (1,880 metric tons), and with a gun not above 5.1-inch (130 mm.) calibre'.[60] Article 16 laid out the total aggregate tonnage each of the three signatories would be allowed in the four covered categories: heavy cruisers, light cruisers, destroyers, and submarines. For destroyers, the allocated tonnages were: 150,000 tons each for the United States and Great Britain, and 105,000 tons for Japan. This gave Japan a 70 per cent allocation of tonnage in the destroyer category compared to the 60 per cent allocation in capital ships and aircraft carriers she had been granted at Washington, a result that, in theory, addressed the Japanese goal of better treatment in this round of negotiations.

An interesting proviso was added in the 4th section of Article 16:

> 4. In the destroyer category not more than sixteen per cent. of the allowed total tonnage shall be employed in vessels of over 1,500 tons (1,524 metric tons) standard displacement. Destroyers completed or under construction on the 1st April, 1930, in excess of this percentage may be retained, but no other destroyers exceeding 1,500 tons (1,524 metric tons) standard displacement shall be constructed or acquired until a reduction to such sixteen per cent. has been effected.[61]

The intent of this clause was obvious. Neither the Americans, who had no destroyers of any size planned or under construction and all of whose older destroyers displaced much less than 1500 tons, nor the British, of whose twenty new destroyers then being built only one exceeded 1500 tons and who possessed only eight older destroyer leaders displacing more than that amount, would be impacted immediately by that restriction. The Japanese, on the other hand, were the obvious target of the clause, as the twenty-four Special Type destroyers them in the works were all over 1500 tons standard displacement. These were similar to, but not exactly the same as, the limits proposed three years earlier in Geneva by the British; the main difference was an increase of 1000 tons for both the maximum of what would be considered a destroyer and for what would constitute a standard destroyer. These new limits were proposed by the British in a policy memorandum outlining their opening position promulgated at the opening of the conference on 7 February 1930.[62]

That same document stated that the British building programme then envisaged would require an aggregate total of 200,000 tons for destroyers, but that could be reduced if the submarine totals for other nations (meaning France and Japan) were reduced. The Americans reacted to that number as might be expected. Hoover took the time before departing on vacation to state his conviction that he believed that figure was too high and he hoped that a number closer to 150,000 tons could be agreed upon.[63] On 28 February, Stimson, who led the American delegation, cabled back to Washington that agreement had been reached with the British to lower the 200,000 ton figure, but that much depended on French and Japanese concurrence.[64] The French, as seen, soon dropped out of these discussions, but the Japanese remained to be convinced. In the end, however, it was the Americans who were the last party reluctantly to agree to a deal. If the Americans and British accepted a 150,000 ton limit for destroyers and if, as the Americans continued to insist, the 5:5:3 ratio negotiated at Washington was still to be applied, then the Japanese would be allowed only 90,000 tons; but they wanted a higher limit. On 4 March, Stimson sent back to Washington his reluctant acquiescence with the Japanese demands:

> They ask 105,000 tons in destroyers as against 150,000 tons for the British and Americans. Ninety thousand for Japan is our maximum. In view of the great preponderance in destroyers which we now have this appears fair.[65]

With this concession, the deal was struck.

The fact that the Japanese accepted this proviso in the treaty was no doubt in part an acknowledgement of the fact that they knew that it made economic sense to begin building a somewhat smaller ship, one that could displace less than 1500 tons, and, perhaps, in part because they most likely were already considering the renunciation of the arms limitation process, something that would not formally occur until 1936. Regardless of the thoughts in the minds of the participants, the bizarre ceremony at which the First London Treaty of Naval Arms Limitation was signed, in part by the five 'powers' and in total by Great Britain, the United States, and Japan took place on 22 April 1930.

Chapter 7

Strange breed – London's children: 1930–1937

THE LONDON TREATY, FOR what it was worth, bound the world's three largest navies to limit the size and aggregate tonnage of destroyers they could build, and impacted the building plans of other nations as well for the next five years. The degree to which a nation was impacted depended on a number of factors including: whether that nation had signed the treaty, especially Part III with its specific restrictions on smaller warship size and numbers (or even if it was one of the nations that was part of the treaty process); where that nation was in its shipbuilding cycle; and fiscal issues, because the worldwide financial downturn had impacted different countries differently, and each country reacted to it differently, but every nation was impacted in some manner. Some reacted by contracting public spending dramatically to reflect the reduced tax revenues and others expanding spending, attempting to 'build' their way out of the depression. The circumstances peculiar to each nation will be addressed in the following sections

More of the same in England, except . . .

At the time of the London Conference, when the Great Depression was just starting and no one could predict its duration or when it would reach its nadir (as it would in early 1933), the British had an aggressive naval building programme under way including, as described in the preceding chapter, twenty destroyers from the 1928 and 1929 programs, establishing the precedent that each year's naval budget would include monies for a new flotilla of destroyers – eight 'standard' destroyers and one leader – to replace the war-built ships reaching obsolescence. The Royal Navy asked

Exercising in the Baltic just before the outbreak of the Second World War, two of the four large destroyers that the Polish navy built after becoming a state again in 1920 can be seen in this image. On the left is ORP *Burza*, one of two ships built in France to an improved *Bourrasque* design, entering service in 1930. On the right is ORP *Grom*, one of two very large destroyers built at J S White (Cowes) between 1935 and 1937. They were, briefly, the fastest destroyers in the world. Of note is the main battery; each of the Groms carried seven Bofors-built 12cm/50 SK M36/P in three twin mounts and one single mount on the forecastle forward. The unusual arrangement of main-battery guns can be made out in this image with a twin-mount superfiring over the single-mount. (Fulvio Venturi)

for a similar programme in 1930, but with the London Conference under way for the first few months of that year and the first full impacts of the depression being felt, MacDonald opted to ask only for a half-flotilla of four destroyers and one leader, which became the 'C' class, differing from the preceding 'B' class in being slightly longer and fitting a single-mount 12pdr(3in)/45 QF HA gun. These ships, including their leader, HMS *Kempenfelt* (I18) never fitted into the Royal Navy's flotilla structure and, so when the opportunity arose in 1937–8, they were transferred to the Canadians, where they were actively involved in the Battle of the Atlantic, only two surviving the war. Eight ships and a leader of identical design were ordered as the 'D' class in 1931.

Still pleased with the basic design, the Royal Navy continued ordering nine-ship annual programs in 1932 and 1933 to an only slightly modified design. These had a somewhat longer hull with a bit more flare and sheer forward. Continuing past practice, they were each given names starting with the letters 'E' and 'F'. Based on feedback from the fleet stating that the 12pdr HA gun was too slow-firing to be an effective anti-aircraft weapon, especially without direction, it was replaced in these ships by a pair of quadruple 0.50in/62 Vickers MGs Mk III, which were equally ineffective, but made a lot more noise. The identical 'G', 'H', and 'I' classes, each of nine ships ordered in each of the following years, differed from the 'E' and 'F' in only minor details.

Nonetheless, with the eyes of Royal Navy planners still mainly rivetted on the Pacific, and on Japan as the most likely next enemy, the emergence of the Fubuki class and the following Special Type destroyers starting in 1928 could not be completely ignored. As happy as the Royal Navy was with their standard destroyer type, it was evident that the Japanese Special Types outclassed them on an individual ship basis, and the need to build a ship capable of operating with the fleet in

the Pacific and taking on any potential enemy counterpart had, by the end of the 1930s, come to the forefront. Other designs have been mentioned as influencing Admiralty thinking in the mid-1930s. Among those are the Yugoslav destroyer *Dubrovnik*, designed and built at Yarrow in Glasgow starting in 1930. Large, at 1880 tons standard displacement, and heavily armed with four Skoda 14cm (5.5in)/56 single-mount guns, this ship was nevertheless very much patterned after earlier British destroyer leaders. Another design that may have been even more influential was that made by J S White for the Polish Navy in 1935. The Poles had previously ordered two destroyers similar to the French Bourrasque class but, looking for a ship to fight the new large destroyers building on either side of them in the Baltic (see below), they turned to British yards, seeking tenders for a much larger leader-type destroyer. They chose the design offered by White's, which specified a ship of over 2100 tons armed with seven 12cm (4.7in)/50 guns. The main battery of the two ships of the Grom class, laid down in July and October 1935, shared the same calibre as the standard British destroyer main-battery gun, but was manufactured in Sweden by Bofors and fired a 90.4 lb fixed round; the British 4.7in/45 Mk XII fitted in their newest destroyers fired a separate round with shell and cartridge having a combined weight of 80.5 lb. The unusual number of main-battery guns in the Groms was arrived at by arranging them in one single mount and three twin mounts. They achieved a high trials speed (39 knots) on a relatively modest 54,000shp by dint of lighter construction than normal British practice, a feature that appears to have caught the Admiralty's attention.[1]

The decision to proceed with the design and construction of a class of large destroyers, initially dubbed 'V' class leaders, was made by the Royal Navy in late 1934. The not-unexpected announcement by the Japanese in December of that year that they were withdrawing from the arms limitation process added to the pressure to consider larger destroyers for use in the Pacific.[2] The instructions generated by the naval staff listed the full range of destroyer duties, emphasising those tasks related to escorting major fleet units, such as scouting, screening, and supporting destroyer flotillas, with less emphasis on individual attack or anti-aircraft defence. This led to a strong emphasis on single-purpose gun armament at the expense of torpedoes. Preliminary sketches dating from March 1935 showed a ship mounting ten 4.7in guns in the main battery and a single quintuple torpedo tube mount on the centreline. The various sketch designs offered different layouts of the guns (and in some cases substituting one or more anti-aircraft weapons for one of the main-battery mounts) and different top speeds, ranging from thirty-five to thirty-seven knots. In all cases, the stated displacement was 1850 tons, in line with London Treaty restrictions. At the beginning of August, the Admiralty signed off on a design with four main-battery twin mounts, a single quadruple torpedo tube mount – reduced from a quintuple when it was learned that the twin 4.7in mounts would be heavier than anticipated – and an as-yet-undetermined number of 2pdr (40mm) Vickers pom-poms. (Part of the reason for choosing

This image shows ORP *Blyskawica*, sister of *Grom*, at a British port soon after they fled the Baltic ahead of the fall of Poland in September 1939. The censored patch on the ship's side was the Royal Navy pennant number (H34) hastily painted on to show the ship's allegiance with the Royal Navy. (For reasons that must have made sense at the time, the British wanted to keep the accession of these Polish assets a secret.) As naval technology improved, among other gains was a shrinkage in the size of boilers; *Blyskawica*'s four Admiralty-type three-drum boilers were small enough that their uptakes could be grouped into a single funnel, allowing deck space for two banks of torpedo tubes and two twin 40mm/60 Bofors anti-aircraft mounts, one of the earliest examples of this soon-to-be-ubiquitous gun. (NHHC)

The Royal Navy continued to build destroyers in annual flotilla-sized programs (eight ships and a leader) through the first half of the 1930s, giving each year's batch names starting with the next letter of the alphabet. Thus, the 1929 program provided for the 'C' class and so on, through the 'I' class of the 1935 program, each class being only an incremental improvement over the preceding, with very little to distinguish them visually or tactically. HMS *Hereward* (H93) of the 'H' class was an exception, because she was used briefly to test a new twin mount for the standard 4.7in/45 gun that had been the main battery in Royal Navy destroyers since being introduced in the Shakespeare class leaders in 1917. The twin mount proved successful, but was too heavy for the 1340 ton standard displacement *Hereward* to bear and was replaced by her normal main battery of two single-mounts forward in April 1937.

HMS *Hereward* is seen here soon after the outbreak of the Second World War, after her reversion to the normal main battery for an 'H' class destroyer, with four single-mount 4.7in/45 QF Mark IXs. Note the RAF roundel painted on the forward slope of her bridge face as an air recognition symbol. That came in handy in this instance, as this photograph was taken by a US Navy patrol aircraft off the Florida coast on 20 December 1939, well within American territorial waters. (NHHC)

the eight-gun rather than the ten-gun design was a concern, well-founded in retrospect, about the ability of Vickers to deliver the required number of 4.7in guns and mountings for the rapidly increasing number of destroyers being ordered.) The first seven ships to the following specification were ordered in March 1936:

HMS *Cossack*: 1938[3]

Displacement (standard):	1854 tons
Length (oa):	377ft
Beam (wl):	36ft 6in
Draft (max):	13ft
Powerplant:	44,000shp; two-shaft Parsons single-reduction geared turbines; three Admiralty three-drum small-tube boilers
Speed:	36kt
Armament:	4 × 21in trainable torpedo tubes (1 × 4; centreline); 4 × MkIX 21in torpedoes; 8 × 4.7in/45 QF MkXII (4 × 2); 4 × 2pdr/39 QF MkVII (1 × 4)

Nine more identical ships were ordered in June 1935; in 1939, both Canada and Australia each agreed to build an eight-ship flotilla of these large destroyers – by now known as the 'Tribals' because all the Royal Navy ships, like the First World War-era Tribals, were named after indigenous warrior groups found in the British Empire. All sixteen of the Royal Navy Tribals were completed in 1938–9. Two of the Australian Tribals were ordered immediately from the Cockatoo Island Yard (Sydney) and were completed in 1942; a third was laid down in 1942 and completed in 1945. Four of the Canadian

The twin-mount 4.7in/45 QF Mark XII was intended to be fitted on a ship such as this one, HMS *Cossack* (L03), seen soon after her completion in June 1938. One of the first of the new, large second Tribal class intended particularly for service in the Pacific against the threat posed by the Japanese, who menaced the widely separated possessions of the British Empire in the Far East. For this, the Admiralty believed they needed a class of destroyers with exceptional speed, range, and firepower. The resulting design was a ship with graceful lines and good gunpower, but one that was poorly balanced, with less than adequate anti-air and anti-submarine capability, and substandard durability due to light construction. (NHHC)

Despite their limitations, the Tribals, of which sixteen were built, were popular in the fleet, and were worked hard during the Second World War. Only four of them survived that conflict. One that did survive was HMS *Ashanti* (F51, later G51) seen here at Hvalfjordur, Iceland, on 4 October 1941. She has the fixed antenna for a Type 286 radar on her foremast and a Type 285 gunnery control radar antenna on the fire control director on her forward superstructure. (NARA)

Tribals were ordered in 1940 and 1941, and delivered in 1942 and 1943; the remaining four were to be built at the Halifax Shipyard. The first two of those were ordered in June 1941 and the second pair in early 1943, but they were not started until 1942 and none of the four Canadian-built Tribals were completed before the end of the war.

Once in service, the Tribals were well received in the fleet, which liked their size and firepower, but also exposed their most obvious weaknesses – the relatively light construction led to some fracturing in their shell plating, especially near the break of the forecastle, which in turn caused some leakage into the hull and into fuel bunkers, and the lack of a dual-purpose main battery, which made them vulnerable to air attack. Those that survived the first years of the war had their after superfiring 'X' main-battery mount replaced by a 4in/45 MkXIX twin-mount. (Some of the late-construction Canadian Tribals were completed with all four main-battery mounts replaced by the 4in gun.) Regardless of their weaknesses, these ships were hard-used in the Second World War and paid a heavy price for such use; of the sixteen Royal Navy Tribals, only four survived the war. Those four were all decidedly 'war-weary' and were disposed of in 1948–9. Of the eleven Tribals built for or by the Commonwealth navies, only one was lost in combat; of the remaining ten, nine saw combat during the Korean War before being gradually paid off during the later-1950s and 1960s. One, HMCS *Haida* (G63), has been preserved as a museum ship at Hamilton, Ontario.

Not nearly as lucky as her flotilla mate, seen just above, HMS *Matabele* (G26), seen also departing Hvalfjordur the same day as *Ashanti*, would not last three months from the date of this image before being sunk by a single torpedo from *U454* (Hackländer) while escorting convoy PQ-8 from Iceland to Murmansk. Note that the gunhouses of the main-battery mounts carried by the Tribals were open at the rear, leaving most of the gun crew exposed to the weather and shell splinters. (NARA)

A hesitant restart across the Atlantic

With so many 'flush-deckers' built between 1918 and 1921, the US Navy was not in a strong position to convince a reluctant Congress to fund new destroyer construction, especially after the economic downturn cut into tax revenues. The General Board of the US Navy, responsible for developing ship requirements, was well aware that these old destroyers were obsolescent, and had been so from the day of their completion, lacking the firepower and range of better contemporaries. Nonetheless, with so many relatively new hulls in the water, it was not until January 1930 that serious work defining the requirements for a new fleet destroyer began.[4]

Here, the Americans were perhaps fortunate that the 'flush-deckers' had been such mediocre warships. Unlike the British, whose late-First World War Modified 'W' class had been so generally considered a highly successful design and so widely admired by other navies, that almost twenty years later successive classes of only slightly modified follow-ons were being built in annual flotilla-sized programmes, the Americans could look back to 1918 and see exactly the ship they did *not* want to repeat. Where the 'flush-deckers' had mounted four 4in/50 Mk 9 single-mounts and twelve 21in torpedo tubes in four sided triple mounts, the proposed new design would carry five 5in guns of a new design and eight 21in torpedo tubes in two quadruple mounts on the centreline, giving this ship more gunpower and a greater torpedo broadside.

The new main-battery gun was the 5in/38 Mk 12,

When the US Navy began building destroyers again in 1932, they knew quite well what they did not want, which was to repeat the 'flush-decker' design built a decade earlier. The Farragut class design settled upon after lengthy discussion was a reasonably well-balanced compromise with an excellent dual-purpose main battery, a large, heavy but advanced-for-its-day, Mk33 tachymetric gun director mounted above the forward superstructure (seen covered in canvas in this view of USS *Dale* (DD 353) dated 28 April 1938), and six torpedo tubes. Note that in an attempt to save topweight, only the forward guns had partial shields. Also note the depth charge racks aft, added after completion. (NARA)

developed from the 5in/51 Mk 9 'wet gun', which used separate projectile and cased powder charge, by the simple expedient of cutting down the length of the barrel.[5] All models were hand loaded, but power rammed at all angles of elevation, which facilitated their use as a dual-purpose weapon. Most mounts elevated from -5° to +85°. The first models of this gun were a pedestal mount design with ammunition supply through the superstructure behind the mount; later versions, starting with those carried in the USS *Gridley* (DD 380) class of FY1935, had a base-ring mount with shell and powder hoists within the mount, which permitted the rate of fire to increase from 12–15rpm to 15–22rpm. Equally critical to the success of this weapon was the employment of the Mk 33 director control system, a tachymetric gunfire control director that incorporated a mechanical, analog, electrically-driven computer – a Mk 10 Ford Rangekeeper – with an attached stable element, all in a huge, open-topped pedestal mount atop the bridge structure. With the integrated 12ft stereoscopic optical rangefinder, this system could track aircraft at speeds up to 275kt, later increased to 320kt, taking approximately 30sec to generate a firing solution including range, elevation, and bearing transmitted by synchronised pointers to each main-battery mount. This combination of gun and DP fire control proved to be highly effective, especially with later director models, such as the Mk 37, which separated the fire control computer from the rangefinder housing, positioning the Ford Rangekeeper and its stable element below decks, allowing a small and lighter control top that was far less cramped.

The Farragut class destroyers served into and through the Second World War, as exemplified by USS *Aylwin* (DD 355), seen here gathering guard mail from *Sargent Bay* (CVE 83) near the Philippines on 5 December 1944 and then breaking away. The visible wartime modifications mainly have to do with electronics systems and anti-aircraft defence. She has an SC-2 air search radar antenna at the foremast head and an SG surface-search radar below that, and a Mk28 parabolic antenna (45in diameter) on the front of the Mk33 director. A pair of twin Bofors 40mm/56 mounts have replaced the No3 main-battery mount abaft the second funnel, while at least six single 20mm/70 Oerlikon mounts can be seen. A pair of depth charge throwers have been added each side abreast the after deckhouse.
(NARA)

Despite being 500 tons heavier than the preceding Farraguts, the Porters had questionable stability, carrying so much topweight on a relatively small hull, and their single-purpose main battery limited their utility as fleet escorts. Therefore, a plan was devised to replace their main battery – and heavy directors – with a more conventional main battery of five dual-purpose 5in/38 guns (carried in two twin mounts and one single-mount) served by a single Mk37 director, at the same time landing the eight reload torpedoes. A lack of time and money prevented this conversion from being carried out in all but one of the Porters. Only *Selfridge*, seen here post-conversion, on 19 April 1945, had those changes made, and that only because she lost her bow back to No2 main-battery mount at the Battle of Vella Lavella in October 1943. (NARA)

Equally important, the US Navy put extra emphasis on improving range with this new design. The old 'flush-deckers' had proved to be disappointingly short-legged, many achieving only 2400nm at economical fifteen knot cruising speed. (In comparison, the Royal Navy's Modified 'W' class could reach 3210nm at fifteen knots, still far from a satisfactory radius, given the distances that would be involved in a war in the Pacific.[6]) There were two basic ways to improve a ship's range; either fuel bunkerage could be increased or fuel efficiency improved. It was the intent of the General Board that both remedies would be pursued. Increasing fuel stowage was the easy part; the new design the board was proposing would be twenty-four feet longer and almost four feet broader than the 'flush-deckers', providing room in the hull for almost double the bunkerage (371 tons vs 600 tons war emergency capacity). Increasing endurance came in part from incremental improvements in turbine design, but mainly from the adoption by the US Navy, almost alone among the major navies at this time, of higher steam operating temperatures and pressures. The new design was to make use of both to get more energy out of the steam being generated by any given amount of fuel; where the Royal Navy had built their first post-war destroyers with engineering plants operating in the range of 290psi and 400°F, the first American post-war class, admittedly not built until five years later, had an engineering plant operating at 400psi and 648°F. These factors combined to give the new design a planned endurance of 7200nm at twelve knots. (Actual range of the USS *Farragut* (DD 348) class, as built, proved to be 6500nm at twelve knots.)

Proposing such a ship and getting them built was always two different things in the United States. Democratic Senator David I Walsh of Massachusetts, longtime Chairman of the Senate Committee on Naval Affairs, left a detailed account of the transition from the austerity of the Harding-Coolidge-Hoover years to the major build-up of naval forces that came under Franklin D Roosevelt, elected president in November 1932. The sea-change in American politics, however, that saw the waning of conservative Republican sentiment, which had held sway steadily since the end of the First World War and which had strongly opposed military expenditures, had begun even before the Great Depression took hold in late 1929, and had accelerated in 1930. In the FY1933 naval appropriation bill, passed in June 1932, just as Congress had begun to swing in favour of the more liberal Democrats, monies were earmarked for five new destroyers (DD 348–352).[7] Three more (DD 353–355) would be funded in the following fiscal year. The specifications to which this first class of post-First World War American destroyers were built were as follows:

In a rare, detailed close-up view, the forward main-battery mounts and the Mk33 gun director, with the Mk28 radar antenna mounted forward, can be seen on the lead ship of the Mahan class destroyers, USS *Mahan* (DD 364) as she is wrapping up a refit at Mare Island Navy Yard, 24 June 1944. The white circles indicate items newly added or changed during the refit. These include a pair of 20mm/70 Oerlikons and the Mk28 antenna in front of the Mk33 director. Of great interest is this look into the Mk33 director, normally covered with canvas when not in use; its enclosure was open-topped, exposing its integrated stereoscopic rangefinder and analog tachymetric fire control computer. (NARA)

USS *Farragut* (DD 348): 1934[8]

Displacement (standard):	1500 tons (design); 1358 tons (actual)
Length (oa):	341ft 3in
Beam (wl):	34ft 3in
Draft (max):	11ft 7in
Powerplant:	42,800shp (design); 40,353shp (actual); two-shaft Parsons single-reduction geared turbines; four Yarrow boilers with superheaters
Speed:	36.5kt (design); 36.6kt (actual)
Armament:	8 × 21in trainable torpedo tubes (2 × 4; centreline); 5 × 5in/38 Mk12 (5 × 1); 4 × 0.50cal/90 M2 Browning MG (4 × 1)

The Farraguts proved to be every bit as good as the US Navy had hoped. The head of BuC&R, RAdm Emory S Land, famously gave a presentation to the General Board in January 1933 in which he described the Farraguts in glowing terms, claiming that, compared to the preceding 'flush-deckers', they increased speed by 10 per cent, doubled metacentric height (meaning improved stability), and significantly increased armament and endurance, plus providing the better seakeeping offered by the raised forecastle deck and vastly improved accommodation for officers and crew. All this was achieved in ships that actually displaced almost 10 per cent less than their designed displacement and only 14 per cent more than the 'flush-deckers' they were replacing.[9] They were indeed so far in advance of their predecessors that American destroyermen gave them and the several similar classes that followed them a couple of nicknames; they were called '1500-tonners' for their general average displacement, but more affectionately were dubbed 'goldplaters' for the comparative lavishness of their crew's accommodations.

Of course, like any other warship design, the Farraguts proved far from perfect in service, nor had their designers been able to anticipate future demands on them. Even when new they were a somewhat unbalanced design, favouring the anti-ship aspect of a destroyer's missions – being well-armed with guns and torpedoes – over anti-air or anti-submarine roles. These other roles were not ignored, simply not emphasised to the same extent. Their main armament was an excellent DP weapon, especially when combined with the Mk33 director, but there was an obvious limit to what five heavy anti-aircraft guns could do, even when well-directed, against a sustained attack, and the four Browning machine guns were a token addition at best. A *Farragut* would be, at best, a useful addition to a layered anti-aircraft defence formation around a more valuable target. Likewise, each of the Farraguts was completed with a QC sonar installed, but no depth charge racks or throwers. Visual evidence indicates that by 1938 depth charge racks had been added aft.

Despite RAdm Land's assertion about the *Farragut*'s high metacentre, from the beginning they were noted for being top-heavy and prone to heel excessively in turns and heavy weather; the addition of topweight as electronics and additional medium anti-aircraft weaponry were fitted before and during the upcoming war only made the problem worse. The No3 main-battery mount was removed during the war to help compensate for the added topweight, but this did not resolve the problem. Three of the eight Farraguts were lost during the war, one by grounding in the Aleutians and two during a typhoon in December 1944. An enquiry into those latter losses concluded that the Farraguts by this point in their evolution suffered from a basic lack of stability. Nonetheless, the remaining ships in the class served out the war on active duty, but were decommissioned almost

The US Navy continued to push naval technology, particularly in the realm of steam engineering; each succeeding class during the 1930s had a more compact and efficient powerplant operating at higher temperature and steam pressures. The savings in the size of boilers allowed the trunking of all the exhaust gases into a single funnel in the destroyers of the Bagley class, such as USS *Helm* (DD 388), seen here from *Tangier* (AV 8) on 8 April 1942, at Nouméa, New Caledonia. Besides trunking four boiler uptakes into a single funnel, this broadside view shows the significant changes from the Farragut class, which included the decision to reduce the main battery by one gun and increase the torpedo battery to four quadruple banks of tubes – two per side. Note that both the forward main-battery guns and the Mk33 director are now fully enclosed. (NARA)

The last class of American inter-war destroyers – before the massive wartime production classes were initiated – were the twelve ships of the Sims class started in 1937. This design was able to reduce the number of boilers by one, which led to less prominent above-deck trunking and allowed the torpedo tubes to be returned to the centreline and the reversion to five main-battery guns. Early war experience proved the need for more anti-aircraft and anti-submarine weaponry, and all ships in the class, including USS *Anderson* (DD 411) seen here at Espiritu Santo on 5 February 1943, lost their No3 main-battery mount in favour of depth charge throwers and multiple 20mm/70 Oerlikon mounts. Note that all but main-battery mount No4 are fully enclosed; that mount was shielded up to the overhead level, which was open, covered only by canvas. The fire control director is now a smaller, lighter Mk37, which achieved its size and weight savings in part by moving the computer elements into the hull. (NARA)

immediately after the end of hostilities and sold off in 1947.

The five successive classes of standard destroyers built for the US Navy comprised fifty ships funded in FY1934–9. Each class was incrementally larger with improvements in detail, but all followed the same basic pattern with a raised forecastle deck forward and five single-mounted 5in/38 guns in the main battery. They were the Mahan (18 ships), Gridley (4), Bagley (6), Benham (10), and Sims (12) classes. They provided the core of the US Navy's destroyer forces in the years just before America's entry into the Second World War and the critical early days of that conflict.

Not only had the Americans not built any new standard destroyers in the years between 1921 and 1933, but likewise there had been no new larger destroyers or small cruisers that could act as flotilla leaders built since the last of the ten *Omaha* (CL 4) class light cruisers was completed in 1925. The Omahas were well liked in the fleet, being fast and good sea boats, but there were simply too few of them to cover the many tasks a small cruiser was required to fulfil in a modern fleet and they were rather large for the destroyer leader task. When the political climate shifted to one more amenable to the modernisation of the fleet in the early 1930s, the US Navy began exploring potential designs for dedicated flotilla leaders to go along with the new destroyers being funded starting in FY1933. While there were many in the US Navy who preferred the idea of light cruisers leading flotillas, and the Americans had no history of building larger destroyers to fulfil the leader role, the limitations to which they had agreed in the Washington and London treaties effectively eliminated the option of building small cruisers to lead their flotillas. The London Treaty did, however, allow them to build 16 per cent of their allocated 150,000 tons of destroyer-sized vessels with displacements greater than 1500 tons, but less than 1850 tons.

Thus, in early 1932, the General Board began by requesting a series of design sketches from Preliminary Design, which responded with three options, all weighing in at 1850 tons, differing in the number and arrangement of 5in/38 main-battery mounts: Scheme 5 had five single mounts; Scheme 10 had six single mounts; and Scheme 11 had six guns in two twin mounts and two single mounts.[10] All of the designs featured main-battery guns in weatherproof, unarmoured shields. In May 1932, the General Board chose Scheme 10 to pursue and asked for a detailed design to be worked up; in the process the option of substituting eight single-purpose guns of the same size and calibre in twin mounts was offered and the General Board accepted this option, in part because this saved enough weight that a full set of eight reload torpedoes could be carried for the two quadruple torpedo tubes. The resulting ships, eight of which were funded along with the eighteen ships of the Mahan class as part of the monies released in President Roosevelt's Executive Order No 6174 of 16 June 1933, were built to the following specifications:

USS *Porter* (DD 356): 1936[11]

Displacement (standard):	1850 tons (design); 1834 tons (actual)
Length (oa):	381ft 1in
Beam (wl):	36ft 11in
Draft (max):	13ft
Powerplant:	50,000shp (design); 47,271shp (actual); two-shaft Parsons single-reduction geared turbines; four Babcock & Wilcox boilers with superheaters
Speed:	37kt (design); 38.19kt (actual)
Armament:	8 × 21in trainable torpedo tubes (2 × 4; centreline); 8 × Mk8 21in torpedoes; 8 × 5in/38 Mk12 (4 × 2); 8 × 1.1in/75 Mk1 (2 × 4); 2 × 0.50cal/90 M2 Browning MG (2 × 1)

When launched, the Porter class leaders looked a lot like the light cruisers they were intended to complement (see Frontispiece); indeed, they looked more like modern light cruisers than did the now-obsolescent Omahas. In particular, their four large twin gunhouses, superfiring forward and aft, their two large tripod masts and the huge, box-like Mk35 gun directors positioned on the forward and after superstructures all tended to make them look more like cruisers than destroyers. Perhaps their most curious feature was their single-purpose main battery, selected because the weight saved over the much heavier dual-purpose mounts of the same weapon allowed the mounting of three more main-battery barrels and two of the newly developed quadruple 1.1in/75 medium anti-aircraft guns. The 1.1in quadruple-mount suffered from numerous teething problems, including excess vibration and a tendency to jam. These problems were resolved over time, but more basic issues, including a poor design that made the inner barrels difficult to access for reloading and extreme complexity that made the gun difficult to repair at sea, made its replacement by the 20mm/70 Oerlikon and the 40mm/56 Bofors starting in 1941 quite popular in the fleet.

From the beginning, the Porters' design was considered to be quite cramped, with little room for growth should any additional equipment or weaponry need to be added. By the time money was made available for the first two of the follow-on Somers class leaders in FY1935, they displaced almost 200 tons more despite sacrificing their tripod masts and one of their Mk35 directors, but they did add a third quadruple bank of torpedo tubes, replacing the eight reload torpedoes, that addition facilitated by the adoption of a single funnel.

From their commissioning, the single-purpose main batteries of these two classes were considered anomalous – they were the only post-1921 American destroyers not to carry a dual-purpose main battery – leading, in May 1940, to a report recommending the replacement of the single-purpose Mk22 mounts with dual-purpose Mk29 mounts; this would have retained the eight-gun main battery, but would have added 30,000 lb per mount, which would have had a ruinous impact on the ships' already questionable stability.[12] A more practical solution was offered in August 1940, which involved the substitution of two dual and two single DP mounts for the four existing dual single-purpose mounts; this change was approved, but never carried out in the rush of events leading to war. Along with most of their sisters that survived the early days of the Second World War, most of these leaders had the No 3 main-battery mount removed along with their 1.1in quadruple-mounts to compensate for the weight of electronics and multiple medium AA mounts added above deck. Only one of these leaders had something close to the proposed August 1940 modification carried out; USS *Selfridge* (DD 357), of the Porter class, losing her bow to a Japanese torpedo at the Battle of Vella Lavella on 6 October 1943, was rebuilt at Mare Island Navy Yard and, in the process, was given one single and two dual DP mounts as her new main battery, along with a single smaller and lighter Mk37 director replacing the two Mk 35s.

Some slight restraint across the Pacific

However reluctant Japan's agreement to the London Treaty of 1930 might have been, compliance with that agreement forced the Japanese to begin looking at a smaller version of the Special Type destroyer that would meet the 1500 ton limit for 'standard' destroyers – the IJN already had more than their

allowed percentage of destroyers exceeding that limit. To achieve this, the twelve ships funded in the 1931 Programme, were to be twenty-nine feet shorter than the original Special Type destroyers, carry one fewer main-battery gun, and be powered by a smaller, lighter engineering plant providing 16 per cent less power, but any saving that these measures might have achieved was countered by the retention of nine torpedo tubes in three triple mounts, one of which had to be mounted higher due to lack of room on the centreline, and of the same full set of torpedo reloads and the same large bridge structure as the Special Type destroyers. The designed specifications for this class of smaller, standard destroyers was as follows:

Hatsuharu class: 1933[13]

Displacement (standard):	1490 tons
Length (oa):	359ft 3in
Beam (wl):	32ft 10in
Draft (max):	9ft 11in
Powerplant:	42,000shp; two-shaft Kampon geared turbines; three Kampon boilers
Speed:	36.5kt
Armament:	9 × 61cm trainable torpedo tubes (3 × 3; centreline); 9 × 61cm Type 90 torpedo reloads; 5 × 12.7cm/50 3rd Year Type (2 × 2; 2 × 1); 2 × 13mm/76 Type 93 MG (2 × 1)

In the end, despite having a stated standard displacement of 1490 tons, the two ships of the Hatsuharu class that were completed to the original design actually displaced approximately 100 tons more than planned, with almost all of that weight high up above the waterline, causing significant instability. If that had not been enough to alarm the Imperial

In the Hatsuharu class, the Imperial Japanese Navy attempted to adhere to the tonnage restrictions imposed by the 1930 London Treaty. Compared to the preceding Special Type destroyers, the Hatsuharus were almost 30ft shorter and, on paper, displaced 260 fewer tons, compensated only by a reduction in one main-battery gun. In reality, only the first two, including *Nenohi* seen here in 1933, were completed before the Tomozuru Incident caused the stability of all Japanese naval vessels to be re-assessed. Note the massive bridge structure and the superfiring gun mount forward, both of which added to this ship's stability issues. (NHHC)

Navy's constructors, the Tomozuru Incident of 12 March 1934, which came soon after the first two ships of the class – *Hatsuharu* and *Nenohi* – had been commissioned in September 1933, caused all recently completed ships to be evaluated for stability and those found deficient, such as the first two Hatsuharus, to be laid up until they could be taken in hand for reconstruction. The remaining ten planned ships in this class presented different problems, depending on how far along they were in the construction cycle. Three had not yet been laid down and three more had been started, but were not far along in construction; work on them was halted and they, along with four more hulls added in the 1934 Programme, were completed to a revised design based in part on the changes made to resolve the stability issues found in the Hatsuharus, as will be described below. Two more ships of this class – *Wakaba* and *Hatsushimo* – had been launched by the time the news of the Tomozuru Incident was absorbed and understood. Work on those two was stopped until a complete redesign could be done. This include moving the single main-battery turret from its superfiring position forward of the bridge to just forward of the after main-battery turret, which in turn required a significant reduction in the size of the after superstructure and the removal of the third,

raised triple torpedo tube mount and the three reloads associated with it.[14] At the same time, the bridge structure and the funnels were both reduced in height, the extended superstructure forward of the bridge where the single-mount turret had been sited was removed, which allowed the bridge to be moved forward several feet, and additional permanent ballast was added, all of which raised the standard displacement to 1700 tons and reduced the top speed to thirty-four knots. The last two ships in this class – *Ariake* and *Yugure* – which had been less far along in their construction, were modified in a similar manner, except that they were fitted with lighter-weight versions of their main-battery gun turrets – Type A_1 single-mounts in the place of Type A single-mounts and Type C twin-mounts in the place of Type B Mod 2 mounts – the weight was saved by restricting the maximum elevation of the 12.7cm guns to 55° instead of 75°.[15] When the first two ships were taken in hand for modification, *Hatsuharu* and *Nenohi* were modified in similar fashion as *Wakaba* and *Hatsushimo*,

The same ship as in the previous photo, *Nenohi*, is seen again, now on 18 August 1937, near Shanghai. The changes are dramatic. The most noticeable is the move of main-battery mount No2 from the forecastle to a position aft, on the main deck just forward of mount No3. The forward deckhouse and most of the aft superstructure have been removed, lowering the height of the after bank of torpedo tubes. Minor reductions have been made in the height and mass of the bridge structure. All of the main-battery gun mounts have been reduced in weight by the lowering in height of the hood over the gunlayer's position. (NHHC)

The next class of Japanese destroyers, the Shiratsuyu class, incorporated all the changes made to improve stability in the Hatsuharus and, additionally, incorporated a smaller bridge structure. The result, exemplified by *Yamakaze* seen here pre-war, was a successful ship, but one that was less capable than the larger Special Types had been, and would be superseded as soon as Japan renounced naval arms limitation restrictions. (USN)

except that it was not possible to reduce the size of the bridge structure as much nor were the funnels reduced as much in height.

The six remaining ships of the 1933 Programme and the four authorised in 1934 were completed to a revised design quite similar to *Ariake* and *Yugure*. Known as the Shiratsuyu class, these ships were designed from the outset to carry their five main-battery guns in the same manner as the revised-Hatsuharus – two forward and three aft – but with a redesigned, more compact, forward superstructure and smaller fire room ventilator covers around the funnels; to make up for the loss of one triple bank of torpedo tubes (and their associated reloads), the Shiratsuyus introduced quadruple torpedo tubes, of which they mounted two with their reloads. The Hatsuharus, in order to make room for reloads for the middle bank of torpedo tubes, had their after funnel offset to starboard; the Shiratsuyus, because they had one fewer bank of torpedo tubes, were able to site both funnels on the centreline. All of the efforts by the IJN to reduce the topweight with this class brought the standard displacement down to 1600 tons.

These sixteen smaller destroyers proved to be quite successful in service; they were worked hard throughout the Second World War, particularly in the South Pacific. None of them survived the war. Despite the Imperial Navy's general satisfaction with these smaller destroyers, their efforts to build an adequate destroyer within the 1500 tons allocated by the London Treaty to standard destroyers had proved unsuccessful, and added one more argument for those naval officers who wanted Japan to renounce the arms limitation regime. Ten more destroyers were to be ordered in the 1934 Programme, but to a larger displacement design that would bring back the firepower and speed of the original Special Type. The December 1934 announcement that Japan was formally withdrawing from the arms limitation process rendered that decision an easy one to make.

Ever bigger challenges across the Ligurian Sea

Having laid down eighteen 2400-ton 'four-pipers' between 1926 and 1931, with which it was quite pleased, the MN nonetheless faced an emerging problem coming from its self-identified rival, Italy. In response to the large and capable French *contre-torpilleurs*, the Italians conceived their *grandi esploratori* (large scouts) in 1926, which were small, fast, lightly armed cruisers, the so-called 'Condottieri' because they would be named after Renaissance warlords, carrying eight 152mm/53 guns in their main battery at 36.5kt, the first of which would be laid down in 1928. As happens in any arms race, the French felt compelled to respond. Although the French believed that a three-ship division of their newest *contre-torpilleurs* could hold their own against one of the new Condottieri, there was a lingering concern they hoped to address in the next iteration of the type.[16] That was the range of their main-battery guns. The Condottieri were very lightly-armoured, to the extent that the 138.6mm/40 Mle 1927 main-battery guns carried by the later classes of *contre-torpilleurs* would have no trouble penetrating their side protection if they could get within range. However, this gun had a maximum range of 18,150yds, while the Condottieri's main battery could reach 31,060yds.[17] To counter this, the CDN demanded a main battery on the next *contre-torpilleurs* that could more nearly match the range of the larger gun carried by the Condottieri. The French response was a longer-barrelled version of the 138.6mm gun that extended the range of the main battery by almost 3000yds, and the adoption of higher steam pressure and temperature to improve speed and endurance, as had been tried out experimentally on the last pair of 'four-pipers', MN *Milan* and *Epervier*.

The six *contre-torpilleurs* of the 1930 Programme of the MN were built to the following specifications:

MN *Le Fantasque*: 1935[18]

Displacement (standard):	2570 tons
Length (oa):	434ft 4in
Beam (wl):	40ft 6in
Draft (max):	16ft 5in
Powerplant:	74,000shp; two-shaft Rateau-Bretagne impulse (3) or Parsons reaction (3) geared turbines; four vertical boilers
Speed:	37kts (design); 40kts (actual)
Armament:	9 × 550mm Mle 1923DT torpedo tubes (3 × 3; 1 × centreline; 2 × sided); 2 × depth charge chutes; 8 × 200kg depth charges (plus 6 reloads) per chute; forty × Breguet B4 500kg mines; 5 × 138.6mm/50 Mle 1929 (5 × 1)); 4 × 37mm/50 Mle 1933 (2 × 2); 4 × 13.2mm/76 Hotchkiss Mle 1929 MG (2 × 2)

Probably the first thing that stands out about the ships in this class is their speed. With their high-pressure, superheated steam turbine systems, every one of the six easily surpassed their designed top speed; all exceeded forty-three knots on trials and could maintain forty knots in service.[19] What is known is that the three ships with Rateau-Bretagne turbines had greater reliability and were somewhat faster than their Parsons-powered sisters.[20] If there can be any criticism of the design, it would be that they are somewhat lightly armed for their large size, although, conversely, this gave these ships excellent reserve stability which allowed those that had the chance the ability to add electronics and anti-aircraft weaponry without much worry.

All six of these ships fought into the Second World War and are seen as being the most successful of French *contre-torpilleurs*.[21] They were found to be somewhat lightly constructed when put to hard use in the winter of 1939–40,

when they combined with the battlecruisers MN *Dunkerque* and *Strasbourg* and three cruisers to form the Force de Raid, tasked with hunting down German commerce raiders in the North Atlantic. After that, the units of this class went on paths that led them to quite different destinies. *Le Triomphant* was damaged during a high-speed sweep of the Skagerrak with two of her sisters in April 1940 and was under repair at Lorient at the time of France's surrender in June. Her port propeller shaft, which had been bent by near-misses by German bombs, was under repair ashore, while her starboard turbine was dismantled for maintenance. By dint of tremendous effort, the propeller shaft and needed engine parts were loaded on board, and *Le Triomphant* was towed from Lorient to Plymouth, where the Royal Navy took over responsibility for her repair and refit. She emerged in November with a 4in/45 QF MkV anti-aircraft gun in place of the after superfiring main-battery mount, and British air search radar on her foremast. She joined the Free French cause, joined via various paths by three of her sisters. (One – *Le Malin* – fought the invasion fleet at Casablanca during Operation Torch in November 1942 and was badly damaged, but was captured and repaired by the Americans and joined the Free French cause at that point.) The two others were war losses – *L'Audacieux* captured by the Germans at Bizerte (Banzart), Tunisia, in a damaged state and later salvaged for parts and the other scuttled at Toulon to prevent capture by the Germans. The four which had joined the Free French side served on in the MN post-war, the last, *Le Malin*, being stricken in 1963.

After the last of the *quatre-tuyaux*, the Marine Nationale came up with a completely new design for its next class of *contre-torpilleurs*. Utilising an advanced high-temperature, high-pressure steam propulsion system, all six ships in the Le Fantasque class – the name ship is seen here on 30 May 1936 near Brest – exceeded their designed speed of 37kt with ease. The only external indication of the new powerplant was the use of two larger funnels in the place of four smaller ones. They carried a more powerful, longer version of the 138.6mm gun, but were considered by many to be undergunned for their size. (NHHC)

The MN planned another class of *contre-torpilleurs* to follow after the Le Fantasques. Their evolution would be delayed by the First London Conference, but only by a year, because, as was seen above, the French opted out of all the substantive clauses of that document, in particular those restricting destroyer tonnage. Nonetheless, as a result of French attendance at the conference and initial uncertainty as to its outcome, and then because of subsequent bi-lateral naval arms reduction talks with the Italians, the French government provided no funding for new *contre-torpilleurs* in their 1931 budget.[22] This 'oversight' was corrected in the following year's budget, which provided monies for one new *contre-torpilleur*, even larger and better armed than the preceding class, specifically intended as a replacement for MN *Amiral Sénès*, which had been launched in 1918. The main difference from the Le Fantasques would be a new enclosed twin mounting for the main battery that would allow it to carry six guns in three mounts in the place of five single-mounts.

Plans for this new twin 138.6mm/50 mount began in 1929. It was called a 'pseudo-turret' because it was not a completely enclosed gunhouse, but rather was open at the back, that opening covered by a canvas screen when the guns were not in use. It was not considered a real turret also because the powder and shell hoists did not rotate with the mount; it had, instead, a complex mechanism that delivered the rounds at a fixed point on the centreline to a rotating drum that tracked the training angle of the mount and delivered them to loading trays behind each breech. The STCN had concluded that a true turret with integrated hoists would be too heavy and too large to fit in a hull the size of the Le Fantasques. After considering, and rejecting, a proposal by the Ordnance Department, that anticipated issues with slow rate of fire from the main battery due to the complex ammunition train could be addressed – by storing a dozen ready-service rounds per gun in the pseudo-turrets and a like amount in the upper handling room – the decision was made to adopt the somewhat smaller and lighter 130mm/45 twin-mount proposed for the planned new *torpilleurs d'escadre* that would become the Le Hardi class (see below).

For a number of reasons, work did not begin immediately on this new ship. One reason was the ongoing naval arms limitation negotiations with Italy, brokered by British Foreign Secretary Arthur Henderson, which finally came to a positive conclusion on 1 March 1931.[23] The agreement included language specifically covering smaller cruisers and destroyers as a single category:

> (b) Cruisers with guns of 6.1-inch (155mm) calibre or less and destroyers: tonnage of new construction to be completed shall not exceed the tonnage which is replaceable in this category before Dec. 31. 1936. Vessels already over-age and vessels becoming over-age during the period of the treaty shall be scrapped on being replaced[,] except in cases where either France or Italy prefers to scrap instead an equivalent tonnage belonging to the category of cruisers with guns of more than 6.1-inch (155-mm) calibre.
>
> It is hereby declared that for the purpose of this arrangement member of the British Commonwealth of Nations and France and Italy do not intend to undertake replacement before Dec. 31, 1936, of any destroyer which will be under sixteen years of age on that date.[24]

One of the *contre-torpilleurs* of the Le Fantasque class, MN *Le Triomphant*, was undergoing repairs for a damaged propeller shaft at Lorient when, with the fall of France imminent in June 1940, she was towed to Plymouth, England to avoid capture by the Germans. There, the Royal Navy took over her repairs, substituting a 4in/45 QF Mark V for the No 4 main-battery gun and installing a Type 290 air search radar atop her foremast. In this condition, she was recommissioned as a unit of the Free French Marine Nationale. As such, she is seen at San Diego, CA, in August 1941, on her way to the South Pacific. (NHHC)

This gave France the legal 'cover' to replace *Amiral Sénès* with a ship of similar or even greater size, but even with such legal niceties out of the way and the money theoretically lined up, the actual release of the funds was delayed by the poor economic condition of the French treasury in 1931.

As far as MN was concerned, that delay was convenient, because the mission to which any new *contre-torpilleurs* would be assigned was being rethought. A new threat, at least as dire as that posed by the RM, was looming in the north. On the same day, 11 March 1931, that the text of the accord was first made public and both governments expressed satisfaction with its terms, the German Reichstag voted initial funds for the second new large cruiser allowed them under terms of the Versailles Treaty, meant to replace the antiquated pre-dreadnought battleships they had been allowed to keep after the First World War.[25] The first of these *panzerschiffe*, allowed to displace 10,000 tons and carry a main battery of 28cm/52guns, had been laid down in 1929 and would launch in May 1931; it was still referred to as 'Cruiser A' or *Ersatz Preussen*, following the German custom of naming a ship under construction after the existing ship it was to replace until it was christened at launch. This ship would be named RM *Deutschland* at launch, later renamed KM *Lützow* in 1939. It was the first of three large cruisers dubbed 'pocket battleships' by the British press. The second of these ships, 'Cruiser B' or *Ersatz Lothringen*, would be launched as RM *Admiral Scheer* in 1934.

The Germans had built a dozen 900-ton torpedo boats in the 1920s as treaty-compliant replacements for old destroyers, which had required no reaction from the French. (They would begin building the first of a series of super destroyers in 1934, as will be described immediately below.) The French, however, did feel compelled to react to the construction of the German *panzerschiffe*, all the more so after the 11 March vote to begin work on the second of the class. Design work had begun on the French response, which would be a fast, well-armed battlecruiser armed with 330mm/50 main-battery guns, as early as 1930, but funding by the constantly-shifting French governments was slow to materialise. Money to build the first of two Dunkerque class battlecruisers would not be approved until March 1932; at the same time money for the construction of the long-delayed new, large *contre-torpilleur* was also allocated, but the start of construction was delayed yet again, this time because the STCN wanted to re-evaluate the design in order to provide a ship specifically suited to scout for *Dunkerque*.

To meet this new requirement, this ship would need a hull large enough and strong enough to operate in the North Atlantic, where it was assumed the enemy's *panzerschiffe* would be active, and where, once it found the enemy, it would be armed with guns of enough power and a torpedo battery of enough size to damage the enemy in the short time that would be available as it used its superior speed and manoeuvrability to escape the enemy's longer-ranging guns.[26] Looking at these requirements, the STCN accepted that reverting to the twin-mounted 138.6mm/50 would be necessary to achieved the needed firepower; the only question was how many to mount. Two options were offered to the EMM. One was a slightly enlarged *Le Fantasque* with three main-battery twin-mounts, somewhat slower because of heavier construction; the other was more enlarged to carry a fourth main battery pseudo-turret. The staff opted for the second option, and with a few more minor changes, the first two of these new *contre-torpilleurs*, money for the second having been approved only in mid-1934, was laid down in December 1934 to the following specifications:

MN *Mogador*: 1938[27]

Displacement (standard):	2880 tons
Length (oa):	451ft 2in
Beam (wl):	41ft 4in
Draft (max):	15ft
Powerplant:	92,000shp; two-shaft Rateau-Bretagne impulse geared turbines; four Indret small-tube vertical boilers
Speed:	39kt (design)
Armament:	10 × 550mm Mle 1923DT torpedo tubes (2 × 3; 2 × 2; sided); 2 × depth charge chutes; 8 × 200kg depth charges (plus 8 reloads) per chute; forty × Breguet B4 500kg mines; 8 × 138.6mm/50 Mle 1934 (4 × 2); 2 × 37mm/50 Mle 1933 (1 × 2); 4 × 13.2mm/76 Hotchkiss Mle 1929 MG (2 × 2)

Both ships were commissioned in 1938 and entered service in 1939. Four more ships of this class were authorised in 1938, but were never laid down due to the outbreak of war. The two that were completed were never as successful nor as well liked in the fleet as the Le Fantasques for a number of reasons. They were designed with a bulbous forefoot, a common feature in larger ships, intended to improve buoyancy forward in ships with long, narrow bows, allowing them to handle rough weather better, riding through rather than plunging into oncoming seas; in the case of a smaller ship like the Mogadors, the forefoot tended to throw a tremendous spray up over the forecastle, making work there hazardous at high speeds even in calm weather. The heavy pseudo-turrets mounted high in the ship and the taller bridge necessary to provide visibility over the gun mounts led to a reduced metacentric height compared to the preceding class, making them less stable in turns and bad weather. The most serious problem faced by the two Mogadors was with their main-battery mounts. The ammunition train – which was a scaled-up version of the system developed for the 130mm guns of the Le Hardi class – was fiendishly complex and prone to regular breakdowns,

The last French *contre-torpilleurs* completed before the Second World War – and the fall of France that ended the line of development started with the Jaguar class of 1923 – were the two ships of the Mogador class of 1936, the lead ship being seen here in late December 1939, when she took Adm Darlan to Portsmouth, UK, for a strategy meeting with Royal Navy leaders. (Four more Mogadors were authorised, but never started due to the outbreak of war.) These ships attempted to make up for the perceived lack of firepower of the Le Fantasque class by increasing the main battery from five to eight by carrying the 138.6mm/50 guns in twin mounts. The guns were mounted in 'pseudo-turrets' which had a large opening at the rear covered by a rolling canvas curtain – as can be seen towards the left in this image – when not in use. Her hull is painted in dark grey, with light grey upperworks. (NHHC)

and, even when it worked correctly, the catapult rammers for the projectiles were so weak they rarely worked at an elevation greater than 10°.[28] Work under way to correct these issues essentially ended after the surrender of France in May 1940; both Mogadors were scuttled at Toulon in November 1942 to prevent their capture by the Germans.

Including the excluded

Two nations that had been major naval powers up to the time of the First World War had, for different reasons, fallen off the charts and had not been part of the naval disarmament discussions. One was Germany, which, as the primary instigator of the war, in the eyes of the victorious Allies, was naturally the nation to receive the harshest punishment at its conclusion. The Treaty of Versailles allowed the Germans to retain, among a very short list of assets, twelve destroyers and a like number of torpedo boats.[29] Not only were they to retain few ships, but the ones they were allowed to keep would be some of the oldest and most worn-out in their fleet.[30] The newest of the twelve destroyers incorporated into Germany's post-war Reichsmarine was RM (ex-SMS) *S23*, launched in March 1913; the oldest had been launched in September 1911. All twelve displaced approximately 570 tons. The treaty allowed for their replacement after fifteen years by new ships displacing 800 tons. This meant that the Weimar Reichsmarine would be legally allowed to replace the relics they inherited with new destroyers beginning in 1927.

When Weimar Germany was allowed to replace its oldest destroyers, it was natural that they built on their experience in the late war, and in particular on the designs of the classes of large destroyer built late in the war, such as the S113 class (see above). The lineal derivation of the design of KM *Georg Thiele* (*Z2*), seen here in 1939, from S113 is obvious with their short, low forecastles with minimal flare and a straight bow with minimal rake. Of course, there were differences: *Georg Thiele* had five main-battery mounts of a smaller calibre compared to S113's four larger guns; the new design displaced approximately 400 tons less on a hull that was more than 40ft longer; and the new ships had much more topweight, with a taller, tower bridge, like contemporary British designs, and taller funnels, all of which adversely affected stability. (NARA)

With those rules firmly in mind, the post-war Reichsmarine turned to the last successful destroyer design the Kaiserliche Marine had produced, in much the same way that the first light cruiser laid down by the Weimar navy in 1921 was a virtual repeat of the late-First World War Cöln class. As much as they might have preferred to repeat the large S113 class destroyers that represented the peak of German wartime destroyer development (and two of which had gone on to lengthy post-war careers in the French and Italian navies as MN *Amiral Sénès* and RN *Premuda* respectively), it would have been difficult to adapt the design of a 2060 ton destroyer to meet an 800 ton limit. Instead, the Reichsmarine was forced to look towards the late-war mobilisation destroyers built starting in 1916, ranging in size from 919 tons to 990 tons, for a more suitable design basis. These ships were armed with three 10.5cm/45 LC/16 single-mounts and six 50cm torpedo tubes in two single-tube and two double-tube sets. The Reichsmarine had wanted to mount a new, larger 12.7cm (5in) gun then being proposed, but the NIACC (Naval Inter-Allied Commission of Control) which oversaw the enforcement of the Versailles Treaty mandates, rejected that request. Six of these new ships were laid down in 1925 with virtually identical specifications, differing mainly in mounting their torpedo tubes in two triple sets and in being fitted with geared turbines. All were named after birds of prey and were known in the fleet collectively as Raubvogel. Six more virtually identical ships were laid down in 1927 and completed in 1928–9; named after hunting animals, they were known as the Raubtier. These twelve small destroyers (classed as torpedo boats by the Reichsmarine) were well liked and served into the Second World War, during which all were lost.

When it came time to consider the next set of replace-

Destroyers of the second group built by the Reichsmarine/Kriegsmarine differed little from the first, adding some keel depth aft to counter a problem with the stern 'digging in' at high speed. The change, not visible above the waterline, overcorrected the problem, causing the bows of ships in this group to be excessively wet. This view of an unidentified Type 34A destroyer, shows it in late-1939 leaving a German port on a mine-laying mission, its decks lined with mines to be dropped off the stern. (NARA)

ments, the limits under which the first post-war destroyers had been designed no longer were a concern. Even before Adolf Hitler became Chancellor of Germany in January 1933, Weimar Germany was veering increasingly towards the right; in July 1932, while Franz von Papen was Chancellor and Kurt von Schleicher was War Minister, Germany formally renounced the Versailles Treaty, allowing the Reichsmarine to look at purely military (and financial) issues while planning their next destroyers.[31] As early as mid-1932, the Reichsmarine had invited bids from private yards in Germany, prohibited by the Versailles Treaty from building warships, for a large destroyer, with no guidance as to specifications. The only responses came from Schichau, which proposed a 1470 ton ship armed with four 12.7cm guns, and Vulcan (Stettin), which proposed a smaller vessel with three 12.7cm guns. For mainly economic reasons, but also due to the considerable political turmoil in the country, nothing further was done towards acquiring more destroyers until after Hitler's rise to power and the farcical Anglo-German Naval Agreement, reached in June 1935, which allowed the Germans to build up to 35 per cent of the tonnage possessed by the Royal Navy (including the navies of all the Commonwealth nations) in all categories of ships.[32] In regard to destroyers, this allowed the Germans to build up to a total tonnage of more than 51,000 tons; given that their current tonnage in the destroyer category totalled approximately 12,000 tons, this agreement gave them more than enough leeway to build as many new, large destroyers as they could. The problems facing the Kriegsmarine (KM) were not limited simply to competition for funding with the other branches of Germany's military, which were scrambling to expand just as eagerly in the heady first years of the Nazi regime, but also competition for manpower, particularly trained seamen to man engine rooms and radio rooms, while shipyards struggled to find skilled workers.[33]

The design eventually produced by the Naval Architects' Office of the KM in 1934 and supplied to the Deutsche Werke (Kiel) – successor to the state-owned Kaiserliche Werft

Kiel – largely based on the Schichau design from 1932, had the following specifications:

KM Type 1934 (*Z1*): 1937[34]

Displacement (standard):	2100 tons[35]
Length (oa):	390ft 5in
Beam (wl):	37ft 1in
Draft (max):	13ft 11in
Powerplant:	70,000shp; two-shaft Deschimag geared turbines; six Wagner boilers
Speed:	36kt (design); 38kts (actual)
Armament:	8 × 53.3cm trainable torpedo tubes (2 × 4); 5 × 12.7cm/45 SK C/34 (5 × 1); 4 × 3.7cm/83 SK C/30 (2 × 2); 6 × 2cm/65 C/30 (6 × 1)

Probably the most noteworthy aspect of this design was its engineering plant. With little experience to call upon, the Germans decided to commit to the installation of high-pressure, high-temperature steam systems that exceeded the parameters of those in use by the US Navy, which had far more experience with such systems. At about the same time that the Americans increased the operating pressure and temperature of the powerplant in their Gridley class destroyers (1935–6) to 565psi (38.5ato) and 371°C (700°F), the Germans built their Type 1934 class equipped with a steam system designed to work at 995psi (67.7ato) and 480°C (896°F).[36] Some later ships of a slightly modified design, the Type 1934A class, used Benson boilers operating at 1600psi (110ato). While their audacity at taking on such an engineering challenge with little experience and a paucity of trained engine room technicians must be admired, their wisdom can be questioned. After the Second World War began, the availability of German destroyers became a significant issue. Between mechanical issues and the normal demands of training and time in port for replenishment and repair, German destroyers averaged a stunningly low 40 per cent availability rate.[37] There is no question that their engineering plants were a major contributing factor.

Beyond any issues with their powerplants, the four Type 1934s were not considered very successful by the fleet, being wet ships and not particularly stable in weather; they also had a poor tactical radius and a tendency to dig their stern in at high speed which further reduced the effectiveness of the rudder.[38] A follow-on class of twelve Type 1934A ships, built to a similar design, attempted to rectify some of the problems by adding a short additional keel section (*totholz* – 'deadwood') aft. This worked almost too well; it buoyed up the stern at higher speeds, forcing the bow to dig in,

KM *Karl Galster* (*Z20*) was a Type 1936 destroyer, seen here from KM *Prinz Eugen* at the beginning of Rheinübung, the operation intended to break the battleship KM *Bismarck* into the Atlantic in May 1941. The later Type 1936s, including *Galster*, featured an *Atlantik* bow, a flared and curved bow intended to resolve the problem of wetness forward suffered by earlier German destroyers. (NARA)

making the forecastle even wetter. Six more similar ships of the Type 1936 class were given a so-called *Atlantik* bow, a curved and flared bow form, that partially addressed the wetness forward. This modified bow was retrofitted to some of the surviving earlier destroyers during refits. The usage rate of these ships was very high at outbreak of war and the loss rate was commensurately high; of the twenty-two destroyers of this type, twelve were lost in the first nine months of the Second World War.

The other nation pointedly excluded from the between-the-wars naval disarmament process was Russia, now become the Soviet Union. The reasons for this were in part self-imposed isolation and in part the result of intentional exclusion by the remaining powers. The way in which Russia had withdrawn from the First World War – undergoing a double revolution that swept the monarchy out of power and installed a socialist regime determined to end its involvement in the war – both angered and terrified the victorious Allies, who believed the Soviets had betrayed a trust and, at the same time, represented a real threat to each nation's internal stability.

For a while after the end of the First World War it seemed possible to ignore the Russians, at least in terms of naval arms discussions. Badly damaged by the Russo-Japanese War of 1905, the First World War and subsequent revolutions, Russia now seemed irrelevant as a naval power. In terms of destroyers, by 1930 the new Soviet regime could muster a total of seventeen ships, all of them design descendants of the pre-war *Novik*, a large destroyer, modern for its time, but laid down at St Petersburg in 1910. The newest of these ships was not completed until 1927, such being the state of Russian heavy industry in the early years of the Soviet Union.

By 1928, the state of Russian industry had declined to a small percentage of its capability before the First World War. On account of that war, which led to the loss of Poland and Finland, the site of much of Russia's pre-war industrial capacity, and the extended civil war, which saw the flight of a great many of the more affluent and more skilled entrepreneurs and workers in the population, Russia needed an infusion of capital and energy to revive its industrial sector. With that in mind, Stalin instituted in 1928 the first of what would eventually be thirteen five-year plans, with the aim of increasing the output of heavy industry by 350 per cent during the plan's five-year tenure. This proved entirely unrealistic, but the actual increase was nonetheless impressive, more than doubling industrial output; just as the American president Roosevelt would do in 1933, when faced with a similar need to stimulate his nation's economy, Stalin saw the building up of Russia's military with aircraft, armoured vehicles and warships as a 'win-win' that would simultaneously facilitate the growth of critical industries and strengthen the nation militarily. Unlike Roosevelt, Stalin was not a 'navy man', naturally inclined to favour the building of warships over any other form of military build-up, so the emphasis of the first plan was not on an immediate and massive investment in warship construction.

If anything, the Soviet approach to the resumption of warship construction can be described as extremely cautious. The First Five-Year Plan included plans for the construction of three large destroyers, but not until the infrastructure existed to support such an effort, meaning the steel mills, shipyards, and gun foundries needed to be in place before construction began. Thus, it was not until 1932, the last year of the First Five-Year Plan, that the three ships were laid down. Soviet records indicate that the design of these ships was inspired by the French *contre-torpilleurs* of the Vauquelin class, the last class of 2400-ton 'four-pipers'. It is not clear whether French naval architects had any direct input in the design of the three ships of 'Projekt 1'; it is most likely that they did not, at least not officially. The specifications to which they were built are as follows:

Leningrad (Projekt 1): 1936[39]

Displacement (standard):	2150 tons
Length (oa):	418ft 4in
Beam (wl):	38ft 5in
Draft (max):	13ft 9in
Powerplant:	66,000shp; three-shaft geared turbines; three watertube boilers
Speed:	36kt
Armament:	8 × 533mm trainable torpedo tubes (2 × 4); 8 × 533mm 53–38 torpedo reloads; 5 × 130mm/50 B13 Pattern 1936 (5 × 1); 2 × 76.2mm/55 34-K Pattern 1935 (2 × 1); 2 × 45mm/46 21-K (2 × 1)

All three Projekt 1 destroyer leaders were laid down in 1932, one at the Zhdanov (Severnaya Verf) yard in Leningrad (St Petersburg) and two at the Marti (UkroBoronProm) yard in Nikolayev (Mykolaiv). Because of the still primitive state of Russian shipbuilding, they took between four and six years to complete. Three more to a slightly modified design were included in the Second Five-Year Plan, that officially began in 1933; as before, one was built at Leningrad and the other two at Black Sea yards. None of them were particularly successful ships, with large, heavy forward superstructures that gave them a low metacentre and inherent instability. Three ships of this class – *Minsk* (Baltic) and *Kharkov* and *Moskva* (Black Sea) – were lost to enemy action during the Second World War. The three remaining units survived the war (and *Minsk* was raised post-war and returned to service), the last being scrapped in 1964.

Finally, by the mid-1930s, the Soviets had rejoined the world community, capped off by gaining membership of the League of Nations in 1934. With this new 'respectability', it was possible to look outside Russia's borders for help in revitalising the shipbuilding industry. Therefore, under the

The large Projekt 1 destroyer *Leningrad* was built at Leningrad (St Petersburg), fittingly enough, as part of the Soviet First Five-Year Plan. This design borrowed heavily from contemporary French *contre-torpilleurs*, but physically resembled them very little. With their large forward superstructure and heavy armament, they were inherently unstable despite their large size. (NARA)

Second Five-Year Plan, the Soviets turned to Fascist Italy, in what surely was the most curious of partnerships, who agreed to build a prototype of an even larger and faster flotilla leader at the OTO-Livorno yard and to provide technical assistance for the construction of twelve more sister ships at Black Sea yards. The Italians had been dealing with the Soviets since 1920, well before the rise of Mussolini, but continued to do so after the rise of the fiercely anti-communist Fascist state, exchanging Italian military technology for Russian natural resources, united by their shared sense of antipathy towards the French, Germans, and, especially, the British.

The design was consistent with Italian practice, with a long, high, flared forecastle with a gracefully rounded forward superstructure; painted a particularly attractive shade of sea blue for her trials in 1938, *Tashkent* was nicknamed the 'Blue Beauty' or the 'Blue Cruiser', because her size, almost 2900 tons, made her look like a small light cruiser. She was turned over to the Russians without her planned main battery of three twin 130mm mounts because they would not be ready until 1940. In the interim, the Russians fitted *Tashkent* with three single 130mm mounts. In the end, only four of the planned sister ships were started, only two of those were launched and none were completed. *Tashkent* was damaged by German bombers near Sevastopol in June 1942. She was towed into Novorossisk, but the flooding of damaged compartments proved uncontrollable and she foundered there on 2 July.

A far more intriguing Russian ship of the time was *Tashkent*, seen here in 1937 during her builder's trials before she received her armament. She was built at OTO-Livorno and painted an attractive shade of azure, which led to her being nicknamed the 'Blue Cruiser'. Her wartime career in the Black Sea Fleet was brief and inglorious.

Chapter 8

Let slip the dogs of war – the floodgates open: 1935–1939

The London Conference of 1930 had attempted to extend the Washington Treaty in time and to cover more types of warships. It had been, at best, a qualified success. The Americans and, to an even greater extent, the British, hoped to try again to negotiate an end to naval arms competition, and, to that end, preparations for the next conference began in late 1933. The British and Americans shared the belief that the First London Conference had been less than optimally successful primarily because the major 'powers' had prepared insufficiently well, failing to reach basic agreement on the major points of contention before the conference convened, assuming that goodwill and shared purpose would suffice to bring resolution to differences. When the harsh reality of incompatible national goals became apparent, the conference had no mechanism for resolving the differences in real time.

For the purposes of this narrative, the Second London Naval Disarmament Conference that convened on 9 December 1935 amounted to a non-event. The resulting treaty never mentioned destroyers as a type, simply carving out a subcategory of 'light surface vessels' that 'do not carry a gun with a calibre exceeding 6.1 in. (155 mm.) and the standard displacement of which does not exceed 3,000 tons (3,048 metric tons)'.[1] The treaty then defined this subcategory of vessels as being deemed over-age sixteen years after completion.[2] Beyond that, the destroyer category is in no way mentioned or limited in any manner.

Of more consequence than the treaty itself was the drama surrounding it. The Japanese attended the conference intent on achieving parity with the two main 'powers' – the United States and Great Britain – ideally by having those nations agree to reduce their navies to the size of Japan's. This was the ultimate 'non-starter', as neither Western 'power' intended to allow Japan parity in naval strength by any means. Once that became clear, the Japanese withdrew from the conference on 15 January 1936, effectively ending any meaningful naval arms restrictions. Almost as an aside, the Italians had boycotted the conference as a protest against the Western, particularly the British, reaction to their invasion of Abyssinia (Ethiopia). The signing of the essentially toothless Second London Naval Treaty by the United States, the United Kingdom, and France on 25 March 1936 was a meaningless *coup de théâtre*.

The Royal Navy returns to normal

The Tribals were well liked in the Royal Navy for their size, which made them good sea boats, and for their powerful single-purpose main battery, but they were too expensive and costly in terms of resources to be the kind of ship that was needed to fill out the multiple flotillas of destroyers that would be required in the war that, by the later-1930s, was clearly looming on the horizon. What was needed was a smaller, more expendable ship that could be built in larger numbers, preferably one with a dual-purpose main battery.[3]

It soon became obvious that there was no way that the fleet could get all that it wanted, at least not in a ship the size of an 'I' class destroyer – 323ft (oa) and 1370 tons. There were several factors that made this impossible. The most critical was the lack of a true dual-purpose gun suitable for use in the main battery, which meant that this ship would either have to be fitted with only four single-mount 4.7in guns like the 'I' class or three of the dual-mounts carried by the Tribals in order to leave space and available buoyancy to allow a quadruple 2pdr (40mm) pom-pom to be fitted – which weighed as much and took up almost as much space as a 4.7in single-mount gun – in order to provide some of the needed anti-aircraft firepower. Some space could be saved by fitting two larger three-drum boilers in the place of the three boilers of the 'I' class, allowing the use of one large funnel in the place of two smaller ones, but even that would not allow all the desired armament to fit into an 'I' class-sized hull. By the time all of the design decisions were made and the drawing received Naval Board approval in March 1937, the new 'standard' destroyer had the following specifications:

'J' Class: 1939[4]

Displacement (standard):	1690 tons
Length (oa):	356ft 6in
Beam (wl):	35ft 8in
Draft (max):	13ft 8in
Powerplant:	40,000shp; two-shaft Parsons geared turbines; two Admiralty three-drum small-tube boilers
Speed:	36kt
Armament:	10 × 21in trainable torpedo tubes (2 × 5; centreline); 10 × Mark IX 21in torpedoes; 6 × 4.7in/45 QF Mark XII (3 × 2); 4 × 2pdr/39 QF Mark VII (1 × 4); 8 × 0.50in/62 MG Mark III (2 × 4)

This design was considered to be highly successful, a good compromise between the handiness and 'expendability' of the smaller 'I' class and the seaworthiness and gunpower of the

HMAS *Nizam* (G38) was manned from her completion at the John Brown yard on Clydebank in December 1940 by an Australian crew and was commissioned into the Royal Australian Navy in January 1941. Except for that minor technicality, she was typical of Royal Navy early-to-mid Second World War-production destroyers, with a main battery of three twin 4.7in/45 QF Mark XII gun mounts – one fewer than carried by the somewhat larger Tribals – with the same issues caused by this being a single-purpose weapon that required the addition of a separate anti-aircraft suite. In this view most likely dating from September 1944, that anti-aircraft defence appears relatively meagre, limited to a quadruple 2pdr/39 pom-pom immediately abaft the funnel and a single 40mm/56.3 OQF Bofors where the after torpedo tubes had been sited. The radars are a curious mix of British and American, with Type 291 air search antenna atop her foremast, a Type 285 (range only) gunnery radar on her fire control director, and an American SG surface-search antenna halfway up her foremast. (SLV – Green Collection)

Tribals. If there was a complaint against the type, it was that they were wet ships; in any heavy sea, they would take water sufficient to interfere with the operation of both forward guns and the manning of the bridge.[5] (The Royal Navy, unlike almost all other navies, insisted that an open bridge was necessary to allow their commanders complete situational awareness.) The 'J' class re-introduced longitudinal-framing to the Royal Navy, which allowed a lighter, yet stronger, structure for the hull. Welding was restricted to non-weight-bearing joints.[6] Between those two measures, the structural problems suffered by the Tribals were effectively resolved.

The adoption of twin mounts allowed the Royal Navy to reduce the annual order of an eight-ship flotilla and larger leader to simply an eight-ship flotilla, with one of those ships having a larger after superstructure to accommodate the flotilla's command staff. The eight ships of the 'J' class were ordered in March 1937, and eight more identical ships of the 'K' class followed a month later. Eight more virtually identical ships of the 'N' class were ordered in January 1940.

On a parallel track, the lack of a true dual-purpose main battery in a fully-enclosed mount, such as was now being fitted on American and Japanese destroyers, led to the development of a new gun and turret: the 4.7in/50 QF Mark XI and a new destroyer class to carry it. This gun and mount were bigger and heavier than the model shipped in the 'J' class and its successors; the new gun mount and its enclosed gunhouse weighed forty-two tons, not counting the rotating understructure, almost 70 per cent more than the gun mounts and rotating structure in the 'I' class. In order to accommodate three turrets in an arrangement similar to that of the 'I' class, this new destroyer, also included in the 1937 Estimates, needed to be larger, by over 200 tons, and bigger, adding six

The 'Q' class of Royal Navy destroyers resulted from the decision to pursue a somewhat smaller destroyer design after the 'N' class. The resulting 'O' and 'P' classes were only marginally smaller and saved very little money or building time, while carrying significantly less firepower. With the 'Q' class of 1940 the Royal Navy reverted to the larger hull of the 'J' class and four single 4.7in/45 QF Mark IX main-battery mounts, which still left them without an adequate dual-purpose main battery. These images show HMAS *Quickmatch* (G92), which was transferred to the Royal Australian Navy on commissioning in 1942. The first image shows her in December 1944 and the second on 16 December 1946. Her electronics suite is the same in both images and is identical to *Nizam*'s seen just above. The later image here shows her anti-aircraft battery in detail: a quadruple 2pdr/39 pom-pom abaft the funnel and three single 40mm/56.3 Bofors mounts further aft. There appear to have been few, if any, changes in her appearance in the two years between these images except that her paint job has been touched up. (SLV – Green Collection)

This image shows HMS *Quilliam* (G09), the 'Q' class leader, probably dating from late-1944. By this point, leaders differed from the 'standard' destroyers that comprised a flotilla only in having an extended after deckhouse, which provided space for the flotilla staff and their communications equipment. This image shows the single main-battery mounts, the quadruple pom-pom abaft the funnel, and multiple, single 20mm/70 Oerlikon mounts atop the after deckhouse. Note the HFDF antenna at the top of the lattice mainmast. (SLV – Green Collection)

feet of length and more than a foot of beam. Due to delays in development of the gun and its mount, the first of the 'L' class was not laid down until March 1939; this class was followed by the identical 'M' class in the 1938 Estimates. Delivery of the new gun mounts was slow, so some of these ships were completed with four twin 4in/45 QF Mark XVI mounts as their main battery. Most ironic is the fact that the new 4.7in twin-mount proved to be a disappointment in service; the gun had a slow training rate and elevation was manually-operated up to a maximum of 40°, all of which made it a suboptimal anti-aircraft weapon.

In 1938 the First Sea Lord, Admiral Roger Backhouse, realised that the evolution of destroyer design leading up to the 'L' class had lost any of the minimal advantages the 'J' class had once offered over the Tribals in terms of affordability and expendability, and that a new 'standard' destroyer design was needed that was smaller than the 'J' class. He envisaged an 'intermediate' class somewhere between the pre-Tribal 'I' class and the 'J' class.[7] After considerable waffling over what main-battery guns could be fitted (and would be available), especially as the prospect of war and the need for large numbers of these ships loomed ever closer, and what constituted an adequate anti-aircraft battery (whether a 4.7in gun with a 40° maximum elevation could be considered a dual-purpose gun), a design was approved that was, in fact, only marginally smaller than the 'J' class, but would nonetheless become the basis for all the 'war emergency' destroyers, with the following specifications:

'O' Class: 1939[8]

Displacement (standard):	1610 tons
Length (oa):	345ft
Beam (wl):	35ft
Draft (max):	13ft 6in
Powerplant:	40,000shp; two-shaft Parsons geared turbines; two Admiralty three-drum small-tube boilers
Speed:	36kt
Armament:	8 × 21in trainable torpedo tubes (2 × 4; centreline); 8 × Mark IX 21in torpedoes; 4 × 4.7in/45 QF Mark IX (4 × 1); 4 × 2pdr/39 QF Mark VII (1 × 4); 4 × 20mm/70 Mark I (4 × 1)

Comparing this ship to the 'J' class, it is easy to see that this hull is marginally smaller, but had the identical powerplant; in order to maintain adequate stability, it was necessary to

By the time the Royal Navy began building the 'T' class destroyers in late 1941, the need for an effective dual-purpose main-battery gun had become pressing, leading to the development of a new mount for the venerable 4.7in/45 QF Mark IX gun with improved loading and greater maximum elevation (55°), necessitating the more raked forward face of the gunhouse seen here. This gun still had manual elevation, which made it less than ideal in the anti-aircraft role, and the hydraulic training mechanism prevented it from rotating freely through more than about one-and-a-half rotations, so 'X' mount was almost always installed facing forward, as in this image of HMS *Terpsichore* (D48), seen from USS *Wasp* (CV 18) in August 1945, so it could freely cover the forward as well as after arcs. Note the Hazemeyer Bofors in the place of the after torpedo tubes. These were twin 40mm/56.3 Mark IV water-cooled guns on a triaxial mount that incorporated a Type 282 radar and an analog fire control computer. Note also the heavy lattice foremast, carrying a Type 276 'cheese' air search antenna and, above that, the light lattice mast topped with an HFDF antenna. The short mainmast appears to be topped by a Type 291 antenna. The ships in the background all appear to be American. (NARA)

reduce the armament by such expedients as the adoption of quadruple torpedo tubes in the place of quintuples and the reversion to the shorter-barrelled 4.7in/45 in single mounts. Eight of the 'O' class were ordered in September 1939, just before war broke out in Europe, and eight more of the identical 'P' class were ordered a month later. The 'Q' class ordered in December 1939 reverted to the larger 'J' class hull, but was otherwise similar.

By the time consideration began for the 'S' class to be ordered in late 1940, the need for an effective dual-purpose main battery had become pressing, and a new single mount for the venerable 4.7in/45 Mark IX gun was ordered into production, one that allowed 55° elevation and full 360° rotation in train; a spring-operated rammer allowed a faster firing rate, but the gun was still hampered by manual elevation. Size crept up slowly as repeat classes were ordered; the 'U' class ordered in June 1941 was four feet longer than the 'S' class and had a somewhat more raked and flared bow in an attempt to keep the forecastle drier. The 'W' class introduced a combined dual-purpose rangefinder/director Mark III. The 'Z' class ordered in February 1942 brought a new main-battery weapon, the 4.5in/45 QF Mark IV, which, in fact, used the same mount as the 4.7in gun carried in the 'S' class and had the same 55° maximum manual elevation; power elevation was not introduced until the 'Ch' class was ordered in September 1942. The 'Z' class also introduced the improved, but larger and heavier, Mark VI dual-purpose fire control director. Starting counting from the 'J' class up through the 'Cr' class ordered in September 1942, one

The Royal Navy's 'W' class destroyers differed little from the preceding several classes, including the 'T' class seen immediately above, mainly in replacing the separate low-angle and high-angle fire control directors which competed for space atop the bridge structure with a single Mark III dual-purpose director. Note also, in this image of HMS *Wizard* (R72), seen on 30 December 1945, the racks for illuminating rockets on the side of 'B' gunhouse ahead of the bridge. (SLV – Green Collection)

hundred and twenty-six 'standard' Royal Navy destroyers were laid down. Of that number, nineteen were not yet commissioned before the end of the Second World War, leaving 107 available to face Great Britain's enemies in that conflict. Of that number, thirty-one were lost in combat or damaged to the extent that they were written off. Many of the later-built ships served in the Royal Navy, or in nine other navies to which they were sold or given, on into the 1970s.

Among the last of the 'standard' war-production destroyers built for the Royal Navy, HMS *Cockade* (R34), seen here on 21 July 1947, has the massive Mark VI dual-purpose director atop the bridge with the twin 'headlights' of the Type 275 radar to either side. Given the added weight of the new director and its radar, every effort was made to save weight high in the ship, including landing the HFDF antenna, obviously missing from atop the foremast. These destroyers carried a new main-battery gun – introduced in the 'Z' class – the 4.5in/45 QF Mark IV now with full-power training and elevation, which added four tons per mount, further exacerbating the overweight problem. (SLV – Green Collection)

Starting too late in France

Having concentrated much of their energies on improving the successive designs of *contre-torpilleurs*, the MN seemed content to build three more tranches of *torpilleurs d'escadre* to fill the need for general-purpose escorts. Finally, in 1931, it became obvious that the newer, faster cruisers being built by the French would need escorts faster and longer-ranging than the *torpilleurs* so far designed. The MN turned to the STCN for a new design that would be at least five knots faster (at thirty-eight knots) and carry 50 per cent greater firepower (at least six 130mm guns), making them the equivalent of the Italian Navigatori class or the Japanese Special Type destroyers. The design took a long time to be worked out, growing from an original estimate of 1300–1400 tons up to, when the design was finally approved in August 1934, a displacement of over 1770 tons.

Even then, the process moved slowly. The first of the first tranche of six – MN *Le Hardi* – was not laid down until May 1936 and the next one not until August. A second tranche of six was authorised starting in 1936 and not started until 1938–9. Work proceeded slowly on all of the ships; only the first six were launched before the outbreak of war and only the first ship had got as far as running her acceptance trials. Eventually ten were launched, but very few completed, and all those were lost during the war; eight were scuttled in their various states of completion at Toulon on 27 November 1942.

This excellent close-up shows the foredeck of the L'Adroit class *torpilleur d'escadre*, MN *Bordelais*, probably in 1939. The open flaps on the face of the gun shield of the superfiring mount, exposing the openings for the aimer's and layer's sights, show just how thin the protection offered by this shielding really was. (NHHC)

The last French *torpilleurs d'escadre* to be completed were four units of the Le Hardi class; the exact number is not totally clear, depending on the definition of completion. MN *Mameluk* was one of those definitely completed, as she is seen here in the Mediterranean sometime in 1940. Note the 'Neutrality' stripes painted on gunhouses No 1 and No 2; those on the forward mount are blue-white-red from the bow. The main-battery gun was the single-purpose 130mm/45 Mle 1935; this gun was available in a dual-purpose version (Mle 1932) and this model was to be mounted on the last eight units of the class, but only three of those were launched and none of those was completed. (Conrad Waters Collection)

In the same fashion as the Japanese Special Type destroyers, the *torpilleurs d'escadre* of the Le Hardi class carried two of their three twin-mounts aft, as can be clearly seen in this image of MN *Epée*, in the Mediterranean in autumn 1940. The image appears to have been taken from the cruiser MN *Dupleix*. Note the 37mm/50 Mle 1933 twin-mounts on the cruiser's fantail. Even destroyers as late as *Epée* had the complex, and often fragile, below-decks depth charge stowage, with a conveyer belt carrying the charges aft to the single exit chute. On 1 April 1941, four ships of the Le Hardi class were renamed to honour destroyers lost in the war to date; *Epée* was renamed *L'Adroit*. (Conrad Waters Collection)

Soldiering on in Rome

Well before placing the last of the Navigatori into service in 1931, the Italians, like the French and the British, realised they needed smaller destroyers to fill out their flotillas. Two years after beginning work on the large *esploratori leggeri* of the Navigatori class, the RM looked back at the 1070-ton Turbine class, which had followed the Sella and Sauro classes, all laid down in 1925, as the design precedent for the next *cacciatorpediniere*. This was the Freccia class laid down in 1929.[9] Designed specifically to escort the fast Zara class cruisers, these destroyers were two knots faster than the Sauro class (at thirty-eight knots), which required 10 per cent more power (44,000shp), which in turn required a hull nearly ten feet longer and two feet broader, increasing standard displacement by 12.6 per cent (to 1205 tons). The single greatest benefit derived from these changes was an increase in range at cruising speed of over 40 per cent (to 4600nm). At the same time, the longer 120mm/50 gun introduced in the Navigatori was retained in two twin mounts. However the most significant recognition feature of the Freccia class was a massive single funnel into which the three boilers were all trunked. The engineering plant proved underpowered and unreliable, never allowing them to maintain over thirty knots in service, but that would not be the biggest problem facing the Freccias. Despite the increased length and beam, they proved to be marginally stable and remained so even after the addition of bilge keels and the shipping of ninety tons of added permanent ballast. Even then, it was necessary to replace spent fuel in bunkers with sea water to maintain stability, a practice universally deplored by engineers as it inevitably led to fuel contamination.

Despite the issues with the Freccias' design, four more near-sisters of the Folgore class followed, different mainly in narrowing the beam by 1ft 6in in an attempt to attain their

The Freccia class *cacciatorpediniere*, RN *Dardo*, is seen, probably in 1935, leaving the inner harbour at Taranto. The Freccias were a throw-back to the earlier, smaller standard destroyers, and suffered from the same stability problems as the earlier ones, caused by trying to stuff too much atop too small a hull. The passage of the fleet through the narrow channel at Taranto, which required opening the swing bridge, always drew a crowd of onlookers. (NHHC)

A view from the starboard quarter showing two Freccia class *caccia-torpediniere*, RN *Strale* (foreground) and *Dardo*, probably in 1932. Of note are the tall above-water structures, including carrying the after main-battery mount atop the after superstructure, all of which led to a low metacentre and poor stability. (NHHC)

This image shows a close-up of the forward half of the *cacciatorpe-diniere*, RN *Fulmine* of the Folgore class. This class, which came immediately after the Freccia class, was made somewhat narrower in an attempt to resolve speed issues with the Freccias, which had never reached their designed top speed. This, if anything, made the stability issues worse. (NHHC)

After the Folgore class, the Regia Marina understood that the problems with their *cacciatorpediniere* could only be resolved by giving the ship a larger hull. This resulted in several classes of larger, faster, and more stable destroyers that more nearly met the needs of the fleet. The best of these was the Soldati class, so named because all its members were named after famous bands of soldiers from history, such as RN *Carabinieri*, seen here sometime during the Second World War. She has what might be a Gufo EC-3 ter radar antenna above the searchlight platform atop her bridge, which would definitively date this image to January 1943 or later. It is most likely dated to the middle of that year or later, because at that time she had her single midships 120mm/50 gun removed along with her after bank of torpedo tubes, replaced by a platform carrying a pair of single-mount 37mm/54 RM 1939 anti-aircraft guns. (NHHC)

intended top speed. This turned out to be a poor decision, as it reduced range and exacerbated the stability problems. Realising that only a major increase in displacement would resolve the problems, the RM, for the four destroyers in its 1931 programme, authorised a size increase to 1615 tons and 350ft (oa), while all other main specifications remained the same. The resulting Maestrale class, initially designated *esploratori leggeri*, was both faster and more stable than the preceding classes, finally providing Rome with a successful, moderate-sized destroyer. The new chief-of-staff of the RM, Adm Domenico Cavagnari, asked in 1934 that an enlarged Maestrale class of 2100 tons be authorised, but Mussolini, like Hitler, failed to understand the importance of naval power and rejected the idea. Four ships of the Oriani class, nearly identical to the Maestrales, followed in 1935–6, in response to the tensions with Great Britain arising from the invasion of Abyssinia. Later, with war clouds looming closer to home, twelve more near-sisters were ordered in 1936. The specifications of these Soldati class *cacciatorpediniere* were as follows:

Soldati class: 1938[10]

Displacement (standard):	1690 tons[11]
Length (oa):	350ft
Beam (wl):	33ft 7in
Draft (max):	11ft 6in
Powerplant:	48,000shp; two-shaft Belluzzo/Parsons geared turbines; three Yarrow vertical small-tube boilers
Speed:	38kt
Armament:	6 × 533mm trainable torpedo tubes (2 × 3; centreline); 4 (or 5) × 120mm/50 Model 1936 (2 × 2; some also with 1 × 1); 12 × 13.2mm Breda (4 × 2; 4 × 1)

These Soldati class ships, so called because all were named after famed bands of warriors from history, would be the newest and best Italian destroyers with which the RM entered the Second World War. Seven more essentially identical ships were laid down in 1940–41, of which five were completed. Of the seventeen total Soldati to enter RM service, nine were lost in action, one was scuttled at the time of Italy's surrender, five were turned over to Allied nations as reparations post-war (three to France and two to Soviet Russia), and two served on in the post-war Italian *Marine Militare*, the last being stricken in 1965.

Going big in Berlin

As if the Type 1936, at 2400 tons and mounting a main battery of five 12.7cm guns, was not big enough, many in the Kriegsmarine wanted a ship armed, like the S113 class of 1918, with 15cm guns, to make up, in part, for the lack of light cruisers in the German navy.[12] The idea was simply to

German shipbuilders had long memories, and none had forgotten the S113 class destroyers built during the First World War, with their 15cm/45 main-battery guns. When it came time to order the seven destroyers which would follow the Type 1936 class, the Kriegsmarine took this opportunity to request an increase in main battery to five 15cm/48 in three single mounts aft and one twin mount forward. To handle the considerable added weight of the 15cm guns compared to the preceding classes' 12.7cm guns, this class added 6ft of length and 8in of breadth. The logic behind siting a twin mount forward is difficult to ascertain, even had it not proved to be much heavier than planned; the entire class was completed with only a single mount forward, because the twin mounts were not ready to install, but, even with this mount, they proved to be unstable and wet forward. This image shows KM *Z24* in 1941, after the FuMO 21 radar mattress antenna was added atop her bridge, but before she received the long-delayed twin main-battery mount forward. (NARA)

substitute the larger, heavier gun, enlarging the hull by six feet in length and eight inches in breadth to compensate for the greater topweight. However, this marginal increase in hull size could hardly make up for the significant difference in weight between the 12.7cm and 15cm single-mounts (22,500 lb vs 35,500 lb per mount). The Naval Architects' Office declared that this would not be a problem, as a lightweight twin mount for the 15cm gun was under development and the two forward single mounts would be replaced by one lightweight twin mount, addressing any stability and wetness issues. However, as often happens in such cases, the designers had overestimated the buoyancy that would be gained by the minor increase in dimensions, and the twin mount was not yet ready when the first of the Type 1936A, as this revised design was called, was being completed in mid-1940, so only a single mount was fitted on the forecastle deck forward where

The Type 1936A (Mob) class was basically a repeat of the Type 1936A with a significantly bolstered anti-aircraft battery. These two images show KM *Z39*, which was completed in August 1943 and turned over to the United States at the end of the war. In the broadside view dated 22 August 1945, taken in Boston MA harbour, the twin main-battery mount forward can be seen along with multiple 3.7cm/69 single- and twin-mounts. Main-battery mount No 2 (*Bruno*) has been landed. The close-up shows a bow-on view of the ship in drydock, with the twin 15cm/48 mount in the foreground, twin 3.7cm/69 mounts ahead of the bridge, and single 2cm/65 mounts abreast the bridge. The large mattress antenna is for a FuMO 24 search radar; at the four cardinal points surrounding the foremast searchlight position are 'figure-of-eight' antennas for an FuMB 4 Sumatra radar detector; above the crow's nest are horn antennas for an FuMB 26 Tunis detector (seen end-on) and the four, paired inverted cones of an FuMB 3 Bali detector; and, at the masthead, the 'pot' antenna of an FuMO 81 Berlin-S microwave surface-search radar. (NHHC)

the twin mount was to go. Unsurprisingly, even with the lower weight of a single-mount gun forward, *Z23*, the first of the class, proved to be marginally stable and very wet forward. In fact, all seven ships of the class were completed with four single-mount 15cm/48 LC/36 guns between September 1940 and November 1941. The first twin-mount LC/38 was not available to be fitted on *Z23* until February 1942. This mount proved to be grossly overweight, massing more than 130,000 lb, which only exacerbated the problems of stability and wetness exhibited by the class. Nonetheless, three others of the class (*Z24*, *Z25*, and *Z29*) had the twin mount fitted in 1942.

Seven more Type 36A destroyers were ordered at the outbreak of the Second World War in September 1939, designated the Type 1936A (Mob) class. Eventually, seven more would be ordered in several increments.[13] The ships built in the sequences *Z31–Z34* and *Z37–Z39* were virtual repeats of the Type 1936A, differing, for example, in having curved, rather than straight-sided, funnel caps. All received the twin main-battery mount in No 1 position forward. Five more were laid down, of which three were completed, to a somewhat modified designed designated Type 1936B (Mob). The primary difference in the three completed ships from their near-sisters was the reversion to a main battery of five 12.7cm/50 SK C/34 single-mount guns. The shipping of a significantly lighter main battery – the weight saving was a total of sixty-two tons of topweight – made a noticeable, and beneficial, difference in the ships' stability and weather handling, but nothing seemed to resolve the persistent issues with the German high-pressure, high-temperature steam systems, which remained marginally reliable through the end of their design run; the only method found to achieve reliable performance was to run the powerplants at less-than-full pressure, a measure that, while completely contrary to regulations, seemed to resolve the reliability problem.[14]

Built to an Italian design, generally resembling the contemporary Oriani class with its large, blocky forward superstructure and single funnel, but differing in favour of the Russian preference for larger guns in single mounts, forty-eight ships of Projekt 7 were ordered as part of Stalin's Second Five Year Plan starting in 1935. Twenty-eight were actually completed to the original design; ten of those were lost during the Second World War. One of the eighteen survivors, probably *Grozyashschiy*, is seen in a Baltic port in 1956. (NHHC)

The people's destroyers

Having laid down the first leaders of the Leningrad class under the First Five-Year Plan (1928–1932), the Russians took advantage of the Second Five-Year Plan (1932–7) to order a large class of 'mainstream' destroyers. Making use of the Italian designers and engineers who had been involved in the development of *Tashkent*, the design was for a 370ft, 1695-ton destroyer, similar in many respects to the contemporary Oriani class being built by OTO-Livorno, which was also building *Tashkent*. The most notable difference was the Russian preference for larger (130mm/50) guns in single mounts superfiring forward and aft. Forty-eight ships of this design, called 'Projekt 7' by the Russians, were ordered from yards at Leningrad, multiple yards at Nikolayev, and at Komsomolsk-on-Amur; thirty-one were laid down between 1935 and 1939. The first did not commission until mid-1938. One got stranded while being towed downstream from Komsomolsk to Vladivostok to be completed and had to be written off, being replaced by another hull laid down at Nikolayev originally intended to be a 'Projekt 7U' ship; ten were lost during the Second World War; and two were completed as modified Projekt 7U ships – meaning that twenty-eight were actually completed as Projekt 7s and eighteen survived the war, at least one lasting in service into the 1960s.

The Projekt 7U (or 'Projekt 7 Improved') design was conceived in 1937 before the first Projekt 7 was completed. The intent was to get more power out of the same hull by adding a fourth boiler. This allowed the adoption of a unit powerplant arrangement where a pair of boilers and a turbine set were followed by a second similar 'unit'; this provided greater survivability, but required rearrangement of the upperworks because the boilers could no longer be served by a single funnel. Twenty ships to this design were started between December 1936 (when the two ships originally ordered as Projekt 7s were laid down) and May 1939 and nineteen completed to this design. All were commissioned between 1940 and August 1942. Nine were lost during the war. The last of the type was sold for scrap in 1966.

All of the Projekt 7 and Projekt 7U ships suffered from several common ailments. The earliest units had very fragile turbine sets prone to blade stripping, but this was corrected in all later units and the early units had their turbines replaced. All ships of these classes suffered from overly light construction leading to chronic structural weakness traceable to the

Above: The Soviets modified the Projekt 7 design to fit a fourth boiler into the same hull, allowing a unit powerplant arrangement; the changed design, which went by the name Projekt 7U – the 'U' short for *Ulushchenny* (Improved) – required a second funnel to serve the now widely separated fire rooms. Otherwise, the ships of this class, such as *Serdity*, seen here at Tallinn, Estonia, in 1941, were similar to the base class. (NHHC

Below: Working from the Projekt 7U design as a basis, the Russians knew they needed a destroyer that was larger and more strongly constructed to handle the weather to be expected in the Arctic and North Atlantic, where the Soviet Navy was expected to operate. The resulting design, known as Projekt 30, carried its four 130mm/50 guns in two twin mounts, allowing more space for anti-aircraft weapons and improving stability. This image shows *Odaryonniy* (43) at a Baltic port in 1956. (NHHC)

design; the Italian naval architects who led the design effort were used to designing ships to operate in the relatively calm Mediterranean; when these ships were subjected on a regular basis to the harsh weather of the North Atlantic, for example, they suffered badly. At least one was lost in a storm and another lost her bow while escorting a Murmansk convoy.[15]

Even before the first of the Projekt 7 units had been completed, the Soviets understood that a new design was

needed for the Third Five-Year Plan (1938–41). The design was started in 1937, working from the Projekt 7U as a basis, but increasing all dimensions in order to gain stability. In another move designed to increase stability, the four 130mm/50 guns were to be carried in a pair of B-2LM twin mounts, one each forward and aft. The plan called for twenty-four units of this design, dubbed 'Projekt 30', and, as far as can be ascertained, all but one of those were laid down before the German invasion of Russia in June 1941. Sources differ as to how many were launched before the invasion disrupted all construction – the actual number is somewhere between two and seven.[16] What seems fairly certain is that only one or two entered service before the end of the war. In general, these ships – a further eleven or twelve were completed post-war – were well liked and served as the basis for the classes of destroyer built by the Russians at the beginning of the Cold War.

Special again in Tokyo

The IJN worked through the structural and stability issues revealed by the successive incidents of 1934 and 1935 that required considerable reconstruction of existing ships and redesign of new ones, often at the cost of capability. Finally, with the late 1934 decision to renounce the arms limitation treaty regime, any need to consider tonnage limits for future destroyers was discarded. Ten destroyers were authorised under the 1934 programme; determined to regain the full capability of the original Special Type destroyers, but with the lessons of the preceding years now very much in mind, it was obvious to the Design Section of the Naval General Staff that the only solution was a ship that was larger in all dimensions. The single most important requirement was the return to a main battery of six 12.7cm/50 guns. To achieve this, a ship of the following specifications was authorised:

***Asashio*: 1937**[17]

Displacement (trials):	2330 tons[18]
Length (oa):	388ft
Beam (wl):	33ft 11in
Draft (max):	12ft 1in
Powerplant:	50,000shp; two-shaft Kampon impulse geared turbines; three Kampon boilers
Speed:	35kt
Armament:	8 × 61cm trainable torpedo tubes (2 × 4; centreline); 8 × 61cm Type 93 torpedo reloads; 6 × 12.7cm/50 3rd Year Type (3 × 2); 4 × 25mm/60 Type 96 MG (2 × 2)

While this class successfully addressed the issues of stability and structural weakness that had afflicted earlier Japanese super destroyer designs, they had new problems that took time to resolve. They proved to have poor manoeuvrability and an

excessively large turning radius, caused by a rudder of insufficient size, but enlarging the rudder and reshaping the stern with a flatter transom underbody resolved that issue and actually resulted in a slight increase in speed.[19] A tougher problem to solve was with blade stripping in the medium-pressure turbines; the problem was ameliorated, but not solved, by strengthening the rotors to reduce vibration. It was not until 1943 that the cause of the sympathetic vibration was discovered and eliminated.[20]

The ten units of the Asashio class were followed by eighteen almost identical ships of the Kagero class; the first fifteen were ordered in 1937, with three more added in 1939.[21] The Asashio and Kagero classes both were fitted with the Type C main-battery mount that had a maximum

The Japanese decision to withdraw from the international arms limitation regime defined in the Washington and London Treaties freed them to return to building destroyers of the size and power of the original Special Type. Increasing displacement by more than 250 tons over the preceding Shiratsuyu class, the destroyers of the Asashio class, as typified by *Yamagumo* seen here on 15 September 1939, retained the smaller bridge of the Shiratsuyus, but returned to the six-gun main-battery armament with superfiring mounts aft, characteristic of the Special Type destroyers.

The destroyers of the Kagero class, such as *Amatsukaze* seen on 17 October 1940, were virtual repeats of the Asashio class, with slightly more powerful engines and a slightly smaller bridge, both changes that improved performance.

Surrendered at the end of the war, the Kagero class *Yukikaze* was one of the few Japanese destroyers to survive the hard use to which they were put. This close-up of her bridge shows the antenna of the Mk 2 Mod 2 surface-search radar, with separate transmitting and receiving 'horns' on the lattice foremast. By the time these radars were ready to install on Japanese destroyers, the period of the great destroyer battles in the South Pacific had passed, so they were never able to prove their utility in battle. (NARA)

elevation of 55°. The Kagero class introduced the Type 93 61cm torpedo. While the same diameter as the previous Type 90 torpedo, the Type 93 was 20in longer and weighed over 200 lb more than the preceding model. The biggest difference between the Type 90 and the Type 93 was the introduction of oxygen in the place of compressed air, which made them faster and more than doubled their range, while carrying a 30 per cent larger warhead. (These new torpedoes – dubbed the 'Long Lance' by Morison in American post-war literature – were so much better than the preceding model that a programme of retrofitting earlier classes to handle this model was undertaken, going as far back as the Fubuki class, that was not completed until 1943.[22]) Starting in 1939, the first of twenty Yugumo class destroyers was ordered; again, these were near-repeats of the preceding classes. They were three feet longer in an attempt to gain a bit of improved hydrodynamic efficiency, and they were equipped with the Type D main-battery mounts, which had a maximum elevation of 75°.

The forty-eight ships of these three classes provided the backbone of the IJN's destroyer forces that fought in the Second World War. Besides providing escort for major fleet units, they made the near-nightly runs of the 'Tokyo Express' in the Solomons in 1942–3. Not surprisingly, the attrition in their ranks was horrific, particularly after the Americans came to dominate the skies where they operated. Only one of their number, *Yukikaze* of the Kagero class, survived to the end of the war, eventually being handed over to the Republic of China (Taiwan) and serving in that navy until 1970.

Quality and quantity in the New World

In the succession of post-war destroyer classes designed for the United States Navy, from the Farragut class of 1934 to the Sims class of 1939, the 'standard' American destroyer had gradually increased in size and capability, but the numbers being built were still modest, given the approaching obsolescence of First World War-vintage flush-deckers and the war looming in Europe. In the six classes built by the Americans in this series, the capabilities gradually increased, particularly with the addition of more anti-submarine weaponry in the form of larger depth charge racks and more throwers, and with more powerful engineering plants to offset an upward creep in standard displacement; where the Farraguts displaced 1358 tons, the Sims displaced 1764 tons on only minimally enlarged hulls. Where the Farraguts had been more than 130 tons below designed displacement, the Sims class ships had come in almost 120 tons over.[23] Not surprisingly, the Sims class exhibited reduced stability, since much of the excess weight was above the waterline. Immediate steps were taken to address the issue. Where they had been designed to carry three banks of quadruple torpedo tubes, one was removed and one of the others relocated to the centreline. That proved insufficient; sixty tons of ballast was added to the keel and splinter plating that had been installed around the director and pilothouse were removed. Any advantage gained by these measures was offset, once the war began, by electronics and anti-aircraft weapons added high in the ship, so that most in the class lost another bank of torpedo tubes and one of their five main-battery mounts during refits.

As had become typical for the American destroyer design process, the generation of the design for the ships that would follow the Sims class began with a set of requirements generated by the General Board. The impetus for the new design came from the Bureau of Steam Engineering (BuEng), which announced in 1937 that advances in boiler design had reached the point that it would be practical to implement a unit powerplant arrangement, increasing the likelihood that the ship could survive a single major casualty amidships, such as a hit by a bomb or torpedo. According to BuEng, such an arrangement could be achieved at an increase of only 50 tons in designed displacement (from 1570 tons to 1620 tons). With that new technology available, contracts for eight destroyers were included in the FY1938 budget; two were to be built to a Gibbs & Cox design by Bath Iron Works (USS *Gleaves* (DD 423) and *Niblack* (DD 424)) and the remaining six to a Bethlehem design at Bethlehem's Quincy yard and three different navy yards (USS *Benson* (DD 421) and

With the approach of war, although America would not be directly involved for two more years, the political and military leadership of the United States was wise enough to understand that their involvement was inevitable and preparation on a massive scale was essential. The most expensive aspect of that preparation was the building up of a navy far beyond the needs of a peacetime force, with a program of building destroyers every bit as large as the 'flush-decker' program during and immediately after the First World War. It began with an improved Sims class design, eight feet longer to allow a unit powerplant arrangement to be installed, which, in turn, required a return to a two-funnel silhouette. Otherwise, the characteristics of the new destroyers were similar to the preceding class, except that the torpedo broadside was increased by a switch to quintuple versus quadruple banks of torpedo tubes. Note that in this view of USS *Meredith* (DD 434) dated June 1941, main-battery mounts Nos 3 and 4 had no shields in order to save topweight. *Meredith* was built to the Gleaves class design; note the round cross-section funnels. She appears to be painted in the rarely seen Ms4 camouflage scheme reserved for destroyers – overall black, with the pole foremast painted white; her hull number is in the reduced-size numbers used with the new camouflage schemes introduced in January 1941, but in an unusual location just forward of the break of the forecastle. *Meredith* has yet to be fitted with any radars. (NARA)

Mayo (DD 422) at Quincy and DD 425–428 at the navy yards). The only significant difference between the designs was in the powerplant; the Gibbs & Cox-designed ships had double-reduction geared turbines with a separate cruising turbine and the Bethlehem-designed ships used single-reduction geared turbines without a cruising turbine. The only external difference was the shape of the funnels; the Gibbs & Cox ships had funnels that were round in cross-section, while the Bethlehems were flat-sided. Except for a ten ton difference in displacement – the Gibbs & Cox-designed ships were heavier – all the initial ships were built to the following specifications:

Benson/Gleaves class (DD 421/423): 1940[24]

Displacement (standard):	1620/1630 tons (design); 1839 tons (actual)
Length (oa):	348ft 4in
Beam (wl):	36ft 2in
Draft (max):	13ft 2in
Powerplant:	50,000shp; two-shaft Bethlehem/Westinghouse geared turbines; four Babcock & Wilcox boilers
Speed:	35kt
Armament:	10 × 21in trainable torpedo tubes (2 × 5; centreline); 5 × 5in/38 Mk12 (5 × 1); 6 × 0.50cal/90 M2 Browning MG (6 × 1)

In the end, both engineering plants proved to be satisfactory in all regards, but BuC&R wished to standardise on a single design and chose the Gibbs & Cox Gleaves class as the standard to which eight more destroyers (DD 429–436) were ordered in the FY1939 programme and a further eight (DD 437–444) in FY1940, ordered in December 1939. (The Gleaves class is sometimes called the Livermore class because the ship on which the US Navy decided to 'standardise' the type was USS *Livermore* (DD 429).) The first three of this class were laid down in May 1938; the last two of this initial series would not be started until March 1940. The first to be commissioned was *Gleaves* in June 1940; the last entered service in July 1941. Only five of this series were lost during the Second World War.

This initial sequence, however, was just the tip of the iceberg, so to speak, as the looming war in Europe was now a reality, leading to decisions that, in the absence of such an exigency, would have been unlikely. It had been the intent of the US Navy to standardise destroyer production on a new, even larger design in FY1941 (to be described immediately below), but the pressure of external events led to the decision to continue producing not only the favoured Gleaves class design, but also the Benson class as well, in order to fully

Seen again a year later, on 23 June 1942, *Meredith* is now in the South Pacific. She has undergone a refit, during which her No 3 main-battery mount was landed and replaced by three 20mm/70 Oerlikons and another similar mount was added atop her pilothouse. Her after superstructure with its searchlight tower has been removed, but the searchlight retained. The protective cover for the aimers serving the after bank of torpedo tubes has been removed, because it is no longer needed with No 3 main-battery gun removed. No 4 main-battery gun has received a half-gunhouse with a canvas top. She now has an SC air search radar antenna on her foremast and the struts in place to hold an FD radar antenna atop her Mk37 fire control director, which has yet to be installed. She is painted in an MS 12 Mod camouflage scheme. (NARA)

utilise established shipyards and expertise. The construction of seventy two additional ships of this class was authorised in four stages between May 1940 and February 1941. Twenty-four were Bensons, while forty-eight were Gleaves. All were laid down between September 1940 and November 1942; the last was completed in August 1943. All of these follow-on ships were completed to what would be the initial wartime standard configuration of four main-battery mounts, the No 3 mount replaced by a 1.1in/75 quadruple AA gun and a single 20mm/70 Oerlikon or, later, a pair of 40mm/56 Bofors in twin mounts, plus some number of single 20mm machine guns. (This standard was retroactively applied to the ships of the original series as well, as they were refitted.) Many would also lose their second quintuple torpedo tube bank to make room for more depth charges and AA guns.

From the beginning, the US Navy envisaged the Benson/Gleaves class as serving in the Atlantic. Of the original twenty-four ships, only four were dispatched to the Pacific after Pearl Harbor, the rest being retained on the east coast.[25] Despite the long planning for war in the Pacific – since the early twentieth century Japan had been presumed to be America's most likely opponent in any future war – and the understandable urge to strike back after Pearl Harbor, 'Europe First' was official US policy, agreed at the Anglo-American staff talks in early 1941, which committed the US Navy to active patrol work in the western Atlantic well before the nation was formally at war.[26] More of the seventy-two Benson/Gleaves follow-ons made their way to the Pacific. The numbers were small at first; probably the next to head west was USS *Farenholt* (DD 491), which joined four Benham class destroyers in DesDiv15 accompanying USS *Wasp* (CV 7) when she transferred from the Atlantic in June 1942.[27] Eventually, almost all of these ships would see service in the Pacific as the need for destroyers in the Atlantic declined. Eleven of the seventy-two follow-on Benson/Gleaves class destroyers were lost during the Second World War; many of those that survived served post-war, often in the navies of allied nations to which they were sold or lent, lasting into the 1970s.

The ship that would go down in history as the most-produced single class of destroyers ever, started out innocently enough as a design study labelled 'Destroyer 1941' initiated by the US Navy's General Board in late 1939.[28] Aware that large numbers of destroyers would be needed to face the war now under way in Europe and to replace the many 'four-pipers' facing mass-obsolescence, the board asked for a design that was both smaller than the Benson/Gleaves class ships then entering production, and carried a more powerful anti-aircraft armament. The Board queried the various bureau and destroyer commanders in the fleet, asking what could be sacrificed to save space; not surprisingly, nearly all respondents asked for more of whatever feature they most desired, be it

speed, protection, gunpower, or torpedo battery. No one seemed interested in sacrificing any of the aspects of existing designs they liked. The end result was much like the proverbial 'mouse designed by a committee', a ship far larger than the General Board had in mind at the beginning of the process. Fortunately for the Americans, the increased sized allowed it to be a ship capable of accepting considerable added topweight in the form of anti-aircraft weapons and electronics without serious problem.

Having listened to all the respondents, the General Board issued its first set of desired characteristics on 17 October 1939 and asked Preliminary Design for sketch designs as soon

USS *Laffey* (DD 459) was a Benson class destroyer – note the flat-sided funnels – seen here on 16 September 1942 at Espiritu Santo during her brief, but eventful, seven-month-long life between her commissioning and demise at the First Naval Battle of Guadalcanal. In this image, she is loaded with survivors from the sinking of USS *Wasp* (CV 7) the day before. This image is of interest because it shows the standard early war fit of a Benson/Gleaves class destroyer, with an FD radar antenna atop her Mk37 director, a single 20mm/70 Oerlikon and a quadruple 1.1in/75 Mk 1 anti-aircraft machine gun aft of the stub mainmast. Note that the No 3 main-battery mount is now fully enclosed. The ship in the background is USS *Juneau* (CL 52), which would be sunk the same day as *Laffey*. (NARA)

USS *Welles* (DD 628) shows the ultimate configuration of the Gleaves type as built. These came out of the yard with four main-battery guns, all fully enclosed, only one bank of torpedo tubes, greatly enhanced anti-aircraft batteries, and multiple depth charge throwers. One of the most noticeable differences was a lowered bridge structure, with no curved surfaces, intended both to save topweight and ease production; the Mk37 director no longer sat on a tall pedestal, but sat directly on the pilothouse. (NARA)

as possible. The desired characteristics included a displacement no greater than 1600 tons, at least four 5in/38 guns, two quintuple banks of torpedo tubes with four reloads, twenty-eight depth charges with two racks and at least two throwers, and at least four .50cal anti-aircraft machine guns. Top speed was to be thirty-six knots and endurance was to be 6500nm at fifteen knots. BuC&R responded with six sketches, all based on the existing Sims class hull, although some would have had a foot of extra beam to compensate for topweight, with various armament combinations. In November, in response to interest in one particular sketch (Scheme 2), which added some protection for the engineering spaces (in the form of 0.25in added STS hull plating), Preliminary Design responded with a new flush-decked design with greater beam, offering improved stability and better seakeeping, dubbed Scheme 7. This was met with general approval and served as the basis for the eventual Fletcher class design.

Nevertheless, it took several more evolutions of the design, as word of early British war experience began to make its way across the Atlantic, and the importance of armour protection for the engineering spaces and stronger anti-aircraft armament became apparent. After further revision, the General Board approved a set of characteristics on 9 January 1940 that included 0.75in STS plating in way of the engineering plant and the bridge, five main-battery mounts, three of them fully-enclosed, and a quadruple 1.1in/75 anti-aircraft machine cannon along with its Mk44 director and four .50cal machine guns. The first two of what would be 175 Fletcher class destroyers – 181 were ordered, but six were cancelled – were laid down on 3 March 1941 to the following specifications:

USS *Fletcher* (DD 445): 1942[29]

Displacement (standard):	2325 tons
Length (oa):	376ft 5in
Beam (wl):	39ft 7in
Draft (max):	13ft 9in
Powerplant:	60,000shp; two-shaft General Electric geared turbines; four Babcock & Wilcox boilers
Speed:	38kt
Armament:	10 × 21in trainable torpedo tubes (2 × 5; centreline); 4 × 21in Mk15 torpedo reloads; 5 × 5in/38 Mk12 (5 × 1); 4 × 1.1in/75 Mk1 AA machine cannon (1 × 4); 4 × 0.50cal/90 M2 Browning MG (4 × 1)

Almost no ships were completed to this configuration, as experience with operating destroyers in both the Atlantic and Pacific before the first of the Fletchers – USS *Nicholas* (DD 449) – was commissioned on 4 June 1942 – taught crucial lessons. One was that all five main-battery guns needed to be fully-enclosed; operating a large, unprotected gun, even one mounted on the after superstructure, was impractical in any weather. Another was that the .50cal MG was no longer an effective anti-aircraft weapon; as soon as 20mm/70 Oerlikons became available, the lighter Brownings were replaced on a one-for-one basis. As soon as 40mm/56 Bofors

The single most-produced class of destroyers, and the one most identified with the US Navy in the Pacific in the Second World War, can be no better represented than by its name ship, USS *Fletcher* (DD 445), seen on 18 July 1942 running builder's trials off New York. A 'clean sheet' design freed from prior constraints, this ship was longer and beamier than preceding classes, and did away with the forecastle deck that had been characteristic of all post-war American designs, opting for the greater longitudinal strength offered by a flush deck and counting on a curved sheer line to provide enough freeboard forward. As built, as seen here, early Fletchers carried five main-battery guns in fully enclosed gunhouses, a 1.1in75 Mk 1 quadruple-mount or a twin 40mm/56 Bofors mount on a raised deckhouse aft, and six single 20mm/70 Oerlikons. (NARA)

An interesting image of an early-production Fletcher class destroyer with an unusual anti-aircraft suite. In this case, USS *Stanly* (DD 478) is seen at Charleston, SC, where she was built, on 16 February 1943. Her light anti-aircraft outfit has been increased to what was then the standard number: seven single 20mm/70 Oerlikons sited one ahead of the bridge, one each side alongside the bridge, and two each side abaft the second funnel. However, *Stanly* was not a normal Fletcher; she was one of six Fletchers that had been selected to receive an aircraft catapult in the place of the after torpedo tube bank, main-battery mount No 3, and the medium anti-aircraft mount sited atop the after deckhouse. However, early experience with the first ships completed with catapults proved that the idea of operating seaplanes off destroyers – as they existed in 1942 – was impractical, and *Stanly* remained tied up alongside for an extra two-and-a-half months at the Charleston NY, while her catapult was removed and the removed torpedo tubes and main-battery mount restored. As the 40mm/56 Bofors mount intended for the after deckhouse had already been sited right aft on her fantail between the depth charge racks, it was decided to place two more single Oerlikons in the space left by her missing after Bofors mount. (NARA)

in twin mounts became available, in late 1942, these replaced the less-reliable 1.1in quadruple-mount, which weighed almost exactly the same.

This became an interim standard configuration that lasted only until it became obvious that yet more anti-aircraft weaponry was needed. As Fletchers came in for refit in 1942, navy yards frequently added a second twin 40mm Bofors mount on the fantail between the depth charge racks and four more single Oerlikon mounts, two to a side, just aft of the second funnel. To compensate for this added topweight, the Fletchers' design was gradually modified by shortening the cylindrical support for the Mk37 main-battery director by six feet, lowering the after superstructure by one deck, removing the torpedo reloads, and reducing depth charge stowage to enough for five patterns from each of the six K-gun throwers (three per side abreast the after superstructure) and eight per rack. The last seventy-two Fletchers, starting with USS *Brownson* (DD 518), were completed with a lower, more open forward superstructure, influenced by Royal Navy practice, in order to both save weight and give commanders

USS *Isherwood* (DD 520) displays a typical mid-war Fletcher class destroyer layout on 27 April 1943, a few days after commissioning at her builder's yard, Bethlehem-Staten Island. The standard anti-aircraft fit now included six 40mm/56 Bofors in twin mounts, one on the deckhouse aft and one each on large emplacements built up abreast the after funnel. Each Bofors mount had its own Mk 51 director; that for the aft mount was in a tub just forward of and above the gun, the ones for the sided mounts abreast the funnel were in the large cantilever platforms on either side of the after funnel that had been the location of the ship's searchlights. Those were moved to new tubs added to the forward funnel. The standard light anti-aircraft fit now was ten single 20mm/70 Oerlikons: three around the bridge, two either side abaft the after funnel, and three on the fantail. (NARA)

better awareness of aerial activity. These ships could be recognised by the 'squared-off' appearance of their bridge, as opposed to the rounded bridge of the earlier ships. Beginning in 1943, Fletchers had the twin Bofors on their fantail replaced by three single Oerlikons and a pair of twin Bofors abreast their after funnel. By mid-1944, many had five twin Bofors mounts (two more added in place of Oerlikons ahead of the bridge), while retaining seven Oerlikons (abaft the second funnel and on the fantail). In a late-war move to further augment anti-aircraft gunfire in response to the kamikaze threat, the twin Bofors abreast the second funnel were replaced by quadruple-mounts and the seven single Oerlikons replaced by six twin-mounts. It is a testament to the inherent stability of the design, that it was only this last

This image of USS *Hall* (DD 583), a later-production Fletcher, seen on 23 March 1945, shows the same kind of simplification and reduction made to the bridge area that had been done with the Benson/Gleaves class. The bridge was lowered and all curved surfaces replaced with flat ones; the tall cylindrical support for the Mk37 director had likewise been deleted. (NARA)

upgrade that required the sacrifice of any of the original offensive weapons fitted; the forward bank of torpedo tubes was landed to compensate for the added topweight.

The last of the 175 Fletchers to be completed was laid down on 27 October 1943 and commissioned on 2 September 1944. Amazingly, even though these ships saw combat in every theatre and in all actions where the US Navy fought, only nineteen were lost. After the war, they continued in service with the US Navy and with many allied navies in some cases into the 1980s. One ship, the Greek *Velos* (D16) – ex-USS *Charrette* (DD 581) – remains technically in commission, although she, along with three former US Navy Fletchers, is a museum ship, open to the public.

Chapter 9

The test of battle, Part 2: 1942–1945

By the beginning of the Second World War super destroyers were ubiquitous, present in each of the major combatant navies, so finding battles in which they played major roles, even primary roles, is not hard. What is more difficult is identifying ones that truly tested the mettle of the type and, at the same time, have not been described in the historical literature to the point of exhaustion. For this reason, the author has chosen to describe in some detail two quite different engagements from the Second World War: one which shows how the hard lesson learned by the Americans in the Solomons – that technology is an enabler, but not a guarantor, of success – applies universally, and another that shows how, even when faced with overwhelming odds, dogged determination can win through.

An old-fashioned drubbing: The Action off Sfax, 16 April 1941

The spring of 1941 was the darkest of seasons for the British. The catastrophic defeat of France a year earlier had repercussions that were still being felt. Not only had the British lost their primary ally, but their primary enemy now had bases on the Atlantic, free from the blockade that had so effectively restricted U-boat access to the open ocean in the First World War. To make matters worse, the often poorly-handled attempts by the British to, at best, persuade the French fleet to come over to the British side or, at a minimum, neutralise it in the West Indies or American ports, had only soured relations with their former ally. The threat of Italian entry into the war was enough to end British merchant sailings through the Mediterranean in May 1940, forcing convoys carrying war materiel to or from the Middle East and South Asia to take the long detour around South Africa. The German invasion of the Balkans in April 1941, which would drive British forces out of Greece and Crete, was just a further blow against British hopes.

Every bit as concerning to the British as the loss of their positions in Greece was the German-Italian counter-offensive across Cyrenaica which began on 24 March 1941, threatening British control of Egypt and the Suez Canal, and ultimately their access to the oil of the Persian Gulf and the many resources of India. The Admiralty was well aware that the Afrika Korps, which was pushing across eastern Libya, was entirely dependent on supplies coming in regular convoys from Italy across to western Libya, specifically to Tripoli (Tarabulus al-Gharb). The commander of the British Mediterranean Fleet, Adm Andrew B Cunningham, based at Alexandria, was under considerable pressure from London to interrupt this steady flow of supplies. Despite serious reservations, Cunningham ordered the establishment of a Striking Force based at Malta intended to disrupt this regular convoy traffic, comprising the four destroyers of the 14th Flotilla – HMS *Jervis* (G00), flagship of Capt (D) Philip J Mack, *Janus* (G53), *Nubian* (G36), and *Mohawk* (G31).[1] These were all new super destroyers – two 'J' class commissioned just before the war began, including the leader *Jervis*, and two Tribals which entered service in 1938. This flotilla arrived at Malta from Suda Bay, Crete, on 10 April 1941.

With the 14th Flotilla in position, they were immediately put to work. Three of the four destroyers in the force had been fitted with Type 286M radar sets while at Alexandria during the past month.[2] While not the most capable of radars, this set was 'better than no radar at all', which was the unfortunate state of the enemy vessels they came to Malta to face.[3] (Curiously, *Jervis*, Capt Mack's flagship, was the one destroyer in the 14th Flotilla that lacked radar.) The British had one other potential advantage, the ULTRA cryptanalysis network based at Bletchley Park, which had just reached the point of making decrypts of Luftwaffe radio traffic in a rapid enough fashion to be able to provide 'actionable' intelligence to fleet units. But even ULTRA was far from foolproof. On the night of 12–13 April, acting on an ULTRA alert that indicated an Italian convoy was at sea off the coast of Tunisia, Mack led his four ships out of Valletta and headed west-south-west to a point near Sfax off the Tunisian coast called the Kerkennah (Qarqanah) Banks, a point where a convoy that was following the Tunisian coast south from Cape Bon (Ras at-Taib) heading towards Tripoli would have to move away from shore to avoid the Kerkennah (Qarqanah) Islands.

The banks themselves are a stretch of shoal water less than five fathoms (30ft/9.14m) deep that reaches 40nm south-east of Cape Kaboudia (Ras Kaboudia), the easternmost point along the Tunisian coast between Cape Bon and Jerba Island (Djerba) at the southern edge of the Gulf of Gabès (halig Qabis), and almost 20nm to the east-north-east of Roumédia (ar-Rumadiyya), the northernmost of the Kerkennah Islands. (The Italian island of Lampedusa lay some 50nm further north-east; it supported a small guard garrison, but no naval or air assets.) A series of four lighted navigational buoys and one lighted mooring buoy marked the ten-fathom (60ft/18.29m) line from just south of Cape Kaboudia, the easternmost point of the banks, a line approximately 27nm long, running approximately north-west to south-east (135°).

A very near-sister to HMS *Jervis* (G00), HMAS *Napier* (G97) is seen here in August 1944. Like *Jervis*, *Napier* was a leader, so she would also have had the extended after deckhouse to accommodate the flotilla staff and their equipment. *Jervis* would not have had the Type 271 'lantern' at the aft end of her bridge, nor the Type 291 antenna atop her foremast. In fact, she had no radar at all, the only one of Capt Mack's destroyers to lack radar. (SLV – Green Collection)

That night, Mack found nothing and, even worse, was harassed by Axis aircraft on the return leg, fortunately with no damage. It had been Mack's intent to make the intercept of the enemy convoy off Sfax soon after midnight, when the night would be deepest, allowing him to capitalise on his possession of radar, and where the distance from Axis aircraft based in Sicily or Sardinia was the greatest, denying the enemy their one big advantage.

The next time Mack led the 14th Flotilla out from Malta, it was on significantly more solid intelligence. An Italian convoy, large by Mediterranean standards, comprising four German transports, carrying approximately 3000 troops, and one Italian cargo ship loaded with munitions, escorted by the Italian 8ª Squadriglia Cacciatorpediniere (8th Destroyer Squadron), led by Cdr Pietro de Cristofaro in the Navigatori class *cacciatorpediniere* RN *Luca Tarigo*, along with the somewhat smaller and newer Folgore class destroyers RN *Baleno* and *Lampo*, had set out from Naples (Napoli) late on 14 April.[4] This was the twentieth convoy carrying troops, vehicles, and supplies for the newly constituted Afrika Korps, a series that had started in February 1941. These ships were carrying some of the last elements of the 15. PzDiv; most had already crossed over to Libya with minimal interference. The steamers were: *Arta* (German, 2452GRT, carrying 194 troops and sixty-two vehicles), *Adana* (German, 4205GRT, carrying 339 troops and 148 vehicles), *Aegina* (German, 2447GRT, carrying 217 troops and sixty-four vehicles), *Iserlohn* (German, 3704GRT, carrying 292 troops and 118 vehicles), and *Sabaudia* (Italian, 1590GRT, carrying German-made munitions).

They were to follow the normal convoy route to Tripoli, which would take them well clear of Palermo at the western tip of Sicily, heading south-south-west as far as Cape Bon, and then hugging the coast heading south, so as to avoid as much as possible British search aircraft based at Malta, until reaching the Kerkennah Banks, where they would normally follow the buoy line and head out into open water again to cross the mouth of the Gulf of Gabès, and then finally follow the Libyan coast to Tripoli. That was the plan, except that, from the beginning, almost nothing went according to plan.

As soon as the convoy left Naples, it ran into uncharacteristically foul weather. This slowed its progress – already a hardly brisk ten knots – and scattered the ships over a wide stretch of water. What little control De Cristofaro had over his convoy was further compromised by a lack of co-ordination with the other two ships in the escort; the ships originally assigned to accompany *Tarigo* as part of the convoy's escort – the somewhat older destroyers RN *Strale* and *Euro* – had been replaced at the last minute by *Baleno* and *Lampo*. Nonetheless, the bad weather was an advantage for the Italians in one sense, in that it kept British aircraft grounded all day on the 14th; however, by the time the convoy reassembled off Cape Bon on the 15th, the band of bad weather had, for the

Seen at high speed near Crete in 1941, HMS *Nubian* (G36) appears just as she would have looked off Sfax in April of that year. (SLV – Green Collection)

most part, moved off to the north; off the Tunisian coast, the weather had calmed to a steady rain, the seas had subsided to short swells, and the convoy had been sighted at 1357 by an RAF Maryland flying out of Malta.[5] The Italians sighted the Maryland and knew they were being tracked; De Cristofaro was also aware of the presence of the 14th Flotilla at Malta and the strong likelihood that it would be out looking for his convoy. He asked his superiors to provide air protection, but such was the state of co-operation between the RM and the Regia Aeronautica, and the paucity of available aircraft, that only two SM 79s were assigned the task, only one actually made it off the runway, and that one turned back, unable to pass through the storm that still was whipping up rain and high winds south of Sicily.[6]

The Maryland's sighting report had been quite accurate, placing the convoy, comprising five merchantmen and three escorts, near Cape Bon on a southerly course at nine knots. With that information, Mack prepared his flotilla and departed Valletta at 1800 on 15 April. The 14th Flotilla, in line-ahead formation ordered *Jervis*, *Janus*, *Nubian*, and *Mohawk*, set off for the Kerkennah Banks at twenty-six knots.[7] His plan was to arrive at buoy No 4 off the south end of the banks well ahead of the convoy's expected arrival and turn towards the north-west, towards the coast of the largest of the Kerkennah Islands, Chergui. The 14th Flotilla arrived at buoy No 4 at 0044 on 16 April; as expected, on arrival he found a clear horizon and made the planned turn to 310°, slowing to twenty knots. At 0100, having reached the path that he expected the convoy would be following, Mack turned his flotilla to north-north-west (333°), the reciprocal of the convoy's presumed course. Ten minutes later, in order to slow its progress and sweep a wider field, the 14th Flotilla began to zigzag along the northerly course. At 0142, it reached the point that Mack had calculated the convoy should be, had it maintained a speed of eight knots, with still no sign of the convoy to be seen. On the presumption that it had been proceeding at a slower speed, perhaps at seven knots, Mack decided to continue on the same zigzag course towards the north-north-west, but at twenty-five knots.

In fact, the British were close to the Tarigo Convoy, which passed on their shoreward (port) side. One source states they passed each other at 0145.[8] It is more likely they passed about twenty-five minutes earlier; the following is from the report of *Nubian*:

> About 0120. Sighted dark shapes on the port bow, which were thought to be land.[9]

It is not possible to be certain of the distance between the two forces when they passed on opposite courses; it might have been as little as a few thousand yards. By this time, the rain had ceased, and the skies had mostly cleared, but remained hazy; the seas were described as 'almost calm'.[10] It would have been nearly two hours past the rising of a waning gibbous moon, which would have been more to the advantage of the Italians, silhouetting the 14th Flotilla against the faint moonlight. What can be certain is that the British were much closer to the convoy than to the Kerkennah Islands, so that if *Nubian*'s lookouts were seeing anything, it would most likely have been the convoy. Regardless, it was written off as unimportant, and Mack's destroyers continued to the north.

The reader will undoubtedly recall that three of the four British destroyers had radar, Type 286M sets with fixed antennas that broadcast a beam covering an arc approximately 60° either side of the bow; range on a target the size of a destroyer or the small merchantmen in the Tarigo Convoy was 4–7nm.[11] However, the reader must remember that radar, surface-search radar in particular, was still in its infancy in early 1941; the Type 286M, besides having a fixed antenna, which meant it could only 'see' what was directly in front of the ship, also had a relatively weak signal (7kW peak power) compared to the Type 271 (and the similar American SG) introduced a year later, which had peak power exceeding 50kW. The antenna unit of the Type 286M consisted of three Yagi 'fishbone' antennas – the central transmitting antenna facing directly forward and the two receiving antennas angled to either side.[12] The intended use was for the ship to be manoeuvred until the signals received by the two receiving antennas were of equal strength, which would theoretically point the bow of the ship at the object causing the reflection. In practice, this slow and cumbersome operation was rarely carried out; radar operators would estimate target direction from the relative strength of signal received by the two receiving antennas. The relatively weak signal strength limited the radar's effective range and its ability to penetrate rain, or even heavy mist. This problem was only made worse by the proximity of land. The rather primitive 'A'-scope (oscilloscope) display further limited the radar's utility, making it hard to distinguish a valid 'return' from the 'scatter' caused by a rough sea surface. Thus even with three functioning radars well within operational range that, on certain of the 'zigs' of the 14th Flotilla's course must have pointed directly at the Tarigo Convoy heading south-eastward closer in towards shore, none of Mack's ships sighted the convoy as they passed, either visually or with radar, nor did any of the Italians detect any sign of the British force passing north-westward.

The Tarigo Convoy had indeed reassembled north of the Tunisian coast during the afternoon of the 15th and now was

Seen at Malta pre-war, with 'Neutrality Bands' painted on the shield of 'B' mount, HMS *Mohawk* (L31, later F31, then G31) certainly appears imposing with her almost cruiser-like silhouette, a trait the Tribals shared with some other super destroyers, particularly those with large gun mounts fore-and-aft and tripod masts, such as the American Porter class; but the Tribals were lightly built, making this apparent solidity somewhat misleading. However, even a much more stoutly built craft of this size would have had a difficult time surviving the two torpedo hits *Mohawk* received five minutes apart off Sfax which caused her rapid demise on 16 April 1941. (NHHC)

in a compact formation, with the four German ships in two columns of two ships each, the munitions-laden freighter in the centre between the columns, in a 'protected' position, and *Tarigo* leading, also between the columns.[13] The two smaller destroyers were on opposite sides, in sweeping positions – *Baleno* off the convoy's starboard side and *Lampo* trailing directly behind, the expected direction from which enemy attack would likely come. As seen from either beam, the convoy presented an overlapping silhouette of no greater than three or four ship-lengths. Just before 0100, it had come within sight of the northernmost of the Kerkennah Banks buoys and by 0130 the leading transports in each column of the convoy – *Arta* and *Aegina* – had passed the buoy.[14] So close was the convoy keeping to the coast that *Aegina* in the starboard column cleared the buoy by a little less than a mile. The convoy was not zigzagging or taking any other evasive precautions, maintaining a steady course (140°) and speed (8–9 knots).

Finally, at 0155, having reached a point seaward of the northernmost of the Kerkennah buoys, Mack figured that even had he grossly overestimated the enemy's speed, the convoy should have been sighted by this point and that clearly something was wrong with his calculations. He figured that one of two possibilities existed to explain the failure to detect the presence of the enemy convoy. Either the Italians had aborted the mission and reversed course back to the north or they had followed a course further inshore than he had presumed.[15] He reasoned that in the first case, the convoy could be anywhere in a broad arc to the north and, therefore, was unlikely to be located before dawn; in the second case, he could try searching closer towards shore with some possibility of success. Therefore, at 0155, Mack ordered a turn to south-west-by-south (214°) to close Kerkennah buoy No 1. Mack's logic was impeccable and the timing of the flotilla's turn fortuitous. No more than three minutes after making that turn, the radar on *Janus*, the second in Mack's line and the first with radar, picked up a return just off the port bow (bearing 170°) at a range of 12,000yds.[16] *Jervis'* report notes sighting ships bearing 170° at a range of 6nm at the same time, without further detail, so it is not possible to determine whether this was simply reporting *Janus'* sighting or was an independent visual sighting from the flagship, although, in this author's opinion, given the visibility, it is more likely the former.

Either way, Mack acted quickly though, at first, without certainty. He immediately swung his line to port, to the south-east (140°), increased speed to twenty-seven knots, and ordered torpedo tubes trained to starboard, apparently intending to attack from behind the convoy.[17] A few minutes later, he thought better of this manoeuvre:

0203	Altered course to 210 degs. to bring enemy between me and the moon.[18]

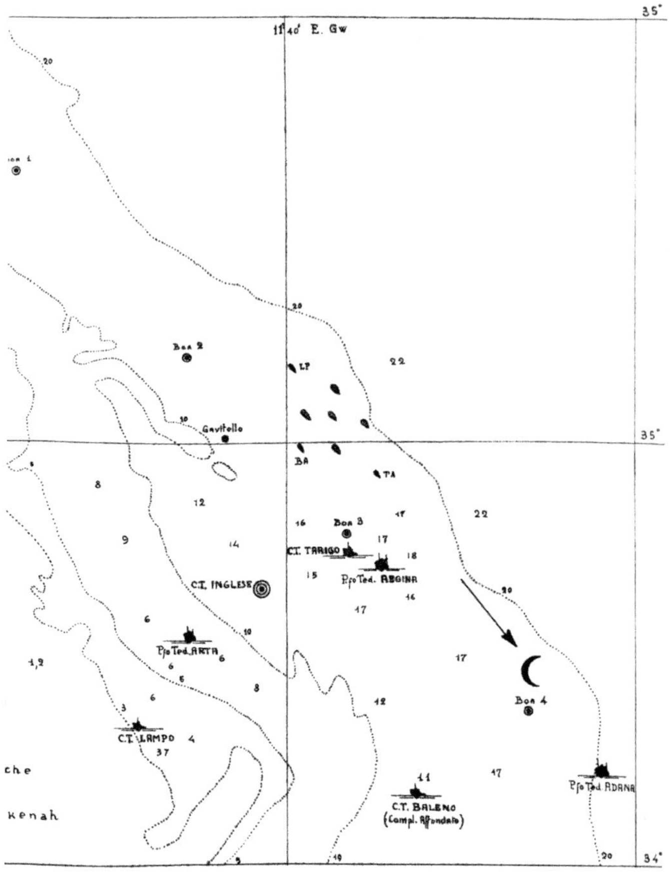

This chart is from the post-action report generated by the Regia Marina for the Action off Sfax, 16–17 April 1941. Water depth is shown in metres, with the line of four lighted buoys – marked 'Boa 1' to 'Boa 4' and one mooring buoy, marked 'Gavitello' – running from north-west to south-east. The Italian convoy is shown in its original formation at the time of the attack, with the escorts marked 'TA' for RN *Luca Tarigo*, 'BA' for RN *Baleno*, and 'LP' for RN *Lampo*. The resting places of seven ships are shown with the markings meaning: 'C.T.' (*cacciatorpediniere* – destroyer); 'C.T. Inglese' (English destroyer); 'P.fo Ted.' (*piroscafo tedescho* – German steamer); and 'Compl. Affondato' (*completamente affondato* – completely submerged). (Ufficio di Stato Maggiore della Regia Marina)

Again, much is unsaid in the narratives, but the presumption must be that with the sky clearing, he hoped that the three-quarter moon, now two hours past moonrise in the east, would silhouette the convoy if he turned his force back towards the shore. With that turn, torpedo tubes had to be reversed to be trained over the port side. However, Mack had no intention of launching torpedoes at this point, even though by 0207 he was able to distinguish five ships in the convoy, because the British Mark IX torpedo had a maximum range of 10,500yds at high speed (36-knot) setting and only 3000 yards more at thirty knots, which meant he would have to get a good deal closer for a chance at an effective torpedo salvo. With that in mind, as he continued to overtake the much slower convoy, Mack steered his line around its western side, ordering a turn to south-by-east (170°) at 0211 and further to south-south-east (160°) at 0213, and then further turns to port by ten degrees at 0214 and 0218.

Finally, with range to the enemy having dropped to 2400yds, *Jervis* opened fire on an unidentified enemy destroyer just south of due east (100°) – almost certainly *Baleno*, which was on the convoy's near side – at 0220. *Janus* joined in two minutes later, as per her narrative:

0220	JERVIS opened fire on a small one funnel destroyer.[19]
0222	Opened fire on this destroyer with an R.D.F. range of 2,400 yards and hit with the opening salvo. This is the only time that an R.D.F. range could be used as later the targets became too numerous and confused.[20]

Jervis noted that by this time the enemy destroyer was being hit by main-battery and pom-pom fire and was returning fire with its Breda machine gun and 'probably 3.9-in with flashless cordite'. She continued firing into *Baleno* until, at 0227, certain that this target was sinking, she checked fire.[21] *Janus* quickly concluded that *Baleno* was 'well hit' and switched her aim to the trailing merchantman, probably the German *Adana*. At a range of 4000yds, she obtained hits on her second salvo and after her fifth salvo saw that the target was on fire. She then switched target, to the German *Aegina*, which was also set on fire. (Italian records indicate that *Adana*, *Aegina*, and *Sabaudia* came under fire at the beginning of the battle and were set on fire by rounds igniting petrol drums carried in their holds.[22]) A torpedo was fired at *Baleno* which missed. After that, *Janus* checked fire briefly, then resumed firing at a target comprising an overlapping group of three merchant ships, obtaining multiple hits. She fired three torpedoes at this target: 'Results not observed from the bridge but tubes' crews claim that one hit was clearly seen.'[23] The time was approximately 0230.

Baleno, as was the case with all the ships in the convoy, was taken by surprise by the sudden gunfire coming out of the darkness. She began a turn towards the enemy but did not immediately return fire because the first salvo of gunfire that hit the ship, mistakenly believed to be 6in (152mm) gunfire from a light cruiser because of its devastating effect, destroyed the bridge and killed the commanding officer, LCdr Giuseppe Arnaud, and all but one of the ship's officers.[24]

> The enemy's first salvo hit precisely on the bridge; it destroyed the fire control director and its base, severing all communications to the guns and rudder controls, killing the commanding officer, who had just given the order to increase speed to full and to head towards the enemy. Also killed in that same explosion were the executive officer, Lt Filippo Ciravegna, the gunnery officer, Lt Enrico Podio, and the navigator, Lt(jg) Oreste Natoli. The Engineering Officer was not aboard. Only Lt(jg) Guglielmo Succi survived amidst the wreckage of the bridge.
>
> Fire broke out immediately in the bridge and wardroom, where paneling with so-called 'fireproof wood' caught fire. BALENO was receiving a crossfire from astern, both port and starboard. The second salvo

The Italian convoy escort at Sfax was led by Cdr Pietro de Cristofaro in the Navigatori class *cacciatorpediniere* RN *Luca Tarigo*, seen here pre war. (NHHC)

> destroyed the communications compartment and boiler No 1, the fourth dislodged the forward engine. In rapid succession, hits were received that damaged steam pipes, the forward bank of torpedo tubes, the naphtha tank and officers' quarters.
>
> The ship rapidly came to a stop, afire from the funnel forward. Meanwhile, machine gun fire swept the deck, disabling many of the torpedomen attempting to man the torpedo tubes. At this point, fear that the fires forward might reach the magazines caused word to abandon ship to spread among some in the crew; a small number of men left the ship at this time, some of those returning to the ship shortly thereafter. Making matters worse, the German steamer AEGINA, totally ablaze, scraped up against BALENO's side. About fifty German soldiers made their way across to BALENO seeking refuge.
>
> Later, when the enemy had ceased firing, the chief gunner managed the flooding of the forward magazines and the making safe of the depth charges; the sole surviving officer organised the fighting of the fires, which were slowly brought under control. Since the ship's momentum had carried her over the shoals, he ordered the anchors dropped in 12m of water. The ship was kept afloat until the 17th by a small electrical pump powered by a single diesel generator, at which time the destroyer VIVALDI and the hospital ship ORLANDO arrived to rescue the survivors.
>
> Some of the survivors stated that BALENO had kept her blue formation lights at her mastheads lit before and during the engagement, giving the enemy an aiming point.[25]

None of the British accounts make any mention of seeing formation lights on any of the Italian ships, much less making use of such as an aiming point.

The two Tribals at the end of Mack's line saw the battle somewhat differently from those further forward. *Nubian* and *Mohawk* both opened fire on the ship they saw bringing up the rear of the convoy, which they identified as a merchant ship, but which, based on the timing of events, probably was the other smaller destroyer *Lampo*. Seeing that target was being hit repeatedly by gunfire from multiple ships, *Mohawk* checked fire after eight salvos and then only fired 'spasmodically for the next few minutes' as targets further ahead were being hit by others in the flotilla and Cdr J W M Eaton did not want to waste ammunition.[26] *Nubian* shifted fire by turns to two overlapping targets further forward, probably *Iserlohn* and *Adana*, and then also checked fire because smoke obscured all targets.[27]

One or more of the British destroyers are known to have engaged the other smaller Italian destroyer, *Lampo*, which was at the rear of the convoy formation. In the attached report by *Lampo*'s captain, the time of this attack is given as 0215, when the convoy had just passed buoy No 2. The firing ship, illuminated by its own gunfire, was described as a two-funnelled warship approximately 2200yds distant, aft of *Lampo*'s port beam, steaming 190° at twenty knots.[28]

> LAMPO was attacked from astern at the same time as was BALENO. At the first sign of enemy fire, the commanding officer, who had only recently taken command of the ship, ordered all men to their battle stations and fire opened on the enemy, full speed on the engines and smoke made, and a hard turn made to the left.
>
> However, her forward main-battery guns could not be brought to bear on the enemy and the after mount was disabled by the third enemy salvo. The first salvo that hit destroyed the forward 20mm machine gun, disabled the gyrocompass and all lighting on the bridge, and wounded the captain, LCdr Enrico Marano, and the gunnery officer, Lt Giovanni Pianigiani.[29]
>
> The same salvo destroyed the fire control director.[30]

It is interesting that the gunfire was seen to come from a ship with two funnels, which makes it more likely that it was *Nubian* or *Mohawk* (or both), since *Jervis* and *Janus* each had only one funnel, and that the salvo came from the port quarter, which could only mean that the two trailing destroyers in Mack's line were well to the east of the first two. Of course, eyewitness accounts of naval combat, particularly at night, must be read with considerable scepticism. At this point in Marano's detailed account, he stated that *Lampo* was strafed by a British aircraft, when in fact, no such aircraft was in the area. Marano described Lt Pianigiani being gravely wounded on the port bridge wing and himself receiving a wound in the chest that rendered him unable to move his left arm. Because the enemy gunfire so far had come in from *Lampo*'s port quarter, Marano had to presume that the threat was primarily from the north-east and he reacted accordingly, ordering his ship to move to the far side of the convoy's left column, which caused *Lampo* to cut between *Adana*, already ablaze and sinking, and *Arta*. The intent of this was to protect the east side of the convoy and reduce the target he was presenting to the enemy.[31]

> From the second salvo, one shell exploded in boiler No 3, while another destroyed the radio compartment.
>
> A fire began under the forward superstructure. The spotting top was likewise engulfed in flame. Another salvo damaged the after main-battery mount and set off the ready-service rounds stored there. . . .
>
> The captain gave the order to turn to the west-southwest (240°) and launch torpedoes. These orders had just been given when it became necessary to shut down the engines, which had just reached 200-

The two other escorts for the convoy were provided by the Folgore class destroyers RN *Baleno* and *Lampo*, the latter seen here, probably while running trials, as she is in light condition, lacking such niceties as gun shields. (Fulvio Venturi)

> 230rpm. The forward engine room had to be abandoned because the explosion of boiler No 3 had damaged the bulkhead between the compartments, and escaping water and steam from the damaged boiler made the compartment untenable. The after turbine was losing power due to ruptured steam lines and a steady loss of steam pressure caused by damaged feed water piping to the remaining boilers. The forward main-battery mount continued sporadic fire with some of the remaining ready-service rounds as access to the forward magazine was disrupted by fire in the forecastle. As the enemy had ceased firing at LAMPO, the decision was made to reserve some of those rounds in case the enemy resumed fire.
>
> The captain and the officers, seeing their ship immobilised and reduced to practical impotence, became concerned that the enemy might attempt to capture LAMPO. The captain therefore ordered that charges be prepared to detonate the after magazine. This concern seemed to recede when the enemy passed out of sight, only to be replaced by worry about the possibility of an explosion of the forward magazine, which could not be flooded due to being inaccessible because of the fire in the forecastle. The captain ordered all surviving crew members into liferafts.
>
> All the officers remained on board except for the midshipmen, a few NCOs, and two stokers. Shortly thereafter, MM1c Bertuzzi and Micheluzzi, risking their lives yet again, managed to flood the forward magazines.
>
> At about this time, the drifting ship grounded in shoal water by her stern. . . . Around noon the next day, the survivors were picked up by the hospital ship ORLANDO.[32]

Lampo, as had been the case with *Baleno*, was seriously damaged and knocked out of the fight before she had a chance to respond in any meaningful manner. More from luck than any other reason, *Lampo* was able to ground herself on the shoals and avoid the fate of the other two escorts.

Then, if possible, things got more confusing. The British line, which had thus far held together, had been heading steadily south-eastward at twenty-seven knots, which would inevitably bring it across the front of the convoy, which was rapidly losing cohesion. At approximately 0231, Mack, in *Jervis*, noted:

> 1 merchant vessel of about 3,000 tons attempted to ram me, but I just crossed his bows in time by going full speed ahead on both engines.[33]

This was almost certainly a case of misidentification, because Cdr J A W Tothill in *Janus* noted:

> 0231 approx. A Navigatori class destroyer passed through the line from port to starboard at high speed, between JERVIS and JANUS. Two torpedoes were fired that missed astern. The 4.7-in. guns could not train fast enough to catch the target but the pom-pom claimed to have hit this ship.[34]

This would seem to make it clear that the ship Mack saw bearing down on *Jervis* was not a merchantman, but rather was *Tarigo* in its position leading the main body of the convoy. That *Tarigo* did indeed cross the British line and turn north on the western, inshore, side of Mack's flotilla can be in little doubt, since an enemy vessel described in the various narratives as a 'large destroyer [which] passed down the line to starboard' (*Jervis*), '1 destroyer (Navigatori class) on opposite course, range about 1,000 yards on starboard beam' (*Nubian*), 'a Navigatori class destroyer ahead steaming north . . .' (*Janus*, this after herself turning north at 0241), and 'an enemy destroyer of the Navigatori class was sighted on the starboard bow, steering an opposite and parallel course at high speed. She was immediately engaged by NUBIAN and MOHAWK, hit and set on fire and was last seen stopped and on fire about a mile astern' (*Mohawk*).[35]

Tarigo, out in front of the convoy, was initially untargeted and, unable to see where the gunfire was coming from, neither could she immediately respond to the attack.[36] De Cristofaro ordered his crew to battle stations, increased speed to full, and ordered a hard turn to starboard to bring his ship about.[37] Almost immediately, two enemy destroyers came into view a few hundred metres on the starboard bow.

The British flotilla had caught the convoy by surprise and had wrought terrible damage in the first few minutes – fatally damaging, if not sinking outright, the destroyer *Baleno* and several of the merchantmen, including *Adana* and *Aegina*. The return fire from the remaining escorts was sporadic and inaccurate; the factors of surprise and poor visibility combined to frustrate any attempts by the three destroyers to fight back. The only comment regarding Italian gunfire up to this point in the battle, at approximately 0245, was in *Janus'* narrative, which noted: 'The Italian gunnery was poor, their tracer could be seen going high and wide.'[38] The after-action report of *Lampo*'s gunnery officer, Lt Pianigiani, stated that his ship had fired at enemy gun flashes and appeared to straddle the target before the guns were silenced; the British reported no hits by Italian gunfire, but did suffer some splinter damage.

At this point there did not seem to be much left of the convoy to attack. *Janus* was, for a while, reduced to firing at targets already seriously damaged:

> 0241 Having altered round to port to the northward, opened fire on a large merchant ship on fire to the eastward.[39]

The target was most likely *Adana* or *Iserlohn*, which were the largest of the transports and trailed the two columns, but, based on what happened next, it is not possible to rule out that it might have been *Sabaudia*:

> 0242 Fired one torpedo at this ship, which blew up at 0243.[40]

This explosion, which utterly demolished the Italian cargo ship, served as a hard point in time that allows the narratives of *Jervis* and *Janus* to be synchronised:

> 0250 An ammunition ship blew up with an enormous explosion; smoke and flames rose to a height of 2,000 feet and JERVIS who was 1,500 yards away was showered with pieces of ammunition, etc., weighing up to 20 lbs.; the sea around appeared as a boiling cauldron.[41]

The seven minute discrepancy in the timing in the two accounts would appear to be a recording error made in one or both ships, as their accounts appeared to have been well-synchronised twenty minutes earlier.

It is entirely possible that *Sabaudia*'s explosion and *Janus*'s actions at this point were entirely unrelated, as it had been noted earlier that *Sabaudia* was set afire in the first moments of the battle and it may have simply taken this long – approximately twenty-three minutes – for the fire to have reached her cargo of munitions; an additional cause is not required to explain *Sabaudia*'s cataclysmic demise. Then, again, the target of at least some of *Janus*'s actions may have been *Iserlohn*, which, according to the Italian record, did not come under fire until some minutes after the first three steamers were hit and fatally damaged. Once targeted, however, *Iserlohn* very quickly succumbed to enemy gunfire and sank in a matter of minutes.[42]

Nubian and *Mohawk* made no note of *Sabaudia*'s cataclysmic demise as they were otherwise occupied. In *Nubian*'s account, after noting the initial sighting of *Tarigo* heading north at 0223, a time that was likely as much as ten minutes behind the clock in *Janus*, the narration stated:

> 0225 Increased speed and altered course toward and then away. Engaged destroyer with pom-pom, 0.5 M.G. and 4.7. Destroyer fired back and there was a short engagement which had to be terminated as MOHAWK passed between NUBIAN and the enemy. Hits were definitely obtained under the bridge and on the hull aft.[43]

Both *Jervis* and *Janus* also related engaging the Italian flagship after she turned north. In *Jervis*' account of the engagement, it happened right after *Tarigo* crossed ahead of *Janus*, before *Sabaudia*'s explosion, and before *Jervis* turned north:

> Fighting lights were switched on. One large destroyer passed down the line to starboard and was heavily engaged, hit with the first salvo and set on fire amidships.

RN *Baleno* is seen in this image the day after the Action off Sfax, 17 April 1941, as she appeared from an Italian reconnaissance aircraft. She appears to be floating in stable condition with no visible fires, but appearances, in this case, are deceiving. Her crew is still aboard, waiting to be taken off by the hospital ship RN *Orlando* and destroyer *Vivaldi*; she was badly damaged and being kept afloat by an electrical pump. Not long after the crew was taken off, the diesel generator failed, and the ship foundered and disappeared in 12m of water. (Enrico Cernuschi Collection)

> 0240 Fired one torpedo at large enemy destroyer, probably obtaining a hit aft.[44]

Nothing in Italian accounts of the action corroborates this account, raising the possibility that Mack was observing hits obtained by *Nubian* and believed them to be his own, a not uncommon occurrence in night battles.

Janus turned north as soon as *Tarigo* crossed ahead of her at approximately 0231. Then, after briefly engaging, and possibly torpedoing, *Sabaudia*, she sighted and set off in pursuit of *Tarigo*. This almost certainly happened after *Nubian*'s brief engagement with *Tarigo*:

> 0246 Sighted a Navigatori class destroyer ahead steaming north and making smoke. Increased to 30 knots and hauled out to port clear of the smoke. As soon as his bow could be seen, fire was opened and 3 hits with the first salvo spotted through his smoke.
>
> 0247 to 0259 Engaged the LUCA TARIGO on various courses at speeds varying between 16 and 30 knots, mostly at fixed sight range.[45] During this time two torpedoes were fired in bridge control and one in local control, all of which missed. . . .
>
> A great many hits were scored on the LUCA TARIGO and by 0259 she was stopped on fire, and had ceased firing. I considered that she was in a sinking condition and withdrew to the eastward to join Captain (D).[46]

The Italian account generally agreed with this timeline. After turning north . . .

> ... TARIGO found herself between two transports already afire and two large enemy destroyers, which had moved into a position on the starboard side of the convoy skirting the Kerkennah buoys unseen.[47]

Given the position and the time of this encounter, these were certainly *Nubian* and *Mohawk*.

> What followed was a brief, intense exchange of fire at a range of a few hundred metres. TARIGO had not yet worked up to full speed. The first salvo of enemy fire hit under the bridge, causing massive casualties; among those casualties was Cdr De Cristofaro, whose right leg was severed. The engine room telegraph was disabled, as was the helm, which swung freely. Despite his grievous wound, the captain remained at his post and repeated the order to attack the enemy. Steering was

> transferred to the after command position. TARIGO was able to get off a few rounds at the enemy, whose second salvo hit the forward fire room, causing boiler No 1 to explode. Machine gun fire swept the deck. Later salvos of enemy fire demolished the after superstructure; one gun of main-battery mount No 3 continued firing, the other jammed. A fire in fire room No 3 caused the bridge to be abandoned. . . . The ship slowed to a stop. The Chief Engineer, mortally wounded and blinded, asked to be helped up to make a final report to De Cristofaro; it is not clear whether this was to report on the condition of the powerplant or simply to die next to the commanding officer.[48]

At that point there was a lull in the action as seen from *Tarigo* that can only be explained by *Janus* turning away after concluding that the enemy flagship had been put out of action. However, *Tarigo* was not quite done . . .

> Taking advantage of this pause, Lt(jg) Ettore BESAGNO, with a scratch crew of torpedomen, many of them wounded, managed to train the one working bank of torpedo tubes and fire three torpedoes, sinking the British destroyer MOHAWK. The actual aiming and firing was done by PO3c (S°C°) Torpedoman MARETTI.
>
> More gunfire began hitting the destroyer. Seeing that TARIGO was beginning to list dangerously, De Cristofaro ordered the crew to abandon ship.[49]

Many of these details are, in fact, not terribly specific, because the losses among the officers and crew of the Italian escorts and steamers were so high. For example, it is known that the Italian commander remained at his post conning *Tarigo* until he lost consciousness from blood loss, but it is not clear exactly when this happened.

It appears that at this point *Nubian* turned away to the east to re-engage the convoy, because the next entry in her narrative, noted as occurring five minutes later, states that she '[c]rossed bow of leading merchant ship', which was almost certainly the transport *Arta*, which had been leading the port column of the convoy.[50] *Mohawk*'s record indicated that she followed *Nubian* around the front of the convoy and last saw *Tarigo* 'stopped and on fire about a mile astern'.[51] It must have been right after *Nubian* and *Mohawk* turned away that *Janus* engaged *Tarigo* and delivered the damage that disabled her.

The battle was far from over. Both sides had more fight in them and damage to be done to their enemy. *Mohawk*, following *Nubian* across the front of what remained of the convoy, turned hard to starboard to avoid *Arta*. This exposed her starboard side to an unexpected danger.

> 5. At 0230, NUBIAN led round to port across the bow of the leading merchantman, who immediately altered course to starboard and tried to ram MOHAWK. This ship appeared to be quite undamaged and on avoiding her and crossing over to the port side, I turned to starboard with the object of engaging and sinking her.
>
> 6. As the ship was still under helm and turning to starboard, a torpedo struck on the starboard side abreast 'Y' gun, blowing away the after part of the ship from just forward of this gun. The ship at once stopped . . .
>
> . . . No ship could be seen in the vicinity which could have fired this torpedo apart from the enemy destroyer who was stopped and on fire over a mile astern of MOHAWK, and I came to the conclusion that it must have been a stray torpedo fired by this vessel, possibly with the object of getting rid of top-weight.[52]

Eaton was correct in identifying *Tarigo* as the source of the torpedo that had disabled his ship, although, with the ingrained disdain the Royal Navy felt for their European rivals, it would have been unthinkable to credit their Italian enemies with the courage and skill to launch a torpedo from a sinking, burning ship a mile away with the accuracy to hit *Mohawk*. Neither did the British yet know the Italians had the technology to build torpedoes that were faster and more accurate than their own.[53] Regardless of Eaton's attitude towards his enemy, he had to confront the fact that his ship was now, at least for the moment, losing way, while continuing to turn to starboard:

> 7. The Engineer Officer then reported to me that although most of the stern had been blown away, the propeller shafts and propellers were still in place and that he would try to move them and get way on the ship.[54]

This report is nothing short of remarkable, if it is to be believed. The short fantail of a Tribal class destroyer began just forward of Y gun mount, the aftermost main-battery mount, directly above the point where the propeller shafts exited the hull. Between the glands where they emerged from the hull and the propellers, just ahead of the rudder, the shafts were supported by heavy struts connecting them to the cutaway underbody of the ship's stern. It is difficult to imagine a set of circumstances in which a torpedo hit *Mohawk*'s fantail and caused it literally to disintegrate in such a manner that left the

Seen from the navigator's cupola on the upper fuselage of a Regia Aeronautica seaplane, the forward section of RN *Lampo* is seen where she grounded after the Action off Sfax, 17 April 1941. Evidence of the fire in the forecastle that disrupted ammunition supply to the forward main-battery mount, seen jammed hard over to starboard, can be seen under the bridge, as can other damage to the superstructure and funnel. (Enrico Cernuschi Collection)

propeller shafts undamaged. Perhaps the Engineer's report, made in haste, was mostly wishful thinking – note the implication that he had not yet attempted turning the propeller shafts – it is difficult to believe that Eaton would put much hope in getting his ship under way again. Regardless, he had very little time to enjoy such hope:

> About five minutes after the first torpedo had struck, a second one struck MOHAWK on the port side approximately on the bulkhead separating Nos 2 and 3 boiler rooms. The ship commenced to settle rapidly on an even keel and I ordered all hands on deck.[55]

It probably would not have been possible to hit *Mohawk* in a more vulnerable location. A hit at that bulkhead opened up perhaps a quarter of the ship's length to the sea; there was simply no chance that a ship of her size, particularly one already damaged as she was, would have sufficient buoyancy in her remaining sections to remain afloat. Note that she was hit on her port side; as the source of this torpedo was certainly *Tarigo*, as had been the first, it indicates that, as the ship was drifting to a stop, it continued to turn to starboard, exposing its port side to the successful efforts of Lt Besagno and his intrepid crew to fight their ship to the end.

> Less than a minute after this order had been given, the ship took a very heavy list to port and settled on her beams ends with the after parts submerged as far as the after end of the torpedo tubes. The order was then given to abandon ship as I considered that it was only a matter of minutes before the ship sank.[56]

This, effectively, brought the engagement to an end. The only ship unaccounted for is *Arta*, which was last seen attempting to ram *Nubian*. She was damaged and set on fire in her brief encounter with *Nubian* and *Mohawk*, but managed to stay under power and afloat long enough to run aground in 20ft of water approximately 2nm north-east of *Lampo*.

Tarigo was dead in the water and, by 0300, had capsized, but seemed unwilling to sink, so Mack ordered *Janus* to complete the job. This was done efficiently at 0333, by putting some rounds into her overturned hull, causing it to slip beneath the surface just south of buoy No 3.

The British scoured the area carefully for survivors from

Mohawk. One hundred and sixty-eight of the crew were rescued; forty-one were lost, the only British casualties in the battle.[57] *Mohawk* sank in extremely shallow water, and the Italians were able to dive on the wreck and remove some charts and papers that proved useful in subsequent operations, particularly the Alexandria raid on 19 December 1941. In a curious twist of fate, *Jervis* was one of the British warships damaged in the attack on Alexandria harbour.

Axis losses are harder to pin down. All eight ships were 'lost', but a number of them were grounded on the Kerkennah Shoals. *Baleno* stayed afloat until the morning of the 17th, when her crew was taken off; after that she almost immediately capsized and sank. *Arta* was boarded ten days later by men from the British submarine HMS *Upholder* (N99), who recovered some random items and fired the ship, leaving her a total loss. *Lampo* was refloated and towed back to Italy where she was repaired, only to be lost off Cape Bon on 30 April 1943. Losses in human lives are equally hard to define exactly. None of the German or Italian survivors were picked up by the British before they departed the scene before dawn on the 16th. As far as can be ascertained, Italian losses were as follows: *Tarigo* – 202 lost; *Baleno* – sixty-nine lost; *Lampo* – 141 lost; and *Sabaudia* – thirty lost. German losses are harder to figure because they included both merchant crew and troops. The four ships were carrying 1042 men and approximately 200 crewmen; approximately 270 German soldiers and seamen were lost in the Tarigo Convoy. Perhaps most importantly, from the point of view of the British, was the fact that in the short term this action did little to interrupt the steady movement of supplies and men from Italy to Africa. It would take more than the loss of one convoy, however complete, to do that.

A glimpse of the future?

Every navy needs a utility warship, a type capable of taking on any number of useful roles with little difficulty. Before steam and iron transformed navies in the nineteenth century, corvettes and sloops filled this role. In the early days of the industrial revolution, these were replaced by small cruisers, but the replacement was not quite satisfactory because even the smallest of cruisers never quite fulfilled all of the requirements of a true utility craft, namely that it be available, adaptable, and expendable. It was that latter requirement which assumed that these craft would be procurable in large enough numbers and at a low enough cost that losses could be tolerated without undue concern. This was one of many reasons why the development of torpedo boats and then torpedo boat destroyers was greeted with such enthusiasm in most navies. Here was a ship that not only could perform its original intended role, launching torpedo attacks and defending against such attacks, but it proved to be an adaptable all-purpose escort vessel for fleets and convoys and could easily be transformed into a high-speed transport, minelayer, minesweeper, seaplane tender, or whatever other role could be filled by a small, fast, capable ship.

Thus it was not at all surprising that when the US Navy found itself facing a new and daunting challenge in the spring of 1945 it was destroyers which were asked to respond. That challenge was a novel and deadly form of air attack devised by the Japanese in response to the overwhelming dominance of American naval air power. Where in the last carrier battles of 1942 the naval air forces of America and Japan had been relatively evenly matched, by the time they faced each other again in June 1944 at the Battle of the Philippine Sea the balance had swung dramatically and permanently in favour of the Americans. By the time the next major invasions occurred in the Allied push across Central Pacific, culminating in the invasion at Lingayen Gulf on Luzon in the Philippines in January 1945, the sense of desperation among the Japanese was such that voices that had been advocating extreme measures such as suicide (*tokko* – special) attacks were now given credence. Japanese pilots had been known to crash damaged aircraft into enemy ships, preferring a 'useful' death to a 'wasted' crash mid-ocean. (These were known as *jibaku*, crashes resulting from a spur-of-the-moment decision by a pilot, as opposed to a planned suicide.) But what was now being proposed, and seriously considered, was the organised flight of aircraft on intentional suicide missions to crash into Allied naval vessels assembled at landing beaches for the invasions advancing ever closer to Japan.[58]

The first organised *tokko* attack – the name kamikaze (divine wind) did not come into common use in Japan until after the war – probably took place on 25 October 1944 off Mindanao in the Philippines.[59] These first halting steps were followed by more intense and organised attacks during the invasion at Lingayen Gulf (Luzon) in January 1945 and off Iwo Jima in February. Those invasions taught lessons to both sides. To the Japanese, it taught that while kamikazes were indeed a powerful weapon, they would not force a change in Allied strategy unless they could be massed in sufficient numbers to overwhelm enemy defences and destroy a larger percentage of the invasion fleets; small attacks at random intervals may achieve the occasional spectacular success, but were not enough to slow significantly the momentum of the Allied advance. To the Allies, it taught (or, better stated, emphatically reinforced) the obvious lesson that nothing is more valuable than timely and accurate warning of an incoming attack.

The two sides each made moves to gain benefit from these lessons learned. On the Japanese side, that meant hoarding aircraft and pilots in anticipation of the expected Allied invasion of Okinawa, the logical next and last 'stepping stone' on the way to Japan itself. For almost the entire month of February, there were no kamikaze attacks on Allied warships and even when Iwo Jima was invaded at the end of the month, the response was relatively muted. Aircraft were being gathered and carefully hidden away on the southernmost 'Home Island' of Kyushu, in caves or in camouflaged revetments in such a manner that, even though carrier sweeps were made over

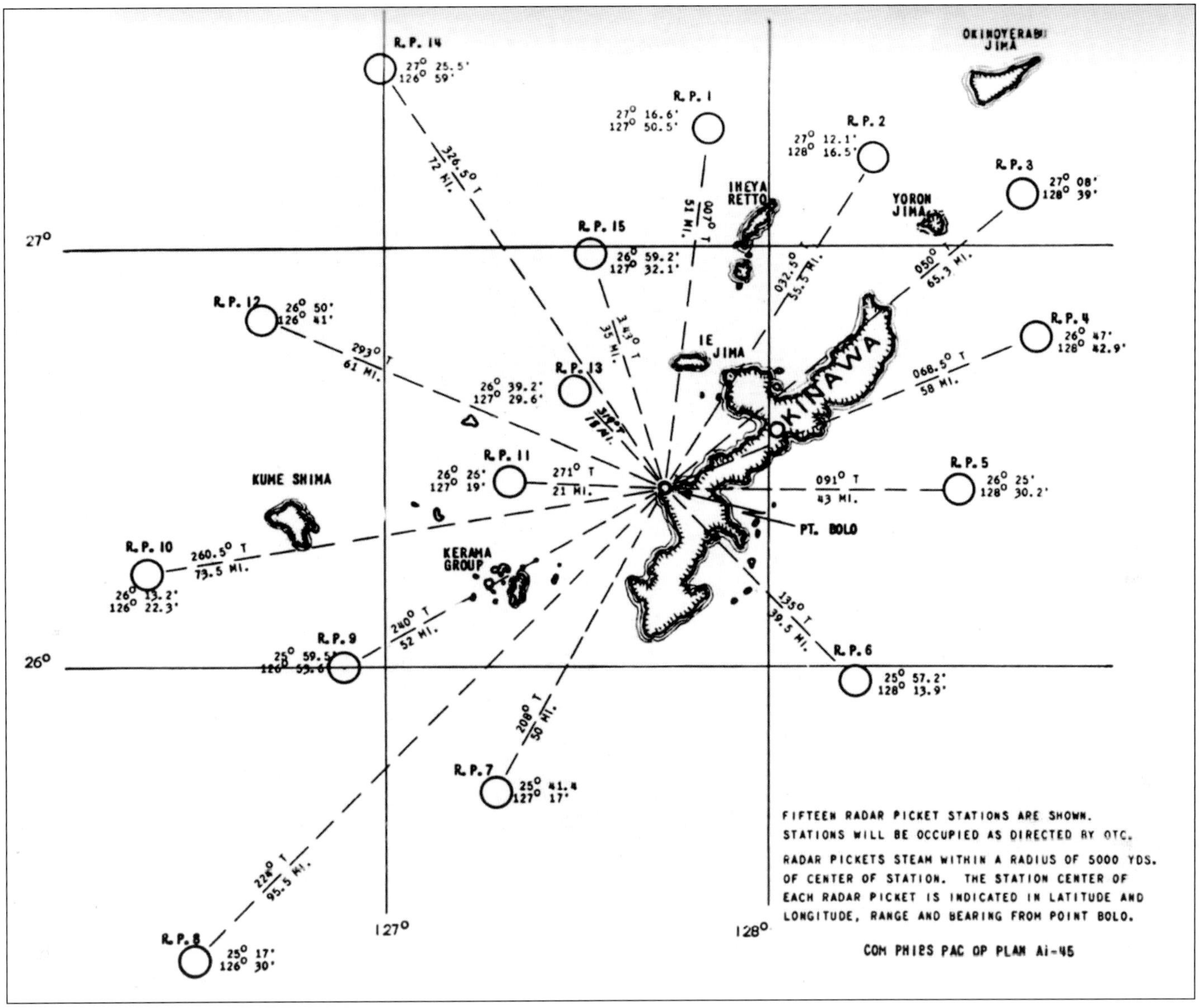

Expecting a sustained kamikaze offensive against the invasion fleet off Okinawa, the Allies established a ring of radar picket stations around a central reference point, code-named Point Bolo. Most of the picket stations were located north or south-west of Okinawa, in the direction of Kyushu or Formosa (Taiwan), respectively, where most nearby Japanese airbases were located. Not all of these picket stations were generally manned; some were never manned. (USN)

Kyushu in mid-March 1945, the leader of the IJN's *tokko* campaign, VAdm Ugaki Matome, held back, awaiting the main blow.[60] The invasion proper began with landings at Hagushi (Toguchi), on the western side of southern Okinawa, but even then Ugaki held back. The triggering event was Operation Ten-Go, the sortie by the last of the IJN's operational surface assets, led by the battleship *Yamato*, which left from Tokuyama anchorage south-west of Hiroshima on the afternoon of 6 April, the aim of which was to beach *Yamato* off the coast of Okinawa in a position to bombard the invasion forces. It was a doomed operation, in no way aided by Ugaki's decision to unleash the first of the major *tokko* attacks on the Allied forces assembled around Okinawa; these major suicide attacks were named Kikusui (Floating Chrysanthemum) operations, for the crest of the famous samurai warrior Kusunoki Masashigi, renowned for his unswerving loyalty to the emperor.

The naval forces off Okinawa were as prepared as possible for the expected attacks. The experience with the still limited kamikaze attacks to date had taught the Allied navies sufficient respect for the potential of this terrible weapon, which had made it obvious that the current system of combat air patrols (CAP) by carrier aircraft, backed up by the heavy, medium, and light anti-aircraft weaponry mounted on the ships of the fleet, was no longer sufficient. Even with the multitude of ever-improving radars carried by the ships of the fleet more was needed.

All the proliferation and improvement in sensors helped, but did not solve the ultimate problem, which was that it was no longer sufficient just to damage an incoming enemy aircraft. In the past, when Japanese attackers carried bombs or torpedoes, attempting to hit enemy ships with those weapons, it was often sufficient to damage an attacking aircraft, spoiling its aim, and thereby rendering it harmless. Now, with the pilot guiding the entire aircraft like a missile, as a weapon that just happened to carry an additional weapon, the aircraft had to be literally destroyed, broken into small fragments, or otherwise rendered incapable of further flight, in order to reduce, if not neutralise, its danger. The only weapons capable of achieving this were heavy AA guns – the 5in/38 gun mounted on all US Navy ships destroyer-sized or larger, but not in sufficient numbers – or massed CAP aircraft. The Japanese very quickly learned how to weaken the overwhelming CAP numbers the Allies could put over their task forces by attacking from multiple directions simultaneously, using radar deception technologies such as chaff ('window') or reflective kites, or attacking at those times of day when the CAP was likely to be weakest, such as at dusk.

The only defence against these tactics was better fighter direction that could more intelligently allocate fighter assets against threats. For this, a key necessity was the earliest possible warning of when and from what direction and at what altitude attacks were developing. To achieve this, the US Navy leaned on a tactic that had been used in one form or another since the beginning of naval warfare, namely the picket destroyer. As early as the Battle of Savo Island in August 1942, the Allies had employed destroyers posted well away from the main body of their naval force with the intent that they should warn of the approach of an anticipated attack. In that battle, they failed in that role, despite being equipped with both air search and fire control radars, because the radars were still new and the operators inexperienced in their use, particularly in conditions such as prevailed that night, with scattered rain squalls and multiple nearby land masses.

It was now two-and-a-half years later and both the radars and the operators were vastly improved. Also improved was the shipboard organisation needed to exploit the information obtained by radar. From the beginning, the planning for the invasion of Okinawa envisaged the establishment of a ring of fifteen radar picket stations around the island, that would be manned as needed to provide early warning of incoming kamikaze raids; the picket stations ranged from 21nm to 95nm from Point Bolo (Zanpa-masaki), a promontory north-west of the landing beaches, chosen as an arbitrary standard reference point for intercept orders.[61] The idea was simply that, by detecting the incoming flights earlier, more time would be allowed for CAP fighters to be vectored into intercept positions well away from the task force. It was in theory an excellent idea and, in practice, it often worked as designed. It only had two problems, neither of which were with the theory, but with the implementation, or rather the means available to the US Navy to implement the theory in April 1945.

The new USS *Laffey* (DD 724) was at RP No 1 because the prior occupant, USS *Cassin Young* (DD 793), had been damaged in a kamikaze raid on 12 April 1945. The damage to *Cassin Young* was not serious, in terms of her survival, but did impair her ability to perform the duties of a radar picket, carrying off her search radar antennas and one yardarm from the foremast, as seen in this image taken the next day at Kerama Retto. (NARA)

Despite the massing of over 1200 naval vessels in TF51, including thirty-nine aircraft carriers of various sizes packed with aircraft, there were simply not enough fighter aircraft over the fleet to be everywhere all the time. (To be precise, the naval forces around Okinawa included eighteen escort carriers (CVEs) carrying a nominal total of 342 fighter aircraft, plus four more CVEs ferrying in 222 Marine fighter aircraft intended to operate from captured land airbases on Okinawa within four days of the invasion, plus it could count on the regular support of TF58, VAdm Marc A Mitscher's Fast Carrier Force comprising eleven large aircraft carriers (CVs) and six light carriers (CVLs) carrying a nominal total of 816 more fighter aircraft.[62]) This was especially true as the Japanese applied their own 'lessons learned' and increasingly attacked at dusk, when they knew that most CAP fighters would be returning to their carriers low on fuel and eager to land aboard before night fell. The other problem was that the Japanese learned very quickly that the picket destroyers (and the small craft such as LSMs and LCSs supporting the destroyers at the picket stations) were a vital link in the Allied defensive plan, and thus became a prime target for kamikazes. Never would that be more so than on 16 April 1945 at RP No 1.

USS *Laffey* (DD 724), the second to bear that name, was still a new ship. She had been commissioned just over a year earlier, in February 1944, one of the new Sumner class destroyers that began to be laid down in May 1943. (These ships, built on a hull of the same length as the preceding Fletcher class, but with increased main-battery armament, will be described in the next chapter.) The first *Laffey* (DD 459) was a Benson class destroyer sunk during the First Naval Battle of Guadalcanal on 13 November 1942, barely six months after being commissioned. The second *Laffey* did not head directly for the Pacific theatre. After working up on the east coast and in the Caribbean, she headed for Plymouth, England, where she supported the Normandy landings. By July, she was back at the Boston Navy Yard, where she received upgrades to her sonar and fire control radars – swapping out the FD radar on the Mk 37 main-battery director for the advanced Mk 12/22 suite – and then proceeded to the Pacific, where she joined TF38, the Fast Carrier Force, in November at Ulithi.[63] She then saw duty supporting Adm Thomas C Kinkaid's Seventh Fleet during the invasion of the Philippines and again supported landing forces at the assaults on Iwo Jima and Okinawa.

The first destroyers were assigned duty at the radar picket stations even before the 1 April 1945 invasion of Okinawa, and the first kamikaze attack on a destroyer at a picket station also came before the invasion. Nevertheless, it took the Japanese a few days to react fully to the invasion; the first, and largest, of the Kikusui attacks, comprising 355 kamikaze aircraft supported by 344 other aircraft came over two days on 6–7 April.[64] The destroyer assigned the station at RP No 1, USS *Bush* (DD 529), was overwhelmed and finally sank the evening of 6 April from accumulated damage; USS *Colhoun* (DD 801), which had come over from RP No 3 to aid in *Bush*'s defence, was similarly overwhelmed and was scuttled shortly after midnight when her salvage was judged to be impossible. More kamikazes attacked the task force supporting the landings on Ie Shima, a small island off the west coast of Okinawa, about 30nm south of RP No 1. Two destroyers there were badly damaged and, while both survived the attacks, neither was ever fully repaired and both were sold for scrap after the war. At other points around Okinawa, one other destroyer and a destroyer escort were likewise rendered constructive total losses, while a fast minesweeper, an LST, and two Victory ships (large civilian-manned steamers), were also sunk.[65] This total of ten ships lost on 6–7 April 1945 was the biggest single-attack result achieved by the 'special attack' offensive. Fortunately for the Allies, the Japanese were never able, in the nine subsequent 'waves' of the Kikusui campaign, to send anywhere near the same number of aircraft against the fleet off Okinawa.

By the time the third Kikusui attack was launched on 15–16 April, only 120 kamikaze aircraft were involved, but that was more than enough to make life miserable for *Laffey*. On 13 April, after spending the preceding night screening a fire support unit off the coast of Okinawa, she was detached and proceeded to Kerama Retto, a small island group off the south-western coast of Okinawa that formed a sheltered lagoon where emergency repairs and replenishment could be performed. There she took on the fighter director team and their equipment from the destroyer USS *Cassin Young* (DD 793). On 6–7 April, *Cassin Young* had the sad responsibility of sinking the mortally wounded *Colhoun* at RP No 1. *Cassin Young* then replenished at Kerama Retto and returned to RP No 1 on 8 April. She was still there on 12 April when she came under sustained kamikaze attack, part of the second Kikusui wave:

> The ship now came to a westerly course, and at 1337 the latter group was sighted: three Vals in a close V formation.[66] At 1338 the U.S.S. PURDY [DD 734] opened fire immediately followed by the U.S.S. CASSIN YOUNG, and the right wing plane was shot down. The leading plane went into a falling leaf dive pulled out at 500 feet, 3000 yards on the starboard bow and swept across the bow in a tight left turn. The rudder was shifted from hard right to hard left and the Val crashed at 1340 25 feet off the port quarter. At 1343 a Val 2000 yards on the port quarter commenced a very low-altitude run on the U.S.S PURDY, was taken under fire and shot down. A fourth Val was taken under fire at 1344 at 5000 yards, position angle 20 as he dove at the ship in a low angle dive on the starboard beam. The rudder was thrown hard right and the plane crashed into and carried away the port yardarm and upper section of the foremast with SG, SC, and numerous radio antennae and exploded in mid air 50 feet from the ship. . . . The last Val was now coming in as did the one before, but on the port beam. The rudder was thrown hard left and at 1348 the plane passed over the ship and crashed on the port quarter.[67]

The two destroyers at RP No 1 were attacked by five aircraft in ten minutes. Neither destroyer was hit squarely in the hull, but the damage to *Cassin Young*, while not in any way threatening her survival, nonetheless proved significant. The explosion of the bomb-laden aircraft in the air just 50ft from the ship had serious consequences. The impact on her crew, many manning anti-aircraft guns in exposed positions, was devastating; while only one crewman was killed, fifty-eight were wounded, many seriously. The damage to the ship was consequential, above and beyond the loss of the sensors atop her foremast:

> A broken salt water line short circuited the main board cutting out all power to the main battery director. A ruptured steam line and loss of ventilation in the forward fireroom and engine room necessitated their evacuation. . . . The U.S.S. CASSIN YOUNG at 1350 turned southward and headed for Kerama Rhetto [sic] as her speed slowly decreased.[68]

The damage to *Cassin Young* rendered her effectively *hors de combat* until her powerplant could be repaired and her radars replaced.[69] Thus *Cassin Young* was at Kerama Retto on 13 April where she transferred the 'fighter director crew, supplies, and ammunition to USS *Laffey* at 0900'.[70] That afternoon, *Laffey*, having completed the transfer of men and materiel, departed Kerama Retto and headed towards 'preferred beaches' at 1517, where she rendezvoused with VAdm Turner's flagship USS *Eldorado* (AGC 11) for a conference of FDOs (Fighter Direction Officers), which allowed *Laffey*'s officers to 'obtain technical assistance with radio equipment'.[71]

The following morning, 14 April, *Laffey* replenished from an oiler and departed for RP No 1, where she relieved the destroyer on duty there at 1612. The next day, 15 April, was the first day of the two-day third wave of the Kikusui campaign. Nonetheless, it was strangely quiet at RP No 1. Except for an afternoon excursion to check out a crashed enemy reconnaissance aircraft – three bodies were found in the water, searched for any useful intelligence items and returned to their watery grave – the daylight hours passed without incident.[72] After the sun set, the skies around *Laffey* became a lot more crowded; radar contact was made with incoming flights of enemy aircraft at 1940, 2030, 2040, 2046, and again at 2220, but none of them came close to *Laffey*'s night station.[73]

Things then remained quiet until 0452 the next morning, when the crew was called to General Quarters as the radars were again filling with contacts, but the screens cleared as the enemy aircraft passed on to the south. At 0600, the first CAP flight of the day reported in – two aircraft directed to orbit at low altitude over the ship, plus two four-aircraft divisions ordered to hold at 8000ft and 10,000ft respectively – the first fighter protection overhead since 1850 the previous day. At 0700, *Laffey* left the night station for the day picket station RP No 1, accompanied by two anti-aircraft-gunfire support ships, *LCS 51* and *LCS 116*. At 0730, a new two-aircraft CAP arrived to relieve the low-altitude pair orbiting over *Laffey*, but

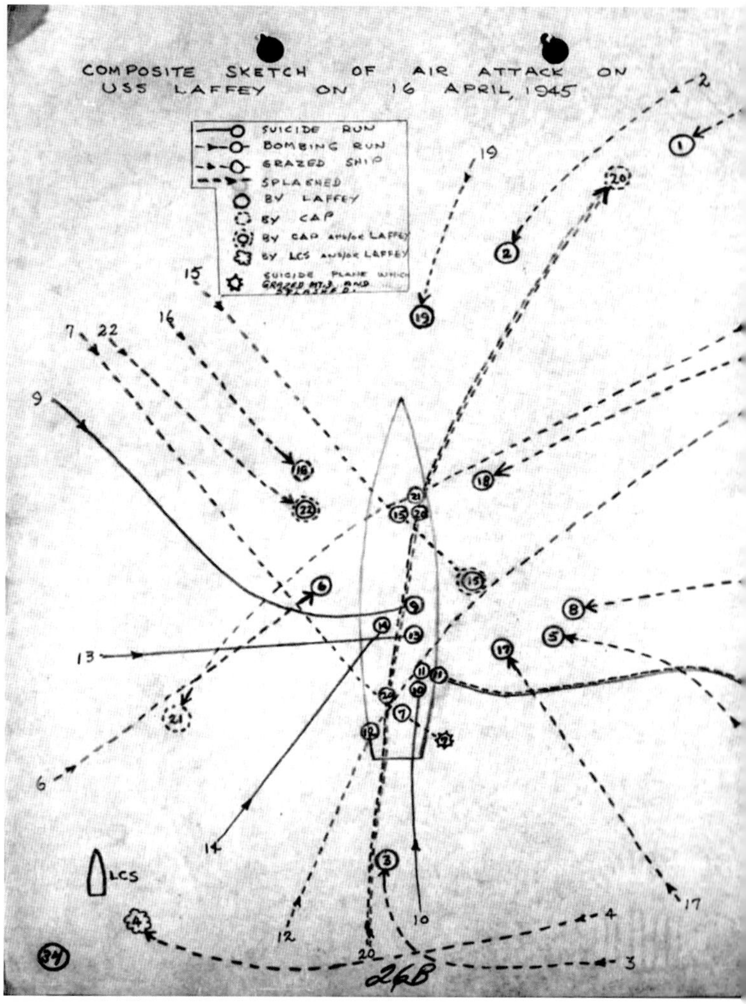

Three days later, it was *Laffey*'s turn to become the target of intense and prolonged attack by kamikazes, at least twenty-two separate attackers over approximately an hour and fifteen minutes. This chart, from *Laffey*'s action report, shows each of the identified attacks, categorised by type and result. Note that most were not specifically identifiable as suicide dives; the Japanese mixed many conventional bombing runs in with the *tokko* (special) attacks. (NARA)

the FDO on board was unable to establish radio contact with either aircraft and neither could the task force FDO in *Eldorado*.[74] This did not portend well for what was about to happen.

At 0735, *Laffey*'s radar picked up a 'bogey' which turned out to be a single Val approaching from the north at an altitude of 5000ft. By 0744 it had approached within six miles and, still unable to raise the low-altitude CAP, *Laffey* opened fire with her main battery. As the destroyer cut a tight circle to port, the aircraft closed following an evasive course that took it around the outside of *Laffey*'s track, closing to within 3500yds, at which point it dropped its bomb and sped away in the same direction from which it had approached, neither side damaged in the encounter. In the process, *Laffey* had the aircraft under fire for three minutes, expending 100 5in rounds, 15 per cent of her store of that munition.[75] The next entry in her report makes her frustration at the turn of events very clear:

0750	Requested relief defensive CAP we could communicate with be sent to replace present one.[76]

She would have no reason to complain about CAP protection during the next attack, which followed hot on the heels of the first. A group of four enemy aircraft was detected on radar coming in from the north-north-west at 20nm. Ten minutes later, they had closed to within 17nm and were being pursued by the CAP. The flight continued on their course, passing *Laffey* and heading off to the west, still pursued by the CAP, having lost three of their number. This minor victory would soon be rendered insignificant.[77]

In this case the CAP had acted as intended, and it would again at the beginning of the next attack; the fact that it ultimately proved inadequate to defend *Laffey*, as did the anti-aircraft gunnery of that ship and her two companions, was no fault of the aircraft, the airmen, the ships, or their crews. No combination of resources available to the US Navy off Okinawa in April 1945 (indeed to any navy of that era) could have been expected to defend against what came next. *Laffey*'s own report puts it best:

0827 to 0947	During this time the ship was under incessant and concentrated attack, with enemy planes so numerous and so well dispersed that organized interception by CAP became impossible after a few attempts. Radar operator reported seeing up to 50 bogies on scope, converging from north, northeast, and northwest. . . . To complicate matters, CAP was in process of being relieved when bogies were closing fast. The CAP undoubtedly intercepted and splashed a large number of planes outside of our gun range, and in several instances the fighter pilots

A photograph does not really do justice to the extent of the damage suffered by *Laffey* on 16 April 1945. As can be seen by comparing this image with the preceding diagram, the attacks came from all points of the compass, but seemed to concentrate their damage on *Laffey*'s aft end. (NHHC)

> waded right into our own AA fire to splash their victims. . . .[78]

After that introduction, the report went on to describe eighty minutes during which at least twenty-two individual kamikaze attacks were identified, enumerated as follows:

1-6 At 0830, four Vals were sighted 8nm to the northeast; they split into two pairs, one of which continued in on the starboard bow and was taken under fire by the two forward main-battery turrets under director control; the first Val was 'splashed' 9000yd from the ship, the second about 3000yd away. The other pair swung around towards the starboard quarter and was taken under fire by the after 5in mount under local control firing VT rounds. One was brought down 3000yd right aft, the other continued across to the port quarter and was brought down with the assistance of one of the accompanying LCSs, near-missing that craft. At the same time, two Judys [Yokosuka D4Y Type 2 Carrier Bombers] came in from opposite sides and were taken under fire by the medium (40mm) and light (20mm) anti-aircraft batteries. The one off the starboard side was dropped well away from the ship, but the one coming in on the port beam did not hit the water until it got much closer. It strafed the deck before splashing, wounding several men, and then splashed close abreast the after funnel. The explosion damaged the Mk12 fire control radar antenna, jamming it in maximum elevation.

7 At 0839, a Val coming in from the port bow was taken under fire by the main-battery and the port side machine guns. It remained intact and struck an open hatch atop the after main-battery mount and then hit the water close aboard the starboard quarter. One man in mount 53 was killed by fragments from the exploding aircraft, otherwise the damage was minimal.[79]

8-9 At 0845, a single Judy was splashed well away on the starboard beam by machine gun fire. Another Judy, coming from the port bow, approached the ship opposed only by the port side machine guns, which proved unable to prevent it from hitting the ship. The aircraft swung around the motor whaleboat abaft the after funnel on the port side and hit machine gun group 23, a pair of single 20mm mounts on the starboard side. There, a wing was broken off, spreading burning avgas over that gun group and the two quadruple 40mm mounts (mounts 43 and 44) nearby on either side of the ship. The 20mm mounts were destroyed and the 40mm mounts knocked out. (Mount 43 would be restored to operation before the engagement was over.) Fire was started in the deckhouse and in the 40mm magazine below that.

10-11 Another Val came in from directly aft about eight or nine minutes later, by now unopposed by anti-aircraft fire, hitting the starboard after corner of main-battery mount 53, demolishing that mount and effectively destroying the three-gun mount 25 right aft as well. A second unidentified aircraft came in from the starboard quarter, dropped a bomb of unknown size that hit two feet inboard of the starboard deck edge aft and the aircraft then crashed into the side of mount 53.

12 At approximately 0911, another unidentified enemy aircraft commenced a steep gliding approach from the starboard quarter, taking advantage of the heavy smoke coming from *Laffey*'s after end. At this point, the ship was in a hard left turn, so the only available AA gun, the after mount of group 24 on the port side on a platform abreast the after funnel, took the attacker under fire briefly with little effect. This aircraft was making a 'conventional'

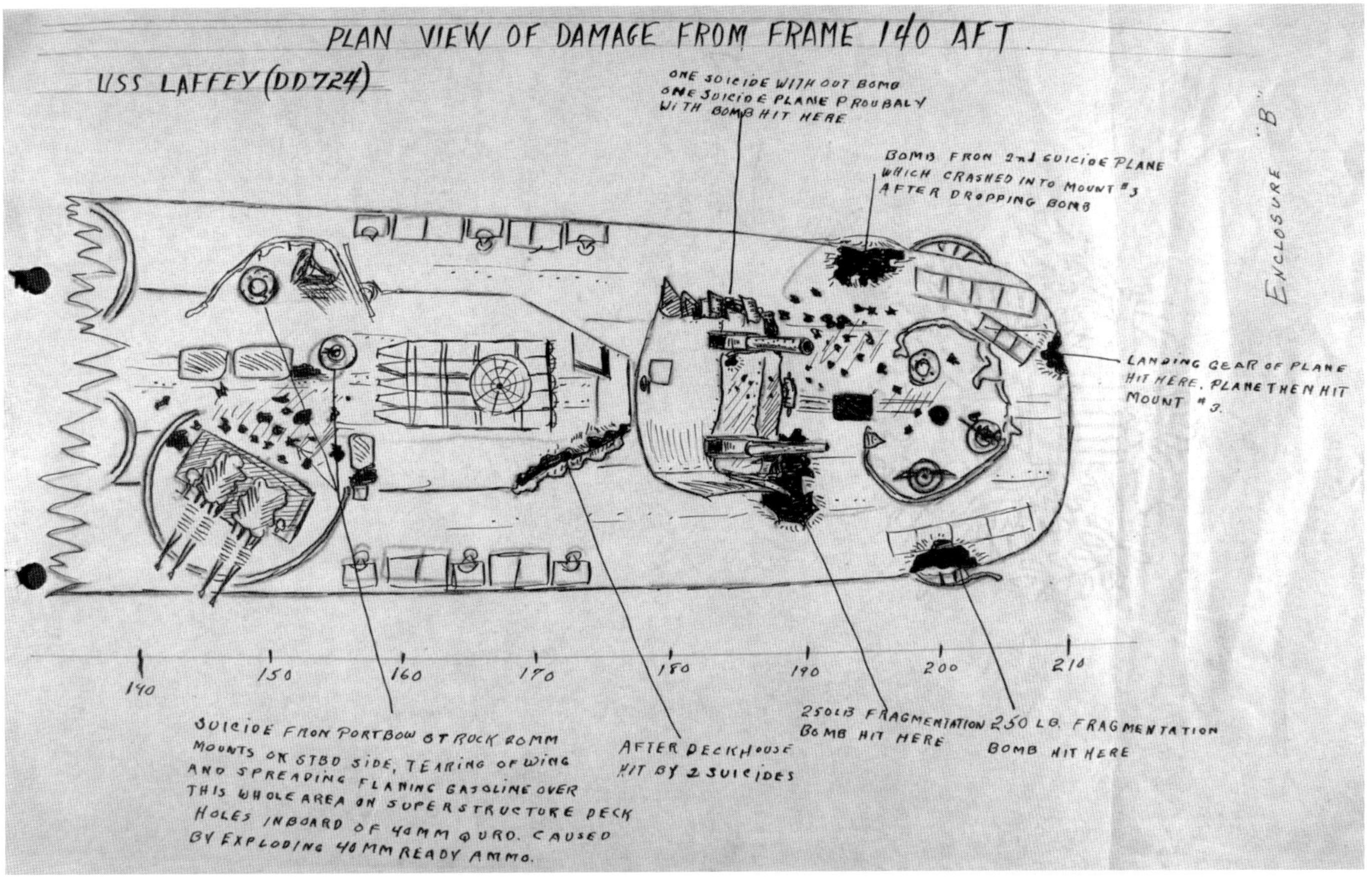

This diagram, also from *Laffey*'s action report, probably does a better job of depicting the damage done to her aft end. Five different aircraft and three bombs hit in this confined area, spreading fire, death, and general destruction, yet the stout structure of the ship and the superb training of the crew managed to limit the damage, fight the fires, tend the wounded, and get the many depth charges in the area overboard before they could detonate. (NARA)

skip-bombing attack, leveling off just above the water. It dropped its bomb too late for it to 'skip' on the water; rather, it hit *Laffey* just at the turn of the deck edge above the port side propeller guard. The bomb detonated in the 20mm magazine; the explosion damaged hydraulic leads in the adjacent steering gear room, jamming *Laffey*'s rudders in hard left position.

13-16 At about 0920, two unidentified aircraft came simultaneously from the port beam and quarter, hitting the after deckhouse between the after bank of torpedo tubes and main-battery mount 53. At the same time, an Oscar [IJA Nakajima Ki43 Type 3 Fighter], which had been engaged in a dogfight with an F4U Corsair of the CAP, broke off from that engagement and dove on *Laffey* from the port bow, followed closely by the Corsair, and under fire from mount 42 and group 22 port side forward. Both aircraft clipped *Laffey*'s foremast as they passed overhead, knocking off her port yardarm, SC radar antenna, and SG wave guide, leaving her with no effective radars. The Oscar crashed off the starboard beam and the Corsair not much farther away. The pilot of the Corsair was seen to have bailed out, but his fate was not ascertained. At the same time, another Judy was shot down by the CAP close aboard on the port bow. It exploded approximately 300yd from the ship, fragments from the bomb it was carrying damaging main-battery mount 52, knocking out its power; the mount was operated manually without difficulty thereafter.

17-19 Perhaps ten minutes later, a Judy approached from the starboard quarter. It was taken under fire by mount 41 and group 21 and, this time, the anti-aircraft fire was effective and splashed the enemy aircraft approximately 800yd off the starboard bow. A few minutes later, an Oscar coming in from the starboard beam was engaged by mount 52 under manual control; a direct hit was achieved on the aircraft's engine at 500yd distance, causing it to disintegrate. At the same time, mount 51 engaged a Val coming in from just off the starboard bow. Using VT rounds, the enemy was splashed 500yd from the ship.

20 At about 0942, a Val came in on a low glide path from just off the port quarter through the thick smoke trail behind the ship. It dropped a bomb that hit just short of the port after corner of mount 53, punching a hole in the main deck. The aircraft then continued forward, clipping the foremast, knocking off the starboard yardarm and kept on flying away from the ship. Taken under fire by group 21 and engaged by CAP fighters, this valiant attacker was finally splashed about 3000yd off *Laffey*'s starboard bow.

21-22 At about 0946, a Val approached in a shallow dive from off the starboard bow, engaged by both forward 5in mounts and the starboard side forward machine guns. It dropped its bomb, which hit and demolished machine gun group 21, and then attempted to escape to the port side, where it was splashed by CAP fighters well clear of the ship. A minute later, a Judy came in from the port bow with CAP fighters right behind. The attacker exploded about 400yd clear of the ship. [80]

Incredibly, the attack ended there with *Laffey* still afloat and under her own power, her rudder jammed hard to port, fires burning out of control aft and threatening the after 5in magazine, with unknown flooding aft. Her two forward main-battery mounts were still operable under manual power and control, along with machine gun mounts 41, 42, and 43, and groups 22 and 24. She had no radar and virtually no radio communications. Every one of the twenty-two aircraft and their pilots that attacked *Laffey* perished in the process, most of them intentionally. Thirty-two men in *Laffey* died that day, with an additional seventy-one wounded, but *Laffey* lived.[81] She attempted to steam to Okinawa under her own power, but no combination of engines could enable her to maintain a steady course. She was towed to an anchorage off Okinawa, where temporary repairs allowed her to leave for Saipan and the American west coast six days later. *Laffey* was fully repaired and served on, including duty off Korea during the conflict there and in the Mediterranean during the 1956 Suez Crisis. She was eventually stricken in 1975, but now serves on as a museum ship at Mount Pleasant, South Carolina, a fitting end for a very tough destroyer.

The gouges in the barrel of the left 5in/38 gun of *Laffey*'s main-battery mount 53 wrenched skyward by the impact of the two kamikaze aircraft that hit the port side of the gunhouse, give testimony to the power of the explosions that devastated this gun mount. (NHHC)

Chapter 10

Building from experience: 1942–1945

As the Second World War progressed, each of the major combatant nations exploited the capabilities of their destroyer types and, undoubtedly, noted those qualities that could be improved. While every one of those navies planned for new types that took advantage of the hard-earned lessons of the war, by no means did all of them have the capability to turn those plans into ships in a timely fashion, much less into numbers of ships sufficient to impact the course of events. Some nations, such as the Italians, wanted to build an improved class, but found themselves unable, amidst the crush of competing demands, to find the resources necessary to complete the ships in time. In Rome's case, the new design was for a ship a quarter larger than the Soldati class, carrying four single-purpose 135mm/45 guns in single mounts. Twenty-four of these Comandanti class *cacciatorpediniere* were authorised in 1942, twenty got so far as being ordered, nine were laid down between December 1942 and August 1943. The final programme, dated 31 August 1943, called for the first six ships to be commissioned by September 1945 and the remainder the following year, but none were completed to the point of being launched before the Italian armistice was announced on 8 September 1943.

Stretching the boundaries in Tokyo

The design for both of the 'wartime' classes built by the IJN pre-dates the entry of Japan into the Second World War, as did the laying down of the first ships of the two classes. Nonetheless, they are considered wartime classes here because their designs were very much influenced by what the Japanese knew of conditions prevailing in the war to date. It should be recalled that for the Japanese, the war had been ongoing since 1937, when the simmering conflict with China broke into open hostilities. While the war between Japan and China was not, to any great extent, a naval war, it nevertheless caused the Japanese to pay close attention to the readiness of the IJN to engage in a war with the major Western naval powers, particularly after 1939. Both classes had their origins before 1939, designed to address two different issues that concerned the Japanese naval leadership in regard to the Special Type destroyers as developed from the Fubuki to the Kagero classes.

Concerned that Japan was not taking advantage of the latest and best steam propulsion technology, in 1938 the Japanese naval staff produced a requirement for a large destroyer, similar in most characteristics to the contemporary Kagero class, but powered by the latest high-pressure, high-temperature steam system driving impulse geared turbines. This was in line with a rethinking of the whole destroyer design philosophy of the IJN, which now proposed that the Japanese navy should deploy three distinct destroyer types: the designation 'Type A' would be applied retroactively to the main line of destroyer development of the Special Type, which would continue to be produced and to evolve to provide general escort and attack capabilities; a new 'Type B' would be built specifically to provide anti-aircraft escort for the independent carrier force (Kido Butai); and another new type, the 'Type C' would provide high-speed surface attack capability against enemy main fleet units.

The proposed Type C destroyer design, ordered under the 1939 Circle Four Programme, called for a single, experimental ship of an enlarged Kagero type, with length increased by twenty-five feet to provide room for the larger powerplant and an almost doubled battery of torpedo tubes. She was built with the following characteristics:

***Shimakaze*: 1942**[1]

Displacement (trials):	3000 tons
Length (wl):	413ft 5in
Beam (wl):	36ft 9in
Draft (max):	13ft 7in
Powerplant:	75,000shp; two shaft Kampon impulse geared turbines; three Kampon boilers
Speed:	39kt
Armament:	15 × 61cm trainable torpedo tubes (3 × 5; centreline); 6 × 12.7cm/50 3rd Year Type (3 × 2); 6 × 25mm/60 Type 96 MG (3 × 2)

The original plan was for *Shimakaze* to carry five reload torpedoes, but all indications are that these (and the ancillary reloading gear) were never fitted. The outbreak of war slowed construction of this ship and caused the plans for sixteen additional ships to the same design to be cancelled; as it was, she was not launched until July 1942 and not completed until May 1943. Once in service, she proved to be highly successful, but the days when she would have been most useful, in the nightly battles in the South Pacific, were already past. She was sunk in an air attack in Ormoc Bay off Leyte Island on 11 November 1944.

The Type B destroyers proposed at the same time as the Type C were also first ordered in the 1939 Circle Four Programme.

Shown running trials in May 1943, *Shimakaze* looked not much different from the Kagero class from which her design was derived, except for the third bank of torpedo tubes on her centreline allowed by the 31ft of added length. Additional power allowed her to achieve 39kt on trials, but, for all her speed and firepower, she was a 'dead end' design; by mid-1943, there was no possible way the Japanese could produce more of such a resource-intensive design. (NHHC)

The requirement grew out of the observation that the naval aviation assets of the IJN, in particular its large aircraft carriers, would be more effective if combined into a single strike force. This idea, which was being pushed in 1939–40 by RAdm Ozawa Jisaburo, would eventually lead to the organisation of the Kido Butai (Mobile Air Squadron) in mid-1941, but it was clearly one that was 'in the air' even before then.[2] It was of sufficient concern to the naval staff that the idea of an anti-aircraft/anti-submarine destroyer to escort an aviation-related task force was included in their planning as the Type B destroyer.

To fulfil this role of dedicated escort for a carrier force, this design called for a true dual-purpose main battery, and the Japanese were able to provide one with the 10cm/65 Type 98 gun.[3] This gun was available in an enclosed twin mount, the shells – 62 lb fixed rounds – coming up from the working chamber below the mount via individual hoists, one for each gun, where they were man-handled into a shell tray, from which they were transferred into a loading tray as part of the recoil process, during which time the fuse was set automatically from range settings entered at a panel on the right side of the mount, and then, once the gun's recoil was complete, rammed into the sliding-block breech by a spring-operated rammer. This mechanism allowed loading over the full elevation arc from 0–90° up to a theoretical maximum of twenty-one rounds per minute. In practice, a good crew could maintain a rate of 15rpm from each barrel. The only complaint against this gun, which proved reliable in service, was that it required relining after about 350–400 rounds. Each 29 lb shell contained 2.1 lb of explosive, making it an effective anti-aircraft munition. To increase effectiveness as an anti-aircraft platform, the design called for main-battery fire control director positions fore and aft, a very unusual feature in a destroyer-sized ship, not seen since the American Porter class.

Only fairly late in the design process was the requirement added for one quadruple bank of torpedo tubes with four reloads, which turned the Type B into a more general-purpose destroyer design. In this form, six of the type were included in the 1939 Programme and ten more in the 1941 Programme, built to the following specifications:

***Akizuki*: 1941**[4]

Displacement (trials):	3430 tons
Length (oa):	440ft 3in
Beam (wl):	38ft 1in
Draft (max):	13ft 7in
Powerplant:	52,000shp; two-shaft Kampon impulse geared turbines; three Kampon boilers
Speed:	33kt
Armament:	4 × 61cm trainable torpedo tubes (1 × 4; centreline); 4 × 61cm Type 93 torpedo reloads; 8 × 10cm/65 Type 98 (4 × 2); 4 × 25mm/60 Type 96 MG (2 × 2)

The other new design to come out of the Imperial Navy's re-evaluation of destroyer requirements in 1938, was the Type B design for a high-speed escort for the navy's aviation assets. The resulting ships turned out to be one of the best general-purpose destroyer designs produced by any nation during the Second World War. The main battery of eight 10cm/65 Type 98 guns was reliable and accurate, suffering only from a tendency to wear rapidly through barrel linings and from a lack of an automatic fuse such as the 'proximity' (VT) fuse developed by the Allies. This image shows *Harutsuki* in 1945, probably at the time of her surrender. (NHHC)

Hanatsuki was surrendered at the end of the war to the Americans. She was partially disarmed, landing her superfiring main-battery mounts and all anti-aircraft guns – but she retained her Mk 2 Mod 2 surface-search radar (foremast) and Type 3 Mk 1 air search radar (mainmast) – for use as a transport carrying Japanese troops back to the homeland from China and other overseas garrisons where they were stranded at war's end. She carries the ship's name in western characters and the national flag (but notably not the 'rising sun' naval ensign), with her name repeated in small hiragana characters on a bulwark alongside her bridge. (NARA)

None of the Akizuki class were ready at the outbreak of the Pacific War. The lead ship did not join the war until June 1942 and her sisters were slow to follow; problems with the availability of shipbuilding materials and labour slowed all new warship production; only three more were completed before the end of the year and only eight more were completed before war's end. In fact, building these ships strained an already over-burdened war economy in Japan and it appears that few Akizukis were completed with both fire control directors, and even in the case of those few that were, the after director was not fully-equipped. This design was very well received in the fleet and thirty-eight more were ordered in 1942, but none of those were even started.

Finally, in 1942, the Admiralty of the Royal Navy insisted that a true dual-purpose main battery be found for destroyers, eventually settling on the 4.5in/45 QF Mark III, which had been developed as a secondary battery for battleships and aircraft carriers (not unlike the 5in/38 twin-mount developed by the US Navy). Because of the size and weight of the twin mount for these guns, it became necessary to increase the size of the then standard destroyers of the 'Cx' classes by 500 tons in displacement and 17ft in length to accommodate the new main battery. Among the first of these new destroyers, called the 'Battle' class because the first group were all named after famous naval battles from British history, was HMS *Armada* (R14), seen here on 31 January 1946, when she arrived at Melbourne to join the British Pacific Fleet. Note the single 4in/40 mount amidships, carried specifically for the purpose of firing starshells, a needless addition, since starshells were available for the main-battery guns. (SLV – Green Collection)

Battle lessons

By mid-1940, the Royal Navy certainly had as much first-hand knowledge of the lessons of war as any navy. By the end of the battles for Norway and France it was clear that a truly satisfactory dual-purpose main battery had not yet been found. The 55° maximum elevation of the available gun mounts was not adequate. The normal dive angle of the German Ju87 Stuka dive bomber was also 55°, but given the natural tendency of a ship to move in three-dimensions (roll, pitch, and yaw), a gun would need a considerably greater maximum elevation to stay on a target approaching from that angle; it was obvious that a maximum elevation limit of at least 75° was required.[5] The one existing candidate weapon that fitted the needed characteristics of 'destroyer-sized' calibre and ample elevation – other than the ubiquitous 4in/45 Mark V, which was considered to be too small to serve as a destroyer's main battery and was used only when other, larger guns were unavailable – was the 4.5in/45 Mark III, a gun and twin mounting which had been developed in the 1930s for use in aircraft carriers and modernised battleships. It had a good combination of elevation, shell size, and power operation, however there were some problems with simply adopting this gun and mount for a destroyer.[6] One problem was that the gun used a fixed round weighing 87 lb; tests with a fixed round for the 4.7in gun weighing 74 lb in the 1920s had shown that gun crews quickly tired after handling such a heavy round and the rate of fire fell off rapidly. Another problem was that the individual gun mounts were significantly heavier than those they were replacing; a 4.7in/45 QF Mark IX twin-mount weighed 25.1 tons, while a 4.5in/45 QF Mark III twin-mount weighed in at 45.6 tons. This severely limited the ability to carry this gun in a destroyer-sized hull. Nevertheless, the need for a true dual-purpose main battery in a destroyer was such that, on 21 October 1940, the Admiralty mandated that all future destroyer designs would carry a true dual-purpose main battery, with at least four of those guns firing forward, and also wanted at least two 40mm mounts, either pom-poms or twin Hazemeyer Bofors.[7] In order for this design to be ordered in the 1942 programme, a decision on the gun suite had to be made quickly. Adding to the demands on the new design was a mandate from the Staff Requirements Director, who demanded greater operating range, up to 4000nm at cruising speed.

The first sketch designs called for four 4.7in guns, in twin mounts superfiring forward, with four twin 4in mounts as an alternative, plus four Hazemeyer Bofors mounts aft. This resulted in a notional design displacing 2250 tons armed with a 4.7in gun in a mount that did not actually exist. This design seemed to please none of the interested parties.[8] Some complained that it was too large to be built two-at-a-time by the yards that normally built destroyers. Others that it was undergunned for its size. After considerable back-and-forth between the interested parties, a grudging agreement was reached on a compromise design featuring two 4.7in mounts forward, spaced well forward from the bridge and well-separated from each other to allow some amount of coverage aft of the beam, one of several possible guns would be placed aft of the funnel amidships to fire starshells, aft of that four twin Bofors mounts and six 20mm Oerlikon guns in single and twin mounts would be arrayed. (Interestingly, all the destroyer officers from the fleet who were asked about the main-battery layout endorsed placing both mounts forward.) The proposed powerplant would produce 50,000shp for a top speed of 31.5 knots. Displacement was estimated at 2280 tons. In this form, the design was approved in October 1941, with the understanding that the Royal Navy would need to continue building smaller 'standard' destroyers, as it would not be possible to build enough of these new, larger type ships to meet the needs of the fleet. It was at this point that a modification of the 4.5in/45 BD mount developed as a dual-purpose gun for HMS *Ark Royal* (91) was proposed in the place of the 4.7in/50, which did not exist in a true DP mount. The 4.5in/45 QF Mark III proved to be an excellent weapon, especially when served with separate rounds. To control this new destroyer main-battery weapon, an equally new fire control suite was to be fitted, combining the Mark VI HA/LA director and the Type 275 centimetric radar. For the first time since experimenting, without notable success, with high-temperature/high-pressure steam with the 'A' class of 1929, the Royal Navy planned to employ a 400psi/700°F powerplant in this class. (Just for comparison, the powerplants of the American Fletcher class, whose characteristics were approved in January 1940, operated at 615psi/850°F.) The first ships of this class, of which twelve were ordered in April 1942 and a further four in August, were all to be given names of famous Royal Navy battles. The initial ships in the Battle class were completed to the following specifications:

The weight of *Armada*'s main-battery mounts was such that only two could be carried and both of those were sited forward so they could share the Mark VI director and its Type 275 radar. Aft, she carried four Hazemeyer twin Bofors mounts each with two 40mm/56.3 guns and an integrated Type 282 radar antenna, electronics, and analog fire control computer. *Armada* is seen at Melbourne on the same day as the preceding image. Only one of the Battle class destroyers arrived in the Pacific in time to participate in the Second World War. (SLV – Green Collection)

A second group of Battle class destroyers was ordered in 1943, none of which were completed in time to fight in the Second World War and most of which were cancelled. These carried the names of famous land battles. Seen here in 1950, at the time of the beginning of the Korean War, is HMS *Dunkirk* (D09); these second group Battle class destroyers differed from the first in being marginally broader, having the American-made Mk37 fire control director, though still mated with a Type 275 radar, and having a single 4.5in/45 mount amidships, in place of the 4in mount. This mount made even less sense than the one it replaced, as it was limited to 55° elevation, making it useful for surface fire only. (Peter Smith)

Battle Class (1st Group): 1943[9]

Displacement (standard):	2315 tons
Length (oa):	379ft
Beam (wl):	40ft 3in
Draft (max):	15ft 2in
Powerplant:	50,000shp; two-shaft Parsons geared turbines; two Admiralty three-drum small-tube boilers
Speed:	35.75kt
Armament:	8 × 21in trainable torpedo tubes (2 × 4; centreline); 4 × 4.5in/45 QF Mark III (2 × 2); 1 × 4in/40 QF Mark XIX; 8 × 40mm/56.3 Bofors OQF Mark IV (4 × 2); 6 × 20mm/70 Oerlikon Mark I (2 × 2; 2 × 1)

The first of the Battle class was laid down in October 1942, and due to be completed in March 1944, but was not actually commissioned until September and even then was not ready to join the fleet because of delays in delivering the new fire control director. HMS *Barfleur* (D61) did not complete her trials until March 1945 and did not join the British Pacific Fleet until June, the only one of her class to see action in the Second World War. All sixteen served on post-war and through the Korean War, into the 1960s and, in two cases, until 1970.

A second group of Battle class destroyers was ordered, differing from the first group mainly in mounting the American Mk37 fire control director in the place of the Mark VI, and

A third group of Battle class destroyers was planned by the Royal Navy, but never started. They were to carry the lightweight 4.5in main-battery mount intended for the second batch, as well as the advanced STAAG mounting for the twin 40mm Bofors with its Type 262 radar, three of which were to be carried aft. While none of the Royal Navy's ships of this type were built, the Royal Australian Navy built two, including HMAS *Anzac* (D59), seen here between 1950 and 1952. These lightweight main-battery mounts were large and boxy, and weighed barely 4 per cent less than the standard mount. (SLV – Green Collection)

would carry a new lighter-weight 4.5in main-battery mount. Four ships to this design were ordered in March 1943, followed in April by orders for fifteen more, with a final four ordered in June. The American fire control system – also fitted to the battleship HMS *Vanguard* (23) and to new-build fleet aircraft carriers – proved very successful, but the new main-battery mount was not ready in time to be fitted to any of the eight ships of this class that were completed in 1946–8; so they carried the same main battery as the first group, except that the 4in starshell gun abaft the funnel was replaced by a single

The massive STAAG mounts for the twin 40mm/56.3 Bofors anti-aircraft guns – STAAG stood for Stabilised Tachymetric Anti-Aircraft Gun – can be seen aft on HMAS *Tobruk* (D37) in 1952. Compare the size of the STAAG mounts with the standard Mark V twin Bofors mount immediately aft of the funnel, which, while it lacked the radar and triaxial stabilization of the STAAG, was simple and reliable, and stayed in service far longer. (SLV – Green Collection)

4.5in/45 QF Mark IV gun, modified from a 4.7in gun mount, that could elevate to a maximum of only 55°, meaning it was effectively a single-purpose, surface-action weapon and most decidedly not a starshell gun.

The remaining ships in the order were cancelled. A slightly modified class with greater beam was proposed but never proceeded with by the Royal Navy. However, two ships of this design were built in Australia, completed in 1951, and became the first to mount the 4.5in/45 QF Mark V, which weighed in at essentially the same weight as the Mark III mount it was meant to replace.

The lightweight 4.5in gun mount was used again by the Royal Navy in the Daring class, designed in early-1945 in an attempt to take advantage of all the lessons of the Second World War. This image, dating from September 1973, shows two ships in Peruvian service. Leading is BAP *Palicios* (D73), the former HMS *Diana* (D126) of the Daring class. The Daring class added 250 tons of displacement and 11ft of length over the Battle class, which gained them room for a third main-battery twin mount carried aft. In the middle distance in this image is BAP *Villar* (D71), the former USS *Benham* (DD 796), a late-production Fletcher class destroyer. (NARA)

Super-Fletchers

For the US Navy, the move to build an improved version of the Fletcher class destroyers started well before the first of those ships was laid down in March 1941. Based on experience with the first units of the Benson/Gleaves classes, which were running trials in mid-1940, there were already concerns being aired about the high profile of these ships, with the instability issues that often accompanied high topweight, and the presence of so much weight forward, which threatened to make them wet in weather. The solution to these problems proffered in June 1940 was to substitute a 5in/38

The Americans were very happy with the Fletcher class, which made up the bulk of their early- and mid-war-production destroyers, but there remained limits to what electronics and anti-aircraft weaponry could be added before stability became seriously compromised. What was needed was a way to maintain existing firepower while freeing up space and available topweight for the inevitable necessary additions. The solution turned out to be remarkably easy – simply replace the five single main-battery mounts of the Fletcher class with three of the twin mounts already developed as a secondary battery for the new battleships and aircraft carriers on the nation's slipways. Thus, one gets the Allen M Sumner class, exemplified by USS *Meredith* (DD 726), seen here working up in Massachusetts Bay on 31 March 1944, barely two weeks after commissioning at Bath Iron Works. It was not quite as simple as replacing five single main-battery mounts with three twin mounts. A Sumner's hull was very similar to a Fletcher's; the new class had just over a foot more beam and also greater draft because the new class displaced just under 300 more tons. They shared the same powerplant which caused an expected drop in top speed and an unexpected decrease in endurance. This class introduced an 'Admiralty-style' enclosed bridge which proved to be very unpopular. (NARA)

twin-mount, such as was already in production for use on the battleships, cruisers, and aircraft carriers then under construction, for the superfiring single 5in guns forward in the Benson/Gleaves design.[10] While this idea was considered at this time, there was strong opposition, particularly from the Bureau of Ordnance (BuOrd), which argued that the twin mount, as it then existed, could not be hand-trained, and thus would become useless if the ship lost power, and that all the current production of twin mounts was assigned to scheduled shipbuilding for the foreseeable future.

Nothing further happened with this idea until September 1941 when, realising that even the new Fletcher class would

likely prove to need increased anti-aircraft firepower, the General Board began investigating ways to improve that aspect of destroyer design. In response to these requests, the newly created Bureau of Ships (BuShips), formed by merging BuEng and BuC&R, responded with six sketch designs, only two of which incorporated single mounts in the main battery. The design that BuShips strongly favoured was Sketch B-II, which replaced the five single mounts of the Fletchers' main battery with three twin mounts, and sacrificed one bank of torpedo

The US Navy did listen to its destroyermen. After the first Sumners reached the fleet, the displeasure with the enclosed 'Admiralty-style' bridge was noted and addressed. This image of USS *DeHaven* (DD 727), the sixth Sumner class destroyer built by Bath Iron Works, seen on 20 July 1944, less than four months after her commissioning, shows the revised bridge, similar to the later bridge design for the Fletcher class, with a small pilothouse and a relatively spacious open bridge around it. (NARA)

tubes to gain a second twin Bofors mount.[11] The General Board was unconvinced and asked for several more design iterations, but ran into stiff resistance from the pertinent bureaus and from destroyer officers in the fleet, who insisted that war experience to date (early 1942) had shown gunpower, particularly anti-aircraft gunpower, to be more important than torpedoes, and continued to favour Sketch B-II. With Adm Ernest King, now in place as both CNO and CominCh, adding his opinion on the side of the bureaus, a final sketch design was generated which retained all the features of Sketch B-II, but added the second bank of torpedo tubes back into the design. All of this was to be accomplished in a hull of the same length as a Fletcher and with the same powerplant as well. The only concession to the added displacement caused by the increased armament, which amounted to over 200 tons, was an increase in beam by 1ft 3in. In order to improve handling, in particular to reduce the tactical radius – a constant complaint against the earlier American destroyer classes concerned their overly large turning circle, which particularly impacted their anti-submarine warfare capabilities – this class was given twin rudders. In this form, Sketch B-VII was approved and of the 117 destroyers ordered on 7 August 1942, sixty-nine were ships of this design ordered to the following specifications:

One of the twelve Sumners converted to fast minelayers, USS *Shannon* (DM 25), is seen at Boston, MA, on 23 September 1944, with a full load of mines lining her rails aft. She never operated as a minelayer, spending her war providing fire support and anti-submarine escort to landing forces and minesweeper units. (NARA)

USS *Allen M Sumner* (DD 692): 1943[12]

Displacement (standard):	2610 tons
Length (oa):	376ft 6in
Beam (wl):	40ft 10in
Draft (max):	14ft 2in
Powerplant:	60,000shp; two-shaft General Electric geared turbines; four Babcock & Wilcox boilers
Speed:	36.5kt (design); 34kt (actual)
Armament:	10 × 21in trainable torpedo tubes (2 × 5; centreline); 6 × 5in/38 Mk12 (3 × 2); 12 × 40mm/56 Bofors Mk1/2 AA machine cannon (2 × 4; 2 × 2); 11 × 20mm/70 Oerlikon Mk2 AA MG (11 × 1)

With the end of the Second World War and the emergence of jet aircraft, it became obvious that even the excellent 40mm Bofors and the ubiquitous 5in/38 were no longer adequate as anti-aircraft weapons, the former firing too light a shell to bring down an aircraft with the mass and momentum of a jet, and the latter too slow to track the much faster targets. The answer was found in the 3in/50 Mk22, which was being used as a main-battery weapon on light escorts. It was deployed in a newly designed, power-operated twin mount, firing a VT-fused round weighing 24 lbs, making it easily man-handled. As seen here on USS *Barton* (DD 722) on 14 June 1960, the twin 3in/50 mount was almost the same size as a quadruple Bofors mount, but as it weighed almost 50 per cent more, the replacement was on a two-for-three basis. The 3in gun proved popular because it was reliable and had excellent range. Note the Mk56 director with its Mk35 3cm band fire control radar antenna to the left. (NHHC)

The first of the Sumner class to be laid down and completed was USS *Barton* (DD 722), commissioned at Bath Iron Works on 30 December 1943. Based on early experience in the South Pacific, it was decided to convert twelve of the Sumners under construction to fast minelayers by the simple expedient of removing the two banks of torpedo tubes, three of the 20mm

The solution to the Sumner class's endurance problem was to lengthen the hull by 14ft, which added fuel bunkerage and, at the same time, made it more hydrodynamically efficient, resulting in the Gearing class, such as USS *Chevalier* (DD 805), seen here. Only a couple of the first Gearings to be completed arrived in time to join the fleet before the end of the Second World War. Twenty-four of the first Gearings were completed as radar pickets, the importance of such craft having been amply demonstrated by the likes of *Laffey*, as seen earlier. Those included *Chevalier*, which still looks much as she would have in the late war, with what appears to be an overall dark tone camouflage – probably a faded Ms21 overall Navy Blue scheme – the small hull number at the bow as commonly displayed during the war, and a wartime radar suite including antennas for an SC-3 and SG on her foremast, a Mk12/22 on her Mk37 director forward, and what appears to be an SP fighter control radar antenna along with a YE homing beacon above it on her tripod mainmast, which replaced her forward bank of torpedo tubes. (NHHC)

guns, and two depth charge throwers to compensate for the weight of two mine rails and 120 mines. As far as can be ascertained, none actually served in this role. The original Sumner class design called for a partly-enclosed 'Admiralty-style' bridge that was cramped and extremely unpopular with American officers. This was redesigned with a small, enclosed pilothouse around the base of the Mk37 director, with the rest of the bridge open; most of the class was completed with this larger bridge design and those early ships built with the small bridge were retrofitted with the new design at their next yard period.

Probably the biggest issue with the Sumner class design, however, was due to the wider hull and greater displacement without any increase in power, which led to an (expected) decrease in top speed and a much greater than expected decrease in endurance. Where the Fletcher class was listed as being able to steam 3480nm at twenty knots, a Sumner could only reach 3240nm, which was 7 per cent less, a considerable difference when considering the constraints that the need to refuel destroyers put on fleet operations. To address this issue, a revised Sumner design was hastily generated and approved by the simple expedient of adding a fourteen-foot section amidships, which allowed an increase in fuel oil bunkerage from 503.9 tons to 740.1 tons, which increased the endurance at twenty knots to 4380nm. The lengthened hull was also more efficient and theoretically should have led to an increase in top speed, but the actual top speed was under thirty-four knots. A total of 116 ships were ordered to this modified Gearing class design, of which ninety-three were actually completed. Most served on into the 1970s.

Gearing class destroyers continued in the radar picket role, even as that role became more greedy in its appetite for space and topweight. This image shows USS *Frank Knox* (DD 742), which was the first of the class to be completed and which joined the 3rd Fleet in June 1945, seen here in a rather embarrassing position, aground on Pratas Reef in the South China Sea in July 1965. *Knox* had received a FRAM II (Fleet Rehabilitation and Modernization) upgrade which included adding the flight deck and hangar for a DASH anti-submarine helicopter drone aft, and the variable depth towed sonar right aft, although it is not clear whether the drone was ever carried. *Knox* had been completed as a radar picket, and, despite the FRAM conversion, retained her now upgraded suite of search and guidance electronics; the large SPS-29 mattress antenna on her foremast for long-range air search; the smaller SPS-10 surface-search antenna above it; a TACAN air navigation antenna atop her mainmast; and the large SPS-8 height-finding radar antenna carried on the DASH hanger. (USN)

Chapter11

A type for the future: 1946

THE END OF THE Second World War ushered in an era of vastly altered global power dynamics, especially in naval affairs. What it did not bring was peace, any more than the end any previous global conflict had done. The remaining 'super powers', if they wanted to retain that status, had to continue investing in their naval forces, as much as they might have wished to spend their money elsewhere. This process had the effect of driving Great Britain and France from the top tier of world powers, because the economics of maintaining a world-class navy on top of the demands of other military obligations and social programmes was judged to be simply too great. That left, for a brief while, the United States and the Soviet Union as the two great world powers. Eventually, the economic burden proved too much for Russia, which has dropped from the top rank, and opened the door for new economic players to become major naval powers as well, specifically China and, perhaps, India.

The strain of post-war economics impacted not only which nations would be naval powers, but how their fleets would be composed. As each individual platform packed in

As had been the case after the First World War, the US Navy found itself at the end of the Second World War with many new destroyers, but there were some important differences, not the least of which being that these destroyers were ones which the navy was quite pleased to retain. Nevertheless, rapid advances in technology made even the best Second World War-vintage warships and weapons systems obsolescent by the mid-1950s, as both submarines and aircraft evolved rapidly into far greater threats than they had been just a decade earlier. Replacing 40mm Bofors anti-aircraft guns with 3in/50 guns was a stop-gap measure at best as jet aircraft kept getting bigger and faster. The Forrest Sherman class built during the late-1950s replaced all the weapons of a Gearing with more modern ones – single 5in/54s replaced twin 5in/38s; twin 3in/50s replaced 40mm/56; and Hedgehogs and ASW torpedoes replaced depth charges – but still more needed to be done to meet the threats. Thus, the Charles F Adams class of the early-1960s, such as USS *Tattnall* (DDG 19) seen here in the Pacific in August 1973. Tattnall lost her aft 5in gun mount, replaced by a one-arm RIM-24 *Tartar* medium-range surface-to-air missile launcher, and her torpedo tubes, replaced by an ASROC (Anti-Submarine ROCket) launcher. (NARA)

Two examples of 1980s destroyer technology, both in somewhat modified form, as seen during Exercise Malabar 2012. On the left is the Indian naval ship *Ranvijay* (D55); she is one of five Rajput class destroyers commissioned in 1988. They were modified Russian Kashin class destroyers built at Nikolayev (Mykolaiv). Her main armament is an eight-cell BrahMos supersonic cruise missile VLS abaft her second pair of funnels and four SS-N-2D Styx anti-ship cruise missiles in the short, inclined tubes forward of the bridge. *Ranvijay* suffered a dockyard fire in 2021, but remains in commission. In the background is USS *Bunker Hill* (CG 52), one of the Ticonderoga class guided missile cruisers built on the hull of a Spruance class destroyer. (The distinction in the US Navy today between a cruiser and a destroyer is one of role and not one of size or weaponry; by all the criteria used in this book, *Bunker Hill* qualifies as a destroyer.) *Bunker Hill* was the first Ticonderoga completed with a VLS rather than a pair of two-arm missile launchers; the VLS system holds 122 missiles, a mix of Standard SAMs, Tomahawk cruise missiles, and ASROCs. The large forward and after superstructures house the AN/SPY-1A/B Aegis phased-array radar system. *Bunker Hill* was scheduled to be decommissioned in 2023, but there was congressional resistance to this move. (USN)

more weaponry and sensors, it became exponentially more expensive, making the maintenance of a large fleet of large ships, such as the United States had built to win the Second World War, entirely impractical. Not only was it required to reduce the number of ships in the fleet, but it was necessary to take a hard look at what types of ship should comprise the fleet going forward. Anyone who had been paying attention to events during that war would draw the obvious conclusion that gun-bearing ships were no longer the key to dominance at sea.

Even the most cursory examination of the war at sea in the Second World War showed that bombs and torpedoes far outweighed shells fired from guns as the cause of damage to ships, and that submarines and aircraft were the most efficient carriers of those weapons. A navy of the future would have to comprise the ships and boats that carried those weapons or could defend against the ships and boats that did. Obviously, the optimal delivery vessels for these weapons were submarines and aircraft carriers. The optimal defender against air attack was either another aircraft or an agile surface platform with

Easily the longest-running class of American destroyers – the first of the Arleigh Burke class was laid down in 1988 – is exemplified by USS *Howard* (DDG 83), a Flight IIA Burke class destroyer, which was commissioned in 2001 and is seen here at RIMPAC16 on 28 July 2016. The Flight IIA ships differed most noticeably from earlier ones in having a hangar aft for two helicopters. These ships also had 96 VLS cells compared to 90 in earlier flights. This ship was one of the first fitted with the longer-barrelled 5in/62 gun. Fourteen Flight III Burkes have been ordered with the first entering service in 2023; plans currently call for as many as forty-two Flight III ships to be built before production ends. (USN)

effective anti-aircraft weaponry; battleships and cruisers had served this role well in the Second World War, but there was nothing they did that could not be provided by multiple destroyers, with the advantage that the defensive firepower was more distributed when spread over multiple smaller platforms and therefore more resilient to attack. Aircraft and submarines were evolving into effective anti-submarine weapons, but there was no question that a destroyer-sized surface ship equipped with adequate sensors and a variety of anti-submarine weaponry, such as depth charges, contact-detonation munitions such as the Hedgehog and its more sophisticated descendants, and rocket-launched anti-submarine torpedoes, doubled as a highly effective anti-submarine platform.

Thus it is easy to understand why, as the larger post-war navies began the melancholy process of paring down their fleets to a more sustainable peacetime level, large platforms requiring massive and expensive manpower commitments, such as battleships and cruisers, were among the first to be 'mothballed'. When the great number of ships of all types built during the Second World War inevitably reached mass-obsolescence in the 1960s, the largest gun-bearing surface combatants any navy built to replace them (with only a few notable exceptions) were destroyers.

Whether a ship was called a cruiser, a frigate, a destroyer leader, a corvette, or a destroyer was really a matter of national tradition and linguistics more than it was one of actual distinction. The Soviets built a large class of large conventionally-gunned light cruisers in the 1950s, but they must be considered anachronistic, the product of pre-war planning and a bureaucracy that could not be derailed. Through the following decades, the world's navies produced surface combatants designated in various manners – ships called destroyers grew as large as 15,000 tons, truly super destroyers, larger than the light cruisers the Soviets built in the 1950s, while a current class of frigates being built in France displaces 6000 tons, nearly three times the size of a Fletcher class destroyer – the only real distinction is whether they are specialised to serve an anti-aircraft or anti-submarine role or, as is more often the case now, general-purpose surface combatants, capable of fighting any of the threats a navy might encounter, their role adjusted by simply altering the mix of missiles in their vertical launch cells. In this form, destroyers, and their somewhat smaller counterparts, whether designated frigates, corvettes, or littoral combat ships, appear likely to be the primary surface combatants in the world's navies for the foreseeable future.

Two experiments by the US Navy that, for the most part, have failed. In the foreground is USS *Zumwalt* (DDG 1000), seen here on 8 December 2016, easily the largest destroyer built by the US Navy, at more than 14,500 tons displacement. Visually, the Zumwalts, of which eventually three have been built, are easy to distinguish, with their characteristic tumblehome hull, tall superstructure, and housings forward for what were to be two single 155mm/62 Advanced Gun System mounts. This gun system has been abandoned due to the prohibitive cost of the intended munitions and will be replaced on the Zumwalts by VLS cells for up to twelve hypersonic cruise missiles to be tested starting in December 2025. This conversion may actually make the Zumwalts significant surface warfare assets. The ship in the background is USS *Independence* (LCS 2), the first of the second design of 'littoral combat ships', conceived by the US Navy as less expensive, more expendable surface combatants, perhaps equivalent to the destroyer escorts of earlier days, although, at approximately 3000 tons displacement, they certainly qualify as destroyers within the definitions used here. Both LCS variants have had mechanical and tactical issues that have prevented their full deployment, and *Independence* has already been decommissioned even as new ships to her design are entering the fleet. (USN)

2
1000

Like every other modern navy, the post-war Royal Navy faced issues of budget restraint, which limited the number of ships it could build and sailors it could put aboard them. This led them, when it came time to start replacing their war-built destroyers, to build a 'high-low' mix of larger and smaller classes, the larger for their capability and the smaller for their economy, though often there has been little to choose between them on either count. In this image, taken on 6 September 2022, HMS *Westminster* (F237), in the foreground, is in the Celtic Sea, that area of water bordered by Ireland in the north-west, Cornwall in the south-east, stretching north to the narrows between Wales and Ireland where St George's Channel begins and as far south as the so-called 'Celtic Margin', where the continental shelf falls off into the North Atlantic deep at about the latitude of Brest. As her pennant number indicates, *Westminster* is considered a frigate, specifically a Type 23 frigate, the newest class in Royal Navy service, although, at 4900 tons, it is hard not to consider her a destroyer. The type can carry eight Harpoon cruise missile tubes, although only two are visible here, and thirty VLS cells behind the bulwark abaft her 114mm/55 gun. She is shepherding the Russian cruiser *Marshal Ustinov* (055), which at 11,500 tons displaces far less than the American Zumwalts, again demonstrating the flexibility (and relative meaningless-ness) of modern ship designations when applied to surface combatants. *Ustinov*, in typical Russian fashion, literally bristles with missile tubes, in this case for the SS-N-12 Sandbox (P-1000 Vulkan) antiship cruise missile, along with 104 VLS cells. She is the sister of *Moskva*, which was sunk in the Black Sea by Ukrainian land-based cruise missiles on 14 April 2022. (MoD)

F237

The current 'high end' surface combatant in the Royal Navy is represented by HMS *Defender* (D36), seen here replenishing from USS *America* (LHA 6) in the Philippine Sea on 20 August 2021. The Type 45 destroyers displacing approximately 8400 tons have been plagued from their service entry with problems with their power systems, on occasion leaving them dead in the water. Since 2020, a program of replacing their auxiliary diesel generators has been under way, which is intended to ameliorate, if not entirely resolve, the problem. Their primary weapon is a forty-eight cell Sylver A50 VLS with Aster SAMs. (USN)

564

Admiral Tributs (564), seen here from USS *Chafee* (DDG 90) in the Sea of Japan on 15 October 2021, is an early example of the Russian Navy's Udaloy class destroyers, dating from the 1980s and 1990s. Despite her chronological age – she was first commissioned in 1985 – *Tributs* has had fewer years of active service than most contemporaries because she underwent a nine-year-long refit between 1994 and 2003. Described by the Russians as an 'anti-submarine frigate', she displaces 6200 tons standard. Her primary armament is a pair of quadruple tubes for SS-N-14 Snowstorm (URPK-4 Metel-U) anti-submarine missiles alongside her bridge and eight VLS cells forward for SA-N-9 Gauntlet (3K95 Kinzhal) SAMs.

A second flight of three ships of the Udaloy class, dubbed Udaloy II in the West, was planned by the Soviet, later Federal Russian, Navy, although only one, *Admiral Chabanenko* (650), seen here, was ever completed. The primary differences from the earlier Udaloys were the substitution of supersonic SS-N-22 Sunburn (P-270 Moskit) missiles in the large cannisters and the newer twin-mount 130mm/70 AK-130 gun. *Chabanenko* has been undergoing refit since 2013; her return to service has been announced regularly since 2019, most recently in 2023. (USN)

The newest class of Russian destroyers (called frigates in Russian documentation, but displacing in excess of 4500 tons, which fully qualifies them as destroyers by the terms used in this book) are the ships of the Projekt 22350M class being built at Severnaya Verf (St Petersburg). The first to enter service was *Admiral Gorshkov* (417), seen here on 10 July 2018. She has thirty-two VLS cells for SAMs immediately abaft the 130mm gun and sixteen VLS cells for larger anti-ship and anti-submarine missiles in the raised deckhouse ahead of the bridge. (MoDR)

At the same time that the Soviets planned the first Udaloy class anti-submarine frigate, they designed a complementary anti-aircraft/anti-surface warfare destroyer dubbed the Projekt 956 Sarych (Buzzard) class, known in the West as the Sovremenniy class after the lead ship. Twenty-five of this type were ordered – twenty-one for the Russians and four improved models for the Chinese People's Liberation Army Navy (PLAN) – of which four of the Russian order were cancelled and of the remaining seventeen, as of the time of this writing, fourteen have been decommissioned, two are in long-term refit, and only one is in active service. Of the four Chinese Sovremenniys, all are still active, including *Fuzhou* (137) seen here, one of the first pair, distinguishable by having two AK-130 gun mounts.

The newest and most potent PLAN destroyer type is represented by the Type 055 *Nanchang* (101) seen transiting out of the Sea of Japan through the Tsugaru Strait on 29 March 2021. Her main armament are the two banks of VLS cells forward and amidships totaling 112 in all, carrying a mix of SAMs, anti-ship and land-attack cruise missiles, and anti-submarine weapons. (MoDJ)

The Japanese Maritime Self-Defense Force (JMSDF) has fostered a robust shipbuilding capability in Japan once again, allowing the construction of ships such as *Shiranui* (120) seen approaching USS *Big Horn* (T-AO 198) for an UNREP in the South China Sea during a FONOP (Freedom of Navigation Operation) on 29 October 2021, a type of operation that has become increasingly important over the last decade as nations such as China have attempted to limit access to formerly openly accessible bodies of water such as parts of the South China Sea and the Taiwan Strait. These operations involve nothing more than openly sailing through waters other nations claim to be restricted. *Shiranui* is a 5100-ton displacement anti-submarine destroyer with phased-array radar and a thirty-two cell VLS system, launched 2019. (USN)

Another nation building up its naval forces in response to the energetic naval expansion taking place in China is South Korea, which has developed several classes of highly regarded destroyer designs, such as *Sejong the Great* (991), seen here at RIMPAC16, 28 July 2016. The weapons and sensors are largely, but not entirely, of American manufacture; for example, while the phased-array radar is the American-made AN/SPY-1D, the 128 VLS cells are split between 80 American-made Mk 41 cells (forty-eight forward and thirty-two aft) and forty-eight Korean-made K-VLS cells aft. The South Korean Navy built two sisters to *Sejong the Great* and, as of the time of this writing, is building the first of three destroyers to an incrementally improved design.(USN)

The Royal Navy's Type 45 destroyer HMS *Defender* sails in the background in company with a ship from another emerging navy; the Indian Navy's *Kolkata* (D63), is seen from USS *Antietam* (CG 54) participating in IFR2016, Bay of Bengal, 9 February 2016. With a standard displacement of 7400 tons, *Kolkata*, her two sisters, and the first two of the derivative Visakhapatnam class, are the largest destroyers operated by the Indian Navy. The high-mounted EL/M-2248 phased-array radar is of Israeli manufacture, and one of two eight cell VLS units holds thirty-two Israeli-Indian-co-developed Barak 8 SAMs, while the other holds thirty-two BrahMos supersonic cruise missiles, a Russian-Indian co-development. (USN)

D63

The British, French, and Italians joined together in the early-1990s on the Common New Generation Frigate (CNGF) program to develop a new anti-air-focused frigate/destroyer. As often is the case with such international developments, differing requirements, in this case having to do with range and the size of the area to be defended by the ships (the British wanting more of both than did the French or the Italians), led to the project splintering; the British went off on their own, developing the Type 45 destroyer which incorporated some features of the CNGF, while the French and Italians continued on with the program (also known as the Horizon class), each building two ships. This image shows *Andrea Doria* (D553) of the Marina Militare (successor to the Regia Marina), which was commissioned in 2007. Although the 'ball-style' radome atop her forward superstructure resembles that of the Type 45s, the radars are different; what they do share is the forty-eight cell VLS of the Sylver A50 system.

A much more successful international program started out as a co-operative venture between the French and the Italians. Called FREMM (Frégate Européenne Multi-Mission), this 6000 ton destroyer has been offered in anti-submarine, anti-air, general-purpose, and land-attack variants, with eleven, two, and eight of the first three variants having been completed for four navies (those of France, Italy, Egypt, and Morocco); only the land-attack variant has had no takers. This image shows the MN *Auvergne* (D654), one of the French anti-submarine variants, as seen in the Arabian Gulf on 19 September 2017. She has a sixteen cell VLS for Aster 15 SAMs and another for MdCN land-attack cruise missile. Abaft the forward superstructure are eight tubes for Exocet MM40 Block 3 anti-ship missiles. Of particular interest is the fact that the US Navy has chosen the Italian FREMM ASW variant as the basis for the FFG 62 class of multi-mission frigates; the plan is to acquire twenty, of which three have been funded by Congress and the first, USS *Constellation* (FFG 62), was laid down on 31 August 2022 at Fincantieri Marinette Marine (Marinette, WI). (DoD)

Sources

Given the research resources available at the beginning of the twenty-first century, many sources once available only at physical archives or from private individuals are now available in cyberspace. (A small percentage of these are available *only* online.) In these cases, I have given the hyperlink to the source rather than the more traditional publisher information. It is an unfortunate characteristic of such sources that they are sometimes more ephemeral than paper-and-ink. When the site that serves the pages is changed or ceases to exist, the effect can be as if every copy of a particular book has instantly vanished. All links listed here were active and available at the time this book was written, and the most recent date accessed is given.

Primary/official sources

The author wishes to acknowledge the effort the British government has made putting the debates of the Houses of Parliament online as far back as 1803. While not absolutely complete, what is there is of great value to any historian interested in the official actions of Great Britain during this period. They may be found at http://hansard.millbanksystems.com/commons/.

—, *Agreement between France, Russia, Great Britain and Italy, Signed at London April 26, 1915*, https://en.wikisource.org/wiki/Treaty_of_London_(1915), last accessed 20 June 2022. (This is the so-called Treaty of London that facilitated Italy's entry into the First World War on the Allied side.)

—, *Attacco da Parte di Forze Leggere Nemiche al Convoglio "Tarigo" Presso le Secche di Kerkenah (16 Aprile 1941/XIX°)*, Notizia 5^ (Rome, Italy: Ufficio di Stato Maggiore della Regia Marina, undated).

Marano, LCdr Enrico, RM, *Rapporto Relativo al Combattimento Sostenuto dai CC.TT. "TARIGO" "BALENO" e "LAMPO" la Notte sul 16 Aprile 1941/XIX°*. (This 12 page report by the senior surviving officer is attached to the above.)

—, *Battle Experience: Radar Pickets and Methods of Combating Suicide Attacks off Okinawa, March-May 1945*, Secret Information Bulletin No 24, US Fleet, Headquarters of the Commander in Chief, Navy Department, Washington, DC, 20 July 1945, https://ibiblio.org/hyperwar/USN/rep/Kamikaze/BatExp-Okinawa/index.html, last accessed 20 January 2023.

—, *Conference on the Limitation of Armament: Address of the President of the United States at the Opening of the Conference at Washington, November 12, 1921, together with the Address of Charles E Hughes, Secretary of State of the United States and American Commissioner, also The Proposal of the United States for the Limitation of Naval Armament*, Senate Document No.77, 67th Congress, 1st Session (Washington, DC: Government Printing Office, 14 November 1921).

—, *Conference on the Limitation of Armament, Washington, November 12, 1921-February 6, 1922* (Washington, DC: Government Printing Office, 1922), https://catalog.hathitrust.org/Record/001756221, last accessed 7 July 2022. (This is the full record of all discussions at the Washington Conference.)

—, *Exchange of Notes between His Majesty's Government in the United Kingdom and the German Government regarding the Limitation of Naval Armaments, 18 June 1935*, http://www.navweaps.com/index_tech/tech-089_Anglo_German_Agreement_1935.htm, last accessed 14 November 2022. (This was the Anglo-German Naval Agreement.)

—, *Handbook for the 10-in R.M.L Gun and Mounting, 1899* (London, England: Harrison and Sons, 1899), https://viewer.slv.vic.gov.au/?entity=IE5359007&file=FL18850719&mode=browse, accessed 7 August 2021.

—, *H M Ships Damaged or Sunk by Enemy Action, 3rd. Sept. 1939 to 2nd. Sept. 1945*, C.B. 4273(52) ((London, England: Director of Naval Construction, Admiralty, 1952). (The copy obtained by this author has the document number C.B. 4273(52) crossed out in pen and replaced with B.R. 1886(2).)

—, *Japanese Naval Guns and Mounts, Article 1 – Mounts under "[18",]* O-47(N)-1 (San Francisco, CA: U.S. Naval Technical Mission to Japan, 16 February 1946), http://bulletpicker.com/pdf/USNTMJ-200E-47N-1.pdf, accessed 15 August 2022.

—, *Limitation and Reduction of Naval Armament (London Naval Treaty)*, https://tile.loc.gov/storage-services/service/ll/lltreaties//lltreaties-ustbv002/lltreaties-ustbv002.pdf, accessed 9 September 2022. (The Treaty was signed in London, England, on 22 April 1930, and was registered by the League of Nations as No. 2608 on 6 February 1931.)

—, *Papers Relating to the Foreign Relations of the United States, 1928* (In Three Volumes) Vol 1 (Washington, DC: Government Printing Office, 1942), https://search.library.wisc.edu/digital/AU5FNC2MGKYRAN8T/pages/ABU6GHEMWFEM7P8A, last accessed 14 September 2022.

—, *Papers Relating to the Foreign Relations of the United States, 1929* (In Three Volumes) Vol 1 (Washington, DC: Government Printing Office, 1943), https://search.library.wisc.edu/digital/AGRQ6ANWL7VORY8W/pages/AVJVDYIPW2OY2U9C, last accessed 18 September 2022.

—, *Papers Relating to the Foreign Relations of the United States, 1930* (In Three Volumes) Vol 1 (Washington, DC: Government Printing Office, 1945), https://search.library.wisc.edu/digital/AYHELVYNH2ZBE786/pages/A7Q74246AE7L7K8G, last accessed 22 September 2022.

—, 'Report of an Action against an Italian Convoy on the Night of the 15th/16th April, 1941,' in *Supplement to the London Gazette of Tuesday, the 11th of May, 1948*, No 38287 (London, England: Her Majesty's Stationery Office, 1948), https://www.thegazette.co.uk/London/issue/38287/supplement/2915, last accessed 13 December 2022. (Besides a brief introduction by Adm Andrew B Cunningham, this report includes 'narratives' by each of the four commanding officers involved on the British side in the Action off Sfax.)

—, *The Statutes at Large of the United States of America from December 1931 to March 1933*, Vol XLVII, Part 1 (Washington, DC: Government Printing Office, 1933), https://tile.loc.gov/storage-services/service/ll/llsl//llsl-c72/llsl-c72.pdf, last accessed 22 October 2022.

—, *The Treaty of Peace with Germany (Treaty of Versailles)*, 1919 For. ReI. (Paris Peace Conference, XIII) 55, 740, 743; Senate document 51, 66th Congress, 1st session.

—, *Treaty for the Limitation of Naval Armament (Second London Naval Treaty)*, London, March 25, 1936, http://www.navweaps.com/index_tech/tech-089_London_Treaty_1936.htm, last accessed 27 September 2022.

—, *Treaty No. 609 – Treaty between the United States of America, the British Empire, France, Italy and Japan, for the Limitation of Naval Armament, Signed at Washington, February 6, 1922.* (The instruments of ratification were deposited at Washington, DC, on 17 August 1923, and the treaty was registered by the League of Nations as No. 609 on 16 April 1924.)

Bevans, Charles I, ed., *Treaties and Other International Agreements of the United States of America, 1776–1949, Vol 2, Multilateral, 1918–1930*, US Department of State Publication 8441, released May 1969.

Cernuschi, Enrico, Alessandro Gazzi, and Michele Maria Gaetani, *Sea Power: The Italian Way* (Rome, Italy: Ufficio Storico della Marina Militare, 2017).

Cunningham, Adm Andrew B, RN, *A Sailor's Odyssey: The Autobiography of Admiral of the Fleet Viscount Cunningham of Hyndhope* (Barnsley, S Yorkshire: Seaforth, 2022).

Medlicott, W N, Douglas Dakin, and M E Lambert, eds., *Documents on British Foreign Policy, 1919–1939, Series IA, Vol IV* (London, England: Her Majesty's Stationery Office, 1971). (This is now available at https://archive.org/details/doc-brit-foreign-policy-1a-s-v-4/page/n7/mode/2up, last accessed on 9 September 2022.)

—, *Documents on British Foreign Policy, 1919–1939, Series IA, Vol V* (London, England: Her Majesty's Stationery Office, 1973). (This is now available at https://archive.org/details/doc-brit-foreign-policy-1a-s-v-5/page/n5/mode/2up, last accessed on 9 September 2022.)

Morison, Samuel Eliot, *History of United States Naval Operations in World War II, Vol. XIV, Victory in the Pacific, 1945* (Boston, MA: Little, Brown & Co., 1960).

Newbolt, Henry, *History of the Great War: Naval Operations, Vol. IV* (London, England: Longmans Green & Co, Ltd, 1928). (This is now available at http://naval-history.net/WW1Book-RN4a.htm, last accessed 28 June 2022.)

—, *History of the Great War: Naval Operations, Vol. V* (London, England: Longmans Green & Co, Ltd, 1931). (This is now available at http://naval-history.net/WW1Book-RN5a.htm, last accessed 5 July 2022.)

Roskill, Capt Stephen W, RN, *The War at Sea, 1939–1945, Vol. I, The Defensive* (London, England: Her Majesty's Stationery Office, 1954).

Ugaki Matome, VAdm, IJN, *Fading Victory: The Diary of Admiral Matome Ugaki, 1941–1945*, trans. Chihaya Masataka (Annapolis, MD: Naval Institute Press, 2008).

Woodward, E L and Rohan Butler, eds., *Documents on British Foreign Policy, 1919–1939, Second Series, Vol I* (London, England: Her Majesty's Stationery Office, 1947). (This is now available at https://archive.org/details/documentsonbriti0001grea/page/n3/mode/2up, last accessed on 22 September 2022.)

Primary/official sources – USN action/damage reports/war histories/war diaries

All of these reports are in RG38, the CNO's Operational and Secret Archives held at the US National Archives and Records Administration (NARA). Specifically, the originals can be found in Entry A1 351: World War II Action & Operation Reports, and have been declassified under Administrative Order NND968133. Both of the records listed below were found by the author in the immensely useful collection of US Navy records available from fold3.com. The author strongly endorses and supports the efforts of organisations like fold3.com that are making historical records available online; the more research that can be done without the expense and environmental impact of flying cross-country each time it is required, the better for all involved.

DD724-1 War Diary for month of August, 1944, Serial 068, DD724/A12, 1 September 1944. (This was the record of activities for USS *Laffey* (DD 724) for the designated period.)

DD724-2 Report of Operations in Support of Landings by U.S. Troops in Kerama Retto – Okinawa Area March 25 to April 22, 1945, Including Action Against Enemy Aircraft on April 16, 1945, Serial 023, DD724/A12, 29 April 1945. (This was the record of activities for USS *Laffey* (DD 724) for the designated period.)

DD793 USS Cassin Young (DD793) War Diary, from 1 April 1945 to 30 April 1945, No Serial, undated.

Secondary sources (author/editor known)

Reference books increasingly are available electronically, which makes them in some ways more useful to a researcher. For example, they can be searched for word matches, making reliance on the author's index less necessary. However, e-books have a huge downside for an author wishing to use them as a citable source; they often lack the original printed book's pagination, so that source citations become somewhat iffy. The author has used a number of e-books as references. When I cite a reference from an e-book, unless the original pagination is displayed in the e-book, I will give the page location given by the e-reader I used – in almost every case the Amazon Kindle app on my Macintosh computer – rather than attempting to intuit the printed version's page number. (I have been able to find no reliable method of translating between a location given by the reader and a page in the print version. Much depends on mutable factors such as screen size, selected display font, etc.) To inform the reader wishing to track down these source citations, each such location reference in the notes will be followed by an asterisk (*) and each source that I have used in e-book form in the following lists will similarly be preceded by an asterisk.

Ando, Elio, 'The Italian Navigatori Class 1926', in *Super Destroyers: The big destroyers built in the 1930s for Britain, France, Germany, Italy, Japan and the United States*, edited by Antony Preston (London, England: Conway Maritime Press, 1978).

Arena, Nino, *Il Radar, La Guerra sui Mari*, Vol 1 (Modena, Italy: STEM Mucchi,1976).

*Brescia, Maurizio, *Mussolini's Navy: A Reference Guide to the Regia Marina, 1930–1945* (Barnsley, S Yorkshire: Seaforth, 2012).

*Brown, David K, *Warrior to Dreadnought: Warship Design and Development, 1860–1905* (Barnsley, S Yorkshire: Seaforth, 2010).

Carlton, David, 'The Anglo-French Compromise on Arms Limitation, 1928', *Journal of British Studies*, Vol 8, No 2, (May 1969), pp 141–62, accessed at https://www.jstor.org/stable/175220 on 7 September 2022.

Chesneau, Roger, ed., *Conway's All the World's Fighting Ships 1922–1946* (London, England: Conway Maritime Press, 1980).

Chesneau, Roger, and Eugene M Kolesnik, eds., *Conway's All the World's Fighting Ships 1860–1905* (London, England: Conway Maritime Press, 1979).

Chihaya Masataka and Abe Yasuo, *IJN Yukikaze/Destroyer/1939–1970*, Warship Profile 22 (Windsor, Berkshire: Profile Publications Ltd, 1972).

Chumbley, Stephen, ed., *Conway's All the World's Fighting Ships 1947–1995* (London, England: Conway Maritime Press, 1995).

Cocker, Maurice, *Destroyers of the Royal Navy, 1893–1981* (Shepperton, Surrey: Ian Allan Ltd, 1981).

Corbett, Julian S, *Some Principles of Maritime Strategy* (London, England: Longmans Green & Co, Ltd, 1911). (This is now available at https://www.gutenberg.org/files/15076/15076-h/15076-h.htm, last accessed 17 April 2022.)

D'Adamo, Cristiano, *Regia Marina Italiana: Sinking of the* Tarigo Convoy, *April 16th, 1941*, https://web.archive.org/web/20060910011054/http://regiamarina.net/engagements/tarigo/tarigo_us.htm, last accessed 12 December 2022.

Evans, David C, and Mark R Peattie, *Kaigun: Strategy, Tactics, and Technology in the Imperial Japanese Navy, 1887–1941* (Annapolis, MD: Naval Institute Press, 1997).

Forsyth, Douglas J, *The Crisis of Liberal Italy: Monetary and Financial Policy, 1914–1922* (Cambridge, Cambridgeshire: Cambridge University Press, 1993).

*Friedman, Norman, *British Destroyers: From the Earliest Days to the Second World War* (Barnsley, S Yorkshire: Seaforth, 2009).

*—, *British Destroyers & Frigates: The Second World War and After* (Barnsley, S Yorkshire: Seaforth, 2008).

Friedman, Norman, *Naval Radar* (Annapolis, MD: Naval Institute Press, 1981).

__, 'The US "1850-Tonners" 1933', in *Super Destroyers: The big destroyers built in the 1930s for Britain, France, Germany, Italy, Japan and the United States*, edited by Antony Preston (London, England: Conway Maritime Press, 1978).

—, *U.S. Destroyers: An Illustrated Design History* (Annapolis, MD: Naval Institute Press, 1982).

Georgescu, Mihai, 'The Romanian Navy's Torpedo Boat *Rindunica*.' *Warship International* No 4 (1987), pp 350–55.

Grey, Randal, ed., *Conway's All the World's Fighting Ships 1906–1921* (London, England: Conway Maritime Press, 1985).

Greene, Jack, and Alessandro Massignani, *The Naval War in the Mediterranean 1940–1943* (London, England: Frontline Books, 2011).

Havelock, Thomas H, FRS, 'The Effect of Shallow Water on Wave Resistance', *Proceedings of the Royal Society A*, Vol 100, Issue 706, (1 February 1922), pp 499–505, accessed at

https://royalsocietypublishing.org/doi/10.1098/rspa.1922.0013, last accessed on 4 January 2022.
Howse, Derek, *Radar at Sea: The Royal Navy in World War 2* (Annapolis, MD: Naval Institute Press, 1993).
Jordan, John, and Jean Moulin, *French Destroyers, Torpilleurs d'Escadre & Contre-Torpilleurs 1922–1956* (Barnsley, S Yorkshire: Seaforth, 2015).
Koop, Gerhard and Klaus-Peter Schmolke, *German Destroyers of World War II*, trans. Geoffrey Brooks (London, England: Lionel Leventhal Ltd, 2003).
Lengerer, Hans, 'Torpedo Boats of the Imperial Japanese Navy, Part I', *Warship International* No 54 (2017), pp 57–73.
—, 'Torpedo Boats of the Imperial Japanese Navy, Part II', *Warship International* No 54 (2017), pp 139–55.
—, 'Torpedo Boats of the Imperial Japanese Navy, Part III', *Warship International* No 54 (2017), pp 293–306.
—, and Lars Ahlberg, *Fubuki-Class Destroyers in the Imperial Japanese Navy during World War II* (Atglen, PA: Schiffer Publishing, Ltd, 2022).
Lyon, David, 'The British Tribals 1935', in *Super Destroyers: The big destroyers built in the 1930s for Britain, France, Germany, Italy, Japan and the United States*, edited by Antony Preston (London, England: Conway Maritime Press, 1978).
Marder, Arthur J, *From the Dreadnought to Scapa Flow, The Royal Navy in the Fisher Era 1904–1919, Volume I: The Road to War 1904–1914* (Barnsley, S Yorkshire: Seaforth, 2013).
—, *From the Dreadnought to Scapa Flow, The Royal Navy in the Fisher Era 1904–1919, Volume IV: 1917: Year of Crisis* (Barnsley, S Yorkshire: Seaforth, 2014).
Mitchell, Alastair, 'The Development of Radar in the Royal Navy 1939–1945, Part 1', *Warship* No 13 (January 1980), pp 2–14.
Mordal, Jacques, 'The French Fantasque Class 1930', in *Super Destroyers: The big destroyers built in the 1930s for Britain, France, Germany, Italy, Japan and the United States*, edited by Antony Preston (London, England: Conway Maritime Press, 1978).
O'Hara, Vincent P, *Struggle for the Middle Sea: The Great Navies at war in the Mediterranean Theater, 1940–1945* (Annapolis, MD: Naval Institute Press, 2009).
—, and Leonard R Heinz, *Clash of Fleets: Naval Battles of the Great War, 1914–18* (Annapolis, MD: Naval Institute Press, 2017).
Olender, Piotr, *Russo-Turkish Naval War: 1877–1878* (Petersfield, Hampshire: MMPBooks, 2017).
*—, *Sino-Japanese Naval War: 1894–1895* (Petersfield, Hampshire: MMPBooks, 2014).
Paine, S C M, *The Sino-Japanese War of 1894–1895: Perceptions, Power, and Primacy* (New York: Cambridge University Press, 2003).
Peattie, Mark R, *Sunburst: The Rise of Japanese Naval Air Power, 1909–1941* (Annapolis, MD: Naval Institute Press, 2001).
Preston, Antony, 'The German Narvik Type 1936', in *Super Destroyers: The big destroyers built in the 1930s for Britain, France, Germany, Italy, Japan and the United States*, edited by Antony Preston (London, England: Conway Maritime Press, 1978).
Price, Scott T, 'A Study of the General Board of the U.S.Navy, 1929–1933.' Masters Thesis, University of Nebraska at Omaha, April 1989, https://www.history.navy.mil/content/history/nhhc/research/library/online-reading-room/title-list-alphabetically/s/study-general-board-usnavy-1929-1933.html#II, last accessed 25 August 2022.
Raven, Alan, 'Aspects of Japanese Warship Design, 1918–1945: No 2: Destroyers', *Warship* No 3 (July 1977), pp 20–25.
Rielly, Robin L, *Kamikazes, Corsairs, and Picket Ships* (Drexel Hill, PA: Casemate, 2008).
Roscoe, Theodore, *United States Destroyer Operations in World War II* (Annapolis, MD: Naval Institute Press, 1953).
Roskill, Capt Stephen W, RN, *Naval Policy Between the Wars I: The Period of Anglo-American Antagonism, 1919–1929* (Barnsley, S Yorkshire: Seaforth Publishing, 1968).
—, *Naval Policy Between the Wars II: The Period of Reluctant Rearmament, 1930–1939* (Barnsley, S Yorkshire: Seaforth Publishing, 1968).
Ruge, VAdm aD Prof Friedrich O, *SM Torpedo Boat B110*, Warship Profile 27 (Windsor, Berkshire: Profile Publications Ltd, 1972).
Scheer, Adm Reinhardt, *Germany's High Seas Fleet in the World War* (London, England: Cassell and Co, 1920).
Sleeman, C W, *Torpedoes and Torpedo Warfare* (Portsmouth, England: Griffin & Co, 1880), accessed at https://archive.org/details/torpedoestorpedo00sleerich/page/n9/mode/2up, last accessed on 30 September 2021.
Sprout, Harold, and Margaret Sprout, *Toward a New Order of Sea Power: American Naval Policy and the World Scene, 1918–1922* (Princeton, NJ: Princeton University Press, 1946).
Stern, Robert C, *Destroyer Battles: Epics of Naval Close Combat* (Barnsley, S Yorkshire: Seaforth, 2008).
—, *Fire from the Sky: Surviving the Kamikaze Threat* (Barnsley, S Yorkshire: Seaforth, 2010).
—, 'Perilous Mission to Guadalcanal', *Naval History* Vol 36, No 5 (October 2022), pp 34–41.
—, *The Battleship Holiday: The Naval Treaties and Capital Ship Design* (Barnsley, S Yorkshire: Seaforth, 2017).
—, *The Modern Cruiser: The evolution of the ships that fought the Second World War* (Barnsley, S Yorkshire: Seaforth, 2020).
—, *The US Navy and the War in Europe* (Barnsley, S Yorkshire: Seaforth, 2012).

Stevens, William Oliver, and Allan Westcott, *A History of Sea Power* (Garden City, NY: Doubleday, Doran & Co, Inc, 1942). (This was the third edition of this work; it was first copyrighted in 1920.)

Taylor, John C, *German Warships of World War I* (Garden City, NY: Doubleday & Co, Inc, 1970).

Walsh, Senator David I, *The Decline and the Renaissance of the Navy, 1922–1944*, presented on 7 June 1944 (legislative day 9 May 1944) (Washington, DC: Government Printing Office, 1944), https://www.ibiblio.org/pha/USN/77-2s202.html, last accessed 1 October 2022.

Watts, Anthony J, 'The Japanese 'Special' Type 1923', in *Super Destroyers: The big destroyers built in the 1930s for Britain, France, Germany, Italy, Japan and the United States*, edited by Antony Preston (London, England: Conway Maritime Press, 1978).

—, and Brian G Gordon, *The Imperial Japanese Navy* (Garden City, NY: Doubleday & Co, Inc, 1971).

Whitley, M J, *Destroyer!: German Destroyers in World War II* (London, England: Lionel Leventhal Ltd, 1983).

—, *Destroyers of World War Two: An International Encyclopedia* (Annapolis, MD: Naval Institute Press, 1988).

Periodicals

The author has made repeated use of the online archives of the *New York Times*, the foremost 'newspaper of record' in the United States.[1] In the US, this function is self-appointed and, with the onset of ubiquitous access to electronic records of government, has become much less necessary than it once was, but the existence of this archive – available by subscription at https://timesmachine.nytimes.com/ – has been invaluable to the author's research.]

Warship International, International Naval Research Organization, PO Box 48, Holden, MA 01520, http://www.warship.org. (This quarterly journal is indispensable for any serious student of naval history.)

Indispensable websites

These are sites I referenced repeatedly during the writing of this and many other books.

Dictionary of American Naval Fighting Ships, http://www.history.navy.mil/danfs/index.html. (The entries in this immense effort vary considerably in detail and completeness. For the most part, though, it is an excellent first reference for any USN ship.)

NavWeaps: Naval Weapons, Naval Technology and Naval Reunions, http://www.navweaps.com/. (Definitely an indispensable site for information on the weapons carried by warships since the late nineteenth century and many related topics.)

Nihon Kaigun, http://www.combinedfleet.com/kaigun.htm. (This site covers the movements of most Japanese warships in the Second World War in great detail.)

World War 2 at Sea: SERVICE HISTORIES of 1,000 ROYAL and DOMINION NAVY WARSHIPS, including British Ships manned by Allied Navies, http://www.naval-history.net/xGM-aContents.htm. (A massively useful site for anyone interested in the Royal Navy in the Second World War.)

Notes

Preface

[1] Robert C Stern, *The Battleship Holiday: The Naval Treaties and Capital Ship Design* (Barnsley, S Yorkshire: Seaforth, 2017); Robert C Stern, *The Modern Cruiser: The Evolution of the Ships that fought the Second World War* (Barnsley, S Yorkshire: Seaforth, 2020); Antony Preston, ed., *Super Destroyers: The big destroyers built in the 1930s for Britain, France, Germany, Italy, Japan and the United States* (London, England: Conway Maritime Press, 1978).

Introduction

[1] 'The Whitehead Torpedoes', The *New York Times*, Vol. XXXIX, No. 12,181, 11 September 1890, 4.
[2] http://www.navweaps.com/Weapons/WTUS_PreWWII.php. The Howell torpedo was an interesting design, powered by a 131 lb steel flywheel spun up to 10,000rpm before launch, which provided a stable weapon that left no telltale trail of air bubbles. The US Navy ended up choosing the Whitehead torpedo for series production in 1890 not because of any dissatisfaction with the Howell torpedo, per se, but because the Howell torpedo was proving difficult to manufacture and appeared to have little potential for improvement in performance. Some sources state that as many as fifty Howell torpedoes were ordered by the USN.
[3] The USN began assigning hull number to ships around 1890. A few American ships mentioned in this text pre-date this period and had no hull number assigned.
[4] The RN used several different pennant number schemes prior to the primary period of interest of this study and again after that period, but the only pennant numbers given here will be those in use during the Second World War.
[5] Kriegsmarine – literally 'war navy' – was the official name of the German navy only from 1935 through the end of the Second World War. Until the surrender of Imperial Germany in 1918, the German navy was called the Kaiserliche Marine ('Imperial Navy') and then until 1935, the Reichsmarine (another way of saying 'imperial navy', as in 'navy of the empire', in the sense of 'national navy').

Chapter 1: Steam launches, torpedo boats, and competing ideas: 1876–1885

[1] 'The Whitehead Torpedoes', The *New York Times*, Vol. XXXIX, No. 12,181, 11 September 1890, 4.
[2] All dates in the discussion of events related to the Russo-Turkish war are given using the common modern calendar, which Russia only adopted in 1918. In contemporary Russian accounts, the dates would have been twelve days earlier.
[3] Mihai Georgescu, 'The Romanian Navy's Torpedo Boat *Rindunica*.' *Warship International* No. 4 (1987), 350–55. Some sources spell this vessel's name *Râundunica*, but the most authoritative source found by this author, the article cited here, spells it *Rindunica* and I have adopted that spelling.
[4] Piotr Olender, *Russo-Turkish Naval War: 1877–1878* (Petersfield, Hampshire: MMPBooks, 2017), 71–76. It is unclear from any of the various accounts of the action on 26 May 1877, whether all (or even any) of the three additional launches that comprised the Russo-Romanian Danube torpedo flotilla were originally Romanian; the names given here are all Russian. The two officers listed here as ensigns held the Russian rank of *michman*, which at this time equated most closely to the Royal Navy rank of sub lieutenant.
[5] C W Sleeman, *Torpedoes and Torpedo Warfare* (Portsmouth, England: Griffin & Co, 1880), 196–98.
[6] The transliteration of Turkish names from this period is iffy at best. Georgescu and Roger Chesneau and Eugene M Kolesnik, eds., *Conway's All the World's Fighting Ships 1860–1905* (London, England: Conway Maritime Press, 1979), 390, give this ship's name as rendered here, differing only by a hyphen, but Olender gives it as *Lüft-ü Celil*. Similar differences exist for many other Ottoman ship's names. I have generally followed the most common spelling when one can be found.
[7] The use of the name Constantinople for the metropolis on the Bosporus that was at this time the capital of the Ottoman Empire is correct; the city was not officially renamed Istanbul until 1930, although that name had been informally used by the city's residents for centuries.
[8] Olender, *Russo-Turkish Naval War*, 129–30.
[9] Ibid, 136–41.
[10] Sleeman, *Torpedoes*, 195.
[11] Olender, *Russo-Turkish Naval War*, 154–57.
[12] Some accounts, such as Sleeman, *Torpedoes*, 202–3, state that *Sinop*'s torpedo was still carried on a towed raft.
[13] The author freely admits that he gave a quite different account of this attack in Robert C Stern, *Destroyer Battles: Epics of Naval Close Combat* (Barnsley, S Yorkshire: Seaforth, 2008), 18–21. Over the intervening years, more information has come to his attention that makes it seem more likely that the sinking of *Intibah* actually did occur, although the accounts still differ greatly and many of the details should be viewed with a fair degree of scepticism.
[14] It is relatively certain *Intibah* was not only unarmoured, but also unarmed (or very lightly armed). It is doubtful she was a naval vessel in any sense; one source (Sleeman, *Torpedoes*, 203) describes the target of this attack as a Turkish revenue steamer, but does not give her name. If that description is correct, *Intibah*'s absence from lists of Ottoman naval vessels can be explained.
[15] Most of the specifications for *Rap* come from Sleeman, *Torpedoes*, 164.
[16] Norman Friedman, *British Destroyers: From the Earliest Days to the Second World War* (Barnsley, S Yorkshire: Seaforth, 2009), 38–40*; David K Brown, *Warrior to Dreadnought: Warship Design and Development, 1860–1905* (Barnsley, S Yorkshire: Seaforth, 1997), 249–51*. See the note at the beginning of the entry for 'Secondary Sources' in the Bibliography to explain the use of the asterisk after the

cited page numbers here and later.
[17] Chesneau and Kolesnik, eds., *Conway's 1860–1905*, 105.
[18] Chesneau and Kolesnik, eds., *Conway's 1860–1905*, 88; Brown, *Warrior to Dreadnought*, 251.
[19] It was a curious characteristic of the RN at this time, quite different from the US Navy, that serving officers were allowed (in fact, at times encouraged) to establish business relations with private firms and to lobby for those firms within the navy.
[20] Chesneau and Kolesnik, eds., *Conway's 1860–1905*, 101.
[21] Brown, *Warrior to Dreadnought*, 251–54*.
[22] Chesneau and Kolesnik, eds., *Conway's 1860–1905*, 101.
[23] Brown, *Warrior to Dreadnought*, 255–57*.
[24] Ibid, 340. Which is not to say the type did not continue to have its supporters. Barnaby, in particular, lobbied hard for the construction of improved-Polyphemus types, but was in a very small minority.
[25] Chesneau and Kolesnik, eds., *Conway's 1860–1905*, 256. This source describes the Blitz class as 'the first true torpedo cruisers'. As will be seen in the next chapter, the Italians had experimented with the idea of the torpedo cruiser almost ten years earlier, but their ambitions were well in advance of the technology and the experiment was not successful.
[26] Ibid.
[27] Ibid, 340; Chesneau and Kolesnik, eds., *Conway's 1860–1905*, 324. These were the four torpedo cruisers of the Condor class, the first of which was laid down in April 1883.
[28] Brown, *Warrior to Dreadnought*, 340–43*; Friedman, *British Destroyers*, 48–53*. The Thornycroft idea, despite being very much along the lines of what would eventually become the first 'torpedo boat destroyers', was rejected by the Admiralty for being excessively protected and too small for fleet work.
[29] Chesneau and Kolesnik, eds., *Conway's 1860–1905*, 80.
[30] Ibid, 324.
[31] Brown, *Warrior to Dreadnought*, 342–7*; Chesneau and Kolesnik, eds., *Conway's 1860–1905*, 88.
[32] Even though Brown, *Warrior to Dreadnought*, 343*, states that the armour in *Rattlesnake* was a protective deck, claiming this proves her lineal descent from the *Scout* torpedo cruiser design, Friedman, *British Destroyers*, 63*, states explicitly that she had no armoured deck and ¾in side armour in way of her engine room only.
[33] Friedman, *British Destroyers*, 64*. The plan had been to build just three of this class at RN yards, but the use of multiple locomotive-type (firetube) boilers and the newly-developed triple-expansion engine in a ship was considered risky and the Admiralty let a contract to the Laird Bros. yard at Birkenhead, the idea being that a commercial yard could build the prototype much faster and the first ship could be finished and put through its trials before the machinery for the three ships from the government yards needed to be ordered. The war scare accelerated the pace of events, causing the three follow-on ships to be laid-down well before *Rattlesnake* was launched, much less tested.
[34] http://hansard.millbanksystems.com/commons/1887/mar/17/decision-of-the-board-upon-sea-going.
[35] Brown, *Warrior to Dreadnought*, 344*.

Chapter 2: Torpedo boat destroyers to lead destroyers: 1884–1915

[1] Chesneau and Kolesnik, eds., *Conway's 1860–1905*, 103; Brown, *Warrior to Dreadnought*, 348*.
[2] Chesneau and Kolesnik, eds., *Conway's 1860–1905*, 327.
[3] Normal displacement is defined as the weight of a ship loaded with all fixed equipment and two-thirds of all moveable items, such as crew, provisions, munitions, fuel, feed water, ballast, etc.
[4] Ibid, 328.
[5] Friedman, *British Destroyers*, 87–88*. It is difficult to tell how much of this origin story is apocryphal. There are claims that Fisher came up with the idea of the destroyer and was even the first to think of the name; these claims are dubious, given the many threads of development under way and the fact that the idea of the torpedo boat destroyer was very much 'in the air' at this time. What is certain is that the arrival of Fisher in the post of Third Sea Lord in February 1892 removed many obstacles in the path of development of the type in the Royal Navy.
[6] Ibid; Chesneau and Kolesnik, eds., *Conway's 1860–1905*, 90. Friedman states that there were six destroyers in this first order, while *Conway's* explicitly states there were five, but then goes on to list six. To further confuse the issue, the quote immediately below from the 1892 Parliamentary debate also states that five 'catchers' were being discussed. Regardless, six were laid down between mid-1892 and mid-1893.
[7] https://hansard.parliament.uk/Commons/1892-03-14/debates/40bf9322-5333-4c37-9abe-e724a6a606d2/Supply%E2%80%94NavyEstimates1892%E2%80%933. It is clear that by no means were all the details of the proposed destroyers released to the public (or to Parliament), because Capt Price, in the elided section of this quote, refers to these planned boats having a speed similar to the torpedo gunboats', some 10kt slower than the designed speed for the new boats; this might account in part for Capt Price's objections and for the First Lord's 'non-answer', assuming the planned speed for these boats was secret.
[8] Chesneau and Kolesnik, eds., *Conway's 1860–1905*, 91; Friedman, *British Destroyers*, 766–67*; 'Fastest Boat Ever Built', The *New York Times*, Vol. XLIII, No. 13,401, 5 August 1894, 20. The *New York Times* report gives *Daring*'s speed at trials as a knot faster than reported by *Conway's* or Friedman; normally, this author would tend to discount a newspaper report against such reliable sources, but all the other specifications given in the article are accurate and a brief outline of destroyer development in Great Britain and Europe is also quite detailed, lending considerable weight to the report.
[9] Thomas H Havelock, FRS, 'The Effect of Shallow Water on Wave Resistance', *Proceedings of the Royal Society A*, Vol 100, Issue706, (1 February 1922), pp 499–505.
[10] Chesneau and Kolesnik, eds., *Conway's 1860–1905*, 87.
[11] Ibid, 104. The Royal Navy continued placing orders for first-class torpedo boats, albeit in much reduced numbers, up until 1902.
[12] Stern, *The Battleship Holiday*, 30.
[13] Chesneau and Kolesnik, eds., *Conway's 1860–1905*, 92; Friedman, *British Destroyers*, 767–69*.
[14] Chesneau and Kolesnik, eds., *Conway's 1860–1905*, 93.
[15] Ibid, 94.

[16] Light displacement measures a ship's weight absent any stores, ballast, feed water, fuel, or crew, including only normal water levels in boilers.
[17] 'Foreign Navies', *New York Times*, Vol. XLV, No. 13,774, 14 October 1895, 5. The *Times* article quoted the British United Service Gazette as saying: 'M. Normand of Havre is to be congratulated on the success of the French sea-going torpedo boat *Forban*, which... establishes herself, for the present, as the fastest boat in existence. The trials,..., were made at the full load displacement of 136 tons, having guns, coal, crew, provisions, as well as electric machinery on board.'
[18] Chesneau and Kolesnik, eds., *Conway's 1860–1905*, 97–98.
[19] Friedman, *British Destroyers*, 139*.
[20] Chesneau and Kolesnik, eds., *Conway's 1860–1905*, 98; Friedman, *British Destroyers*, 139*, 775–76*.
[21] These involved the use of cruising turbines, an entirely separate set of 'stages' attached to one or more propeller shafts, independent of the main turbines, designed to work more efficiently at slower speeds.
[22] Friedman, *British Destroyers*, 255*.
[23] Chesneau and Kolesnik, eds., *Conway's 1860–1905*, 99; Friedman, *British Destroyers*, 139, 778–79*.
[24] Friedman, *British Destroyers*, 236*.
[25] Ibid, 238*; Randal Gray, ed., *Conway's All the World's Fighting Ships 1906–1921* (London, England: Conway Maritime Press, 1985), 73–74. Unlike the move to turbine powerplants, this decision would be second-guessed. As late as the Beagle class destroyers laid down in 1909–10, the Royal Navy reverted to coal for fuel because of worries about the cost and reliability of the oil supply.
[26] Gray, ed., *Conway's 1906–1921*, 72; Friedman, *British Destroyers*, 240–41*, 266–71*.
[27] Gray, ed., *Conway's 1906–1921*, 71–72; Friedman, *British Destroyers*, 783–84*.
[28] Interestingly, sources differ greatly on whether the Tribals were adequate sea boats. Grey, ed., *Conway's 1906–1921*, 72, states emphatically that they 'were not regarded as good seaboats', while Friedman, *British Destroyers*, 274*, states: 'Except for *Mohawk*, the Tribals were considered excellent sea boats'. However, Friedman then goes on to list several instances where COs described them as being 'dry,' but not when compared to the Rivers.
[29] Gray, ed., *Conway's 1906–1921*, 73; Friedman, *British Destroyers*, 253–55*.
[30] Friedman, *British Destroyers*, 255*.
[31] Gray, ed., *Conway's 1906–1921*, 73; Friedman, *British Destroyers*, 780–81*.
[32] Friedman, *British Destroyers*, 259–61*; 'Destroyer Makes 33.7 Knots,' *New York Times*, Vol. LVIII, No. 18,849, 2 September 1909, 1. The *Times* article documents the speed set by the USN destroyer USS *Flusser* (DD20) on trials, then notes in passing that this was slower than *Swift*, which had established a record speed of 'nearly 36 knots'.
[33] Arthur J Marder, *From the Dreadnought to Scapa Flow, The Royal Navy in the Fisher Era 1904–1919, Volume I: The Road to War 1904–1914* (Barnsley, S Yorkshire: Seaforth, 2013), 23–25.
[34] https://hansard.parliament.uk/Commons/1905-03-06/debates/19305d49-cb3d-4797-a8e3-b04c37e73618/Supply(NavyEstimates). The reference to the issue of the scouts and of their utility will be taken up below in this chapter.
[35] These Japanese boats were the Akatsuki class built by Yarrow, which will be discussed below in this chapter.
[36] https://hansard.parliament.uk/Commons/1905-03-13/debates/11dab2c0-3e27-408c-bafc-761996e62439/NavyEstimates1905%E2%80%936
[37] Ibid.
[38] https://hansard.parliament.uk/Commons/1905-03-13/debates/106d1ece-d6c6-4c5b-8542-2fd1e18ed0ac/NavyEstimates1905%E2%80%936
[39] Marder, *From the Dreadnought to Scapa Flow, Volume I*, 122–23.
[40] Gray, ed., *Conway's 1906–1921*, 164; VAdm aD Prof Friedrich O Ruge, *SM Torpedo Boat B110*, Warship Profile 27 (Windsor, Berkshire: Profile Publications Ltd, 1972), 50.
[41] Gray, ed., *Conway's 1906–1921*, 73–74.
[42] Gray, ed., *Conway's 1906–1921*, 73; Friedman, *British Destroyers*, 789–90*.
[43] Gray, ed., *Conway's 1906–1921*, 74; Friedman, *British Destroyers*, 790*.
[44] Chesneau and Kolesnik, eds., *Conway's 1860–1905*, 326.
[45] John Jordan and Jean Moulin, *French Destroyers, Torpilleurs d'Escadre & Contre-Torpilleurs 1922–1956* (Barnsley, S Yorkshire: Seaforth, 2015), 9.
[46] Gray, ed., *Conway's 1906–1921*, 203. The boats in this class were built by seven different yards with four different turbine types and four different boiler types, so the dimensions and displacements varied considerably; an average for all specifications for the class is given.
[47] Jordan and Moulin, *French Destroyers*, 13–14.
[48] Ibid, 14–15; Gray, ed., *Conway's 1906–1921*, 204. These were only four of a dozen boats ordered by the Argentines in 1910, four each from France, Germany and Great Britain; the story of the eight other boats will be told below.
[49] Chesneau and Kolesnik, eds., *Conway's 1860–1905*, 264–65.
[50] Gray, ed., *Conway's 1906–1921*, 167.
[51] Ibid, 164; Ruge, *SM Torpedo Boat B110*, 49.
[52] Gray, ed., *Conway's 1906–1921*, 168.
[53] Enrico Cernuschi, Alessandro Gazzi, and Michele Maria Gaetani, *Sea Power: The Italian Way* (Rome, Italy: Ufficio Storico della Marina Militare, 2017), 85.
[54] Ibid, 93.
[55] Ibid, 89.
[56] Chesneau and Kolesnik, eds., *Conway's 1860–1905*, 355.
[57] Cernuschi, Gazzi, and Gaetani, *Sea Power: The Italian Way*, 94.
[58] Gray, ed., *Conway's 1906–1921*, 265.
[59] Great appreciation to Enrico Cernuschi for this detailed developmental history of the Mirabello class.
[60] David C Evans and Mark R Peattie, *Kaigun: Strategy, Tactics, and Technology in the Imperial Japanese Navy, 1887–1941* (Annapolis, MD: Naval Institute Press, 1997), 12, 24. In 1870, an imperial decree declared the Royal Navy to be the pattern upon which the new IJN would be modelled. Mahan's most important work, *The Influence of Sea Power upon History*, was translated and published in Japan by the naval officers' professional association in 1896.
[61] Ibid, 15.

[62] Chesneau and Kolesnik, eds., *Conway's 1860–1905*, 237.
[63] Hans Lengerer, 'Torpedo Boats of the Imperial Japanese Navy, Part I,' *Warship International* No 54 (2017), 60.
[64] Ibid, 65–66.
[65] Ibid, 237–38; Evans and Peattie, *Kaigun*, 17; Anthony J Watts and Brian G Gordon, *The Imperial Japanese Navy* (Garden City, NY: Doubleday & Co, Inc, 1971), 221–22.
[66] Hans Lengerer, 'Torpedo Boats of the Imperial Japanese Navy, Part II,' *Warship International* No 54 (2017), 140.
[67] Other sources call this Jungch'eng Bay.
[68] Ibid, 151.
[69] The name of the Japanese admiral in charge was alternately rendered in western languages as Ito Sukeyuki or Ito Yuko. Such alternate transliterations of Japanese names are common due to the multiple possible vocalizations of many kanji characters.
[70] This account of the Naval Battle of Wei-hai-wei is drawn primarily from Lengerer, 'Torpedo Boats of the Imperial Japanese Navy, Part II,' 152–53 & Hans Lengerer, 'Torpedo Boats of the Imperial Japanese Navy, Part III,' *Warship International* No 54 (2017), 293–94, with additional details from Piotr Olender, *Sino-Japanese Naval War: 1894–1895* (Petersfield, Hampshire: MMPBooks, 2014), 2748–2826*.
[71] The order in which events were reported to have happened, specifically in which Japanese torpedo boats attacked, on 5 February, differs greatly between accounts. The reader should keep in mind that these events in all likelihood happened more or less simultaneously and any attempt to 'organize' them into a coherent narrative, necessary to write any account of the battle, introduces an implied sense of sequence where none existed.
[72] Gyroscopic guidance, which would facilitate the first major increase in the effective range of torpedoes by allowing them to maintain a steady, aimed course, was just coming into use at this time.
[73] Lengerer, 'Torpedo Boats of the Imperial Japanese Navy, Part II,' 153n46. The best source seems to indicate that *Ting Yuen* had only one hole in her side. While that does not prove she only received one hit, it is strongly suggestive of that case.
[74] Olender, *Sino-Japanese Naval War: 1894–1895*, 2788*. *TB-9* stayed afloat long enough to be boarded by the Chinese at dawn, but sank soon after. After the surrender of the Chinese, *TB-9* was raised and resumed service in the IJN, remaining in commission until 1909.
[75] Strangely, the normally authoritative Evans and Peattie, *Kaigun*, 46–47, gives an account of the Battle of Wei-hai-wei completely at odds with all other accounts this author has found, reporting that the second night's attack was an utter failure.
[76] William Oliver Stevens and Allan Westcott, *A History of Sea Power* (Garden City, NY: Doubleday, Doran & Co, Inc, 1942), 263–69. This authoritative American work, first published in 1920, but clearly reflecting contemporary opinion, covers the Yalu River battle over seven pages and discusses its lessons in detail, but makes no mention at all of Wei-hai-wei.
[77] Watts and Gordon, *The Imperial Japanese Navy*, 228–29.
[78] Only in the strictest legalistic sense was this a premature, surprise attack. Both sides had been exchanging 'notes' of increasingly hostile tone for months in a dispute over control of the northern half of Korea, and the Japanese presented Russia with an ultimatum which had expired unanswered on 4 February 1904, and had broken diplomatic relations on the 6th, the same day the Russians sent troops into the disputed region, so the declaration of war was something of a formality, cf., Stern, *Destroyer Battles*, 25. Certainly, VAdm Starck commanding the Russian Pacific Fleet had every reason to expect that war was imminent, though he was likely unaware that Russian and Japanese forces were already engaged with each other near Chemulpo (Inchon).
[79] The larger ships in the Russian fleet would have had torpedo nets that could have been deployed, but all evidence indicates that they were not.
[80] The first Corbett quote is from Evans and Peattie, *Kaigun*, 97, quoting from Julian S Corbett, *Maritime Operations in the Russo-Japanese War, 1904–05, Vol 1* (Annapolis, MD: Naval Institute Press, 1994), 99. The second is from Julian S Corbett, *Some Principles of Maritime Strategy* (London, England: Longman, Green & Co, Ltd, 1911), 229–30.
[81] Evans and Peattie, *Kaigun*, 130–31.
[82] Watts and Gordon, *The Imperial Japanese Navy*, 246–47; Gray, ed., *Conway's 1906–1921*, 241.
[83] Watts and Gordon, *The Imperial Japanese Navy*, 248–49; Gray, ed., *Conway's 1906–1921*, 242.
[84] Chesneau and Kolesnik, eds., *Conway's 1860–1905*, 210.
[85] Gray, ed., *Conway's 1906–1921*, 309.
[86] Ibid.
[87] Chesneau and Kolesnik, eds., *Conway's 1860–1905*, 157.
[88] In 1899, the Clydebank yard would be bought by John Brown & Co, under which name it would prosper.
[89] Norman Friedman, *U.S. Destroyers: An Illustrated Design History*, (Annapolis, MD: Naval Institute Press, 1982), 11–12. This includes a long quote from a report by a war plans board headed by TR.
[90] https://www.spanamwar.com/Furorbustemente.htm. The reader should recall that at this time, the effective range of torpedoes was still 2000yds at best.
[91] Friedman, *U.S. Destroyers*, 11. At this time, and even when FDR held the position during the Wilson administration, the Assistant Secretary effectively ran the US Navy because the Secretary of the Navy position was a political appointment given out by the president to cronies and big donors.
[92] Ibid, 14–15.
[93] Chesneau and Kolesnik, eds., *Conway's 1860–1905*, 157; Ibid, 392.
[94] Gray, ed., *Conway's 1906–1921*, 122; Friedman, *U.S. Destroyers*, 398.
[95] Friedman, *British Destroyers*, 109–11*.
[96] Chesneau and Kolesnik, eds., *Conway's 1860–1905*, 264; John C Taylor, *German Warships of World War I* (Garden City, NY: Doubleday & Co, Inc, 1970), 86.
[97] Gray, ed., *Conway's 1906–1921*, 73. In the RN, 'Captain (D)' was the position of the officer in command of a formation of destroyers.
[98] Ibid, 77.
[99] According to Friedman, *British Destroyers*, 336*, the initial design was submitted on 24 September 1913 and received Admiralty Board approval on 16 October, what must have been record speed.
[100] Chesneau and Kolesnik, eds., *Conway's 1860–1905*, 264; Friedman, *British Destroyers*, 795–96*.

Chapter 3: The test of battle, Part 1: 1917

[1] Gray, ed., *Conway's 1906–1921*, 169; Ruge, *SM Torpedo Boat B110*, 50.

[2] Gray, ed., *Conway's 1906–1921*, 169.

[3] Gray, ed., *Conway's 1906–1921*, 266.

[4] Much of this account comes from Vincent P O'Hara and Leonard R Heinz, *Clash of Fleets: Naval Battles of the Great War, 1914–18* (Annapolis, MD: Naval Institute Press, 2017), 262–70, supplemented by Henry Newbolt, *History of the Great War: Naval Operations, Vol. IV* (London, England: Longmans Green & Co, Ltd, 1928), 297–305. Critical details were added by Enrico Cernuschi.

[5] O'Hara and Heinz, *Clash of Fleets*, 242–44; Ruge, *SM Torpedo Boat B110*, 66–67; Arthur J Marder, *From the Dreadnought to Scapa Flow, The Royal Navy in the Fisher Era 1904–1919, Volume IV: 1917: Year of Crisis* (Barnsley, S Yorkshire: Seaforth, 2014), 311–15; Adm Reinhardt Scheer, *Germany's High Seas Fleet in the World War* (London, England: Cassell and Co, 1920), 312–14; Henry Newbolt, *History of the Great War: Naval Operations, Vol. V* (London, England: Longmans Green & Co, Ltd, 1931), 184–94.

[6] Gray, ed., *Conway's 1906–1921*, 79.

[7] See the account at Robert C Stern, *The Modern Cruiser: The evolution of the ships that fought the Second World War* (Barnsley, S Yorkshire: Seaforth, 2020), 64–67.

[8] This quote is from Marder, *Dreadnought to Scapa Flow, Volume IV: 1917: Year of Crisis*, 312, which cites Newbolt, *History of the Great War: Naval Operations, Vol. V*, 189.

Chapter 4: The immediate precursors: 1917–1918

[1] Most of the following about HMS *Anzac* comes from Friedman, *British Destroyers*, 392–93*.

[2] Gray, ed., *Conway's 1906–1921*, 82; Friedman, *British Destroyers*, 800–1*.

[3] Friedman, *British Destroyers*, 402*.

[4] Gray, ed., *Conway's 1906–1921*, 83–84; Friedman, *British Destroyers*, 789–99*.

[5] Gray, ed., *Conway's 1906–1921*, 243–44.

[6] Kampon was a contraction for *kansei honbu*, which stood for the IJN's Technical Division, whose task was to gather industrial technologies from other nations and producing designs that could be utilised in shipbuilding.

[7] Ibid., 170.

[8] Jordan and Moulin, *French Destroyers*, 19, 98; http://www.navweaps.com/Weapons/WNGER_59-45_tbts.php accessed 12 June 2022.

[9] Jordan and Moulin, *French Destroyers*, 16.

[10] Gray, ed., *Conway's 1906–1921*, 123; Friedman, *U.S. Destroyers*, 37.

[11] Gray, ed., *Conway's 1906–1921*, 124; Friedman, *U.S. Destroyers*, 401.

[12] Friedman, *U.S. Destroyers*, 40.

[13] Ibid.

Chapter 5: The first of the true super destroyers – Washington swings & misses: 1922–1928

[1] Much of the account of the Leone class comes from Maurizio Brescia, *Mussolini's Navy: A Reference Guide to the* Regia Marina, *1930–1945* (Barnsley, S Yorkshire: Seaforth, 2012), 2595*.

[2] 'British Coal Stock Far Below Needs', *New York Times*, Vol. LXVII, No. 22,142, 8 September 1918, p 6.

[3] 'British to Consult Allies on Trade Preferences', *New York Times*, Vol. LXVII, No. 22,111, 8 August 1918, p 1. In this article, it is reported that George N Curzon, then Government leader in the House of Lords, stated that the subject had not yet been discussed by the Imperial War Cabinet.

[4] https://www.nationalarchives.gov.uk/first-world-war/a-global-view/europe/italy/ accessed 20 June 2022.

[5] 'Text of President Wilson's Speech', *New York Times*, Vol. LXVII, No. 21,900, 9 January 1918, p 1.

[6] Ibid.

[7] Cernuschi, Gazzi, and Gaetani, *Sea Power: The Italian Way*, 121–22. This was by no means the only instance in which the British had used economic pressure to gain desired goals in Italy. After signing the Treaty of London in April 1915, the Italians had declared war on the Austrians the following month, but were able to delay declaring war on Germany, with which they shared no border. The other Entente powers, in particular the British, resented this delay and, finally, in August 1916, when the Italians, desperately short of coal, asked for an increase in the amount being shipped from the United Kingdom, the British used the occasion to force the Italians to agree to declare war on Germany before increasing their coal allocation. Cf., Douglas J Forsyth, *The Crisis of Liberal Italy: Monetary and Financial Policy, 1914–1922* (Cambridge, Cambridgeshire: Cambridge University Press, 1993), 167.

[8] Gray, ed., *Conway's 1906-1921*, 267.

[9] This discussion of wartime designs comes from Jordan and Moulin, *French Destroyers*, 14–19.

[10] M J Whitley, *Destroyers of World War Two: An International Encyclopedia* (Annapolis, MD: Naval Institute Press, 1988), 35.

[11] Jordan and Moulin, *French Destroyers*, 18n14. The 138.6mm gun proposed here had the same bore diameter as the earlier proposed 14cm gun. Right around 1920, the French began designating their guns in millimetres rather than centimetres.

[12] For greater coverage of the origins and events surrounding the Washington Naval Conference, cf., Stern, *The Battleship Holiday*, 98–107, and Stern, *The Modern Cruiser*, 85–92.

[13] —, *Conference on the Limitation of Armament: Address of the President of the United States at the Opening of the Conference at Washington, November 12, 1921, together with the Address of Charles E Hughes, Secretary of State of the United States and American Commissioner, also The Proposal of the United States for the Limitation of Naval Armament*, Senate Document No.77, 67th Congress, 1st Session (Washington, DC: Government Printing Office, 14 November 1921), 24–25.

[14] It probably was not clear to most who heard Hughes' proposal that day, but the measurement of displacement used by him and the Americans at this time, that would become the default measurement system henceforward, was standard displacement. This was defined as a completed ship, fully-manned and equipped for sea, with all stores, provisions, drinking water, and munitions needed for a sea voyage, but

without fuel or boiler reserve feed water, cf., —, *Treaty No. 609 – Treaty between the United States of America, the British Empire, France, Italy and Japan, for the Limitation of Naval Armament, Signed at Washington, February 6, 1922*, Ch 2, Pt 4.

[15] 'Navy Experts Defend Rival Claims', *New York Times*, Vol. LXXI, No. 23,314, 23 November 1921, p 1.

[16] —, *Conference on the Limitation of Armament, Washington, November 12, 1921-February 6, 1922* (Washington, DC: Government Printing Office, 1922), 86.

[17] Ibid, 454–58.

[18] Ibid, 458.

[19] The word Briand used for torpedo boats was '*torpilleurs*', which was understood to refer to standard destroyers in the 1000–1500 ton range being built by many nations towards the end of the war.

[20] Ibid, 460.

[21] Ibid, 462.

[22] Ibid.

[23] 'French Troubled Over Hughes Notes', *New York Times*, Vol. LXXI, No. 23,344, 23 December 1921, p 1. It would be an understatement to say there was considerable anger across the political spectrum in Paris at what was seen as American attempts to influence French naval policymaking.

[24] —, *Conference on the Limitation of Armament, Washington, November 12, 1921-February 6, 1922*, 466.

[25] —, *Conference on the Limitation of Armament, Washington, November 12, 1921-February 6, 1922*, 570.

[26] Ibid, 574.

[27] Ibid, 576.

[28] Ibid, 578.

[29] Harold Sprout and Margaret Sprout, *Toward a New Order of Sea Power: American Naval Policy and the World Scene, 1918–1922* (Princeton, NJ: Princeton University Press, 1946), 210–12.

[30] Roger Chesneau, ed., *Conway's All the World's Fighting Ships 1922–1946* (London, England: Conway Maritime Press, 1980), 267.

[31] This and much else in this section comes from Jordan and Moulin, *French Destroyers*, 20–40.

[32] The Italian destroyers were named *Leone*, *Pantera*, *Tigre*, *Leopardo*, and *Lince*, the last two being cancelled; the six ships of the French Jaguar class included *Panthère*, *Tigre*, *Léopard*, and *Lynx*. This clearly was not a co-incidence.

[33] Ibid, 22; Chesneau, ed., *Conway's 1922–1946*, 267.

[34] At the time the 130mm/40 Mle 1919 was designed, the French had not yet had experience with the German sliding-block breech guns mounted on the destroyers they acquired as reparations after WWI. The Krupp-designed sliding-block breech mechanism did not provide quite as tight a seal as the Welin interrupted screw, and thus required the powder be cased in a metal cartridge, but had the advantage of being significantly simpler and faster to operate.

[35] The deletion of the depth charge throwers was something of a 'no-brainer'. Not only were the Thornycroft throwers poorly sited in their original location in way of the forward funnel, but the Jaguars, while they were designed with spaces for the installation of sonar and its necessary electronics, were completed without that equipment, making them unlikely ASW platforms.

[36] Jordan and Moulin, *French Destroyers*, 77n4. In French, 'four-piper' is *quatre-tuyaux*.

[37] Hans Lengerer and Lars Ahlberg, *Fubuki-Class Destroyers in the Imperial Japanese Navy during World War II* (Atglen, PA: Schiffer Publishing, Ltd, 2022), 13; Anthony J Watts, 'The Japanese "Special" Type 1923', in *Super Destroyers: The big destroyers built in the 1930s for Britain, France, Germany, Italy, Japan and the United States*, edited by Antony Preston (London, England: Conway Maritime Press, 1978), 3.

[38] http://www.navweaps.com/Weapons/WTJAP_WWII.php#61_cm_%2824%22%29_Type_93_%281933%29_Model_1%2C_Mod_1%2C_Mod_2_and_Mod_3, accessed 13 August 2022. The 'Long Lance' name was a post-war naming convention most likely invented by American historian Samuel Eliot Morison; contemporary Japanese sources refer to it as the 'oxygen torpedo'.

[39] Watts, 'The Japanese "Special" Type 1923', in *Super Destroyers*, 7. So extraordinary was the Treasury order for the four additional destroyers, that the four hulls in the Treasury order were laid down before the second of the five ships in the regular programme.

[40] Japanese emperors have a given name and a regnal name that is chosen during his lifetime, but becomes the name by which he and his reign are known after his death. Thus, the first emperor was given the name Mutsuhito at birth, choosing the name Meiji for his reign. His successor was born Yoshihito and became the Taisho Emperor upon his death. His heir was Hirohito, who became the Showa Emperor when he died in 1989.

[41] Chesneau, ed., *Conway's 1922–1946*, 193; Watts, 'The Japanese "Special" Type 1923', in *Super Destroyers*, 8; Lengerer and Ahlberg, *Fubuki-Class Destroyers*, 20. It should be noted that at the time of ordering and laying down the first ships of this type, Japanese destroyers were given only numbers. Until October 1912, Japanese destroyers had been given names – 1st-class destroyers were named after weather phenomena and 2nd-class after natural objects like trees or flowers – but at that time it was decided that any new destroyers would simply receive numbers. This started with the ship that was to have been named *Kiyokaze*, laid down in December 1921, which instead became *Destroyer No 1*. Following this sequence, the first of the Special Type destroyers was to have been named *Destroyer No 39*, but on 1 August 1928, the decision was made to begin naming destroyers again and to give names to all destroyers given numbers in the interim. Thus *Destroyer No 1* (ex-*Kiyokaze*) became *Kamikaze*, and *Destroyer No 39* became *Fubuki*.

[42] The various displacements reported for the completed first subclass Fubukis range as high as 1870 tons (cf., Watts, 'The Japanese "Special" Type 1923', in *Super Destroyers*, 9) and sometimes even more, but rarely are these reported displacements described as standard; more often they are given as 'trials displacement', which is a relatively meaningless term referring to whatever condition a ship was in when it ran its acceptance trials, which was often a relatively 'light ship' state in terms of loading and manning. Later Special Type subclasses displaced more due to the fitting of heavier gun mounts and enclosed control position for the torpedo tubes.

[43] http://www.navweaps.com/Weapons/WNJAP_5-50_3ns.php#mountnote17. While this note explicitly states that the reduction in elevation by 20° was for the purpose of weight reduction, in the main

body of the article, the weight of the Type C twin turret is shown as being somewhat greater than that of the Type B.

Chapter 6: Trying to hold back the flood: 1927–1930

[1] "President's Message and Memorandum", *New York Times*, Vol. LXXVI, No. 25,220, 11 February 1927, 16.
[2] Ibid.
[3] Ibid.
[4] Capt Stephen W Roskill, RN, *Naval Policy Between the Wars I: The Period of Anglo-American Antagonism, 1919–1929* (Barnsley, S Yorkshire: Seaforth Publishing, 2016), 499–501; Scott T Price, 'A Study of the General Board of the U.S.Navy, 1929-1933.' Masters Thesis, University of Nebraska at Omaha, April 1989, 31.
[5] 'France Can Block Arms Conference and She Intends To', *New York Times*, Vol. LXXVI, No. 25,223, 14 February 1927, 1.
[6] 'France Says "No" to Our Arms Proposal; Briand Prefers Methods of League; Talk Now of Three-Power Conference', *New York Times*, Vol. LXXVI, No. 25,225, 16 February 1927, 2.
[7] 'France Can Block Arms Conference and She Intends To', *New York Times*, Vol. LXXVI, No. 25,223, 14 February 1927, 1.
[8] 'Washington Thinks Door Is Left Open', *New York Times*, Vol. LXXVI, No. 25,225, 16 February 1927, 1.
[9] The conference's discussions, which almost entirely concerned cruisers, are covered in some detail in Stern, *The Modern Cruiser*, 136–143.
[10] 'Text of American Naval Limitation Proposal', *New York Times*, Vol. LXXVI, No. 25,350, 21 June 1927, 2. Gibson was not well-known to the British before this conference, but he was then the American Ambassador to Belgium, had for a number of years been the primary American spokesman at League of Nations functions, and had a reputation as a tough, but fair, negotiator.
[11] Roskill, *Naval Policy I*, 499–500.
[12] 'America Urges 5-5-3 Small Ship Ratio at Naval Limitation Conference; Britain and Japan Have Counterplans', *New York Times*, Vol. LXXVI, No. 25,350, 21 June 1927, 1. The subhead for this lead article reads: "British Offer a Surprise".
[13] 'Britain Asks Cuts in Ships and Guns', *New York Times*, Vol. LXXVI, No. 25,350, 21 June 1927, 3.
[14] 'Japan for Caution in Naval Changes', *New York Times*, Vol. LXXVI, No. 25,350, 21 June 1927, 3.
[15] Elio Ando, 'The Italian Navigatori Class 1926', in *Super Destroyers: The big destroyers built in the 1930s for Britain, France, Germany, Italy, Japan and the United States*, edited by Antony Preston (London, England: Conway Maritime Press, 1978), 15–16.
[16] Ibid, 16; Chesneau, ed., *Conway's 1922–1946*, 299.
[17] The designed power output of the engineering plant was 55,000hp using the Italian scale, which equalled 54,241shp in the more common international scale.
[18] Ando, 'The Italian Navigatori Class 1926', in *Super Destroyers*, 15.
[19] Ibid, 17–18.
[20] David Carlton, 'The Anglo-French Compromise on Arms Limitation, 1928', *Journal of British Studies*, Vol 8, No 2, (May 1969), pp 144–49.
[21] W N Medlicott, Douglas Dakin, and M E Lambert, eds., *Documents on British Foreign Policy, 1919–1939, Series IA, Vol IV* (London, England: Her Majesty's Stationery Office, 1971), 594–95.
[22] Ibid, 597. This proposal is in an enclosure to a letter from Chamberlain to Briand dated 10 March 1928; curiously, it is written in French. A somewhat different version, different in a few details but not in intent, is noted in Carlton, 'The Anglo-French Compromise on Arms Limitation, 1928', 145n10, which is cited as being from a Foreign Office memo written by Alexander Cadogan, dated 1 May 1928.
[23] Carlton, 'The Anglo-French Compromise on Arms Limitation, 1928', 146.
[24] W N Medlicott, Douglas Dakin, and M E Lambert, eds., *Documents on British Foreign Policy, 1919–1939, Series IA, Vol V* (London, England: Her Majesty's Stationery Office, 1973), 679–80.
[25] Ibid, 680. M Boncour was Joseph Paul-Boncour, a deputy in the Assemblée Nationale, who was head of that body's Foreign Affairs Committee and a League of Nations delegate.
[26] Ibid. In this passage, Madden was specifically talking about cruiser strength, but it is clear from context that his argument applied to all smaller warships, including destroyers.
[27] The '3rd Bureau' would have been the Operations Division of the Naval Staff, tasked with planning for future conflicts.
[28] Medlicott, Dakin, and Lambert, eds., *Documents on British Foreign Policy, 1919–1939, Series IA, Vol V*, 681.
[29] Ibid, 682.
[30] Ibid, 719–20.
[31] Ibid, 725–26.
[32] Carlton, 'The Anglo-French Compromise on Arms Limitation, 1928', 150. This is from VAdm Kelly's unpublished memoirs, cited by Carlton.
[33] Medlicott, Dakin, and Lambert, eds., *Documents on British Foreign Policy, 1919–1939, Series IA, Vol V*, 769–70.
[34] Ibid, 775–76.
[35] Ibid, 780. The identical text exists in the American records at —, *Papers Relating to the Foreign Relations of the United States, 1928* (In Three Volumes) Vol 1 (Washington, DC: Government Printing Office, 1942), 265.
[36] Ibid, 781–82.
[37] Ibid, 790. In this message from Robert McNeill (1st Baron Cushendun), the British representative at the League of Nations, to Henry G Chilton, Minister to the United States, dated 10 August 1928, McNeill stated that concern about an early convocation of the Preliminary Commission had dissipated, and he should inform the Americans that they could 'defer replying until there has been time for a full consideration of the new proposal in all its bearings'.
[38] Ibid, 782n4.
[39] Ibid, 788.
[40] Ibid, 781.
[41] —, *Papers Relating to the Foreign Relations of the United States, 1928 Vol 1*, 266.
[42] Ibid, 267. That agreement was signed in Paris on 27 August 1928.
[43] Ibid, 267–68.
[44] Ibid, 283.
[45] —, *Papers Relating to the Foreign Relations of the United States, 1929*

(In Three Volumes) Vol 1 (Washington, DC: Government Printing Office, 1943), 94.
[46] Ibid, 92.
[47] Ibid.
[48] Capt Stephen W Roskill, RN, *Naval Policy Between the Wars II: The Period of Reluctant Rearmament, 1930–1939* (Barnsley, S Yorkshire: Seaforth Publishing, 1968), 39–43.
[49] Ibid, 41. Although the American reaction at the time to the idea of 10,000-ton 6in-armed cruisers was strongly negative, it ended up being incorporated in the London Treaty of 1930 and led to the USS *Brooklyn* (CL 40) class of seven ships and the near-sisters of the USS *St Louis* (CL 49) subclass of two ships.
[50] Curiously, Dawes shared the prize with Austen Chamberlain, who won for the Locarno Treaties of 1925.
[51] A number of very (or not so very) distinguished individuals have accomplished two out of those three feats, including (with apologies to anyone left off this list): Theodore Roosevelt, Woodrow Wilson, Jimmy Carter, Al Gore, Barack Obama, Sonny Bono, and Bob Dylan.
[52] —, *Papers Relating to the Foreign Relations of the United States, 1929*, 148–49.
[53] Ibid, 150.
[54] These numbers and those for the Royal Navy come from Roskill, *Naval Policy Between the Wars II*, 41.
[55] Friedman, *British Destroyers*, 471–73*.
[56] —, *Papers Relating to the Foreign Relations of the United States, 1929*, 243. The author was able to find no early reference to the 150,000 ton limit proposed for destroyers in either the British or American diplomatic correspondence for 1929.
[57] Ibid, 254. The same letter appears in Woodward, E L and Rohan Butler, eds., *Documents on British Foreign Policy, 1919–1939, Second Series, Vol I* (London, England: Her Majesty's Stationery Office, 1947), 94–97.
[58] Ibid, 255.
[59] The reader is advised to look at Stern, *The Modern Cruiser*, 167–70, for a concise summary of the treaty negotiations.
[60] Charles I Bevans, ed., *Treaties and Other International Agreements of the United States of America, 1776–1949, Vol 2, Multilateral, 1918–1930*, US Department of State Publication 8441, released May 1969, 1068.
[61] Ibid, 1068–69.
[62] Woodward and Butler, eds., *Documents on British Foreign Policy, 1919–1939, Second Series, Vol I*, 208. It is not clear to whom this document was distributed; as it covered in some the detail the negotiating position to be taken by the British delegation at the First London Conference, presumably it was intended for solely that delegation and perhaps for Commonwealth delegates as well.
[63] —, *Papers Relating to the Foreign Relations of the United States, 1930* (In Three Volumes) Vol 1 (Washington, DC: Government Printing Office, 1945), 22–23.
[64] Ibid, 32.
[65] Ibid, 44.

Chapter 7: Strange breed – London's children: 1930–1937
[1] David Lyon, 'The British Tribals 1935', in *Super Destroyers: The big destroyers built in the 1930s for Britain, France, Germany, Italy, Japan and the United States*, edited by Antony Preston (London, England: Conway Maritime Press, 1978), 50.
[2] 'Text of the Statements on End of the Naval Treaty', *New York Times*, Vol. LXXXIV, No. 28,099, 30 December 1934, p 14. It should be noted that this was NOT early termination of Japan's participation in the Washington Treaty restrictions. The Japanese announced that they would end their participation in that treaty on 31 December 1936, which was the date the treaty was set to expire; what it would do is prevent the automatic renewal of the treaty's terms. Also, at the same time Japan stated their strong desire to continue the naval disarmament process, but on a more equitable basis than enacted in the Washington Treaty. (See next chapter.)
[3] Chesneau, ed., *Conway's 1922–1946*, 40; Ibid, 61; Whitley, *Destroyers of World War Two*, 114.
[4] Friedman, *U.S. Destroyers*, 79–81.
[5] http://www.navweaps.com/Weapons/WNUS_5-38_mk12.php, last accessed 2 October 2022.
[6] Friedman, *British Destroyers*, 798*.
[7] —, *The Statutes at Large of the United States of America from December 1931 to March 1933, Vol XLVII, Part 1* (Washington, DC: Government Printing Office, 1933), 441.
[8] Friedman, *U.S. Destroyers*, 403; Chesneau, ed., *Conway's 1922–1946*, 125.
[9] Friedman, *U.S. Destroyers*, 81.
[10] Ibid, 84.
[11] Friedman, *U.S. Destroyers*, 404; Chesneau, ed., *Conway's 1922–1946*, 125.
[12] Friedman, Norman, 'The US "1850-Tonners" 1933', in *Super Destroyers: The big destroyers built in the 1930s for Britain, France, Germany, Italy, Japan and the United States*, edited by Antony Preston (London, England: Conway Maritime Press, 1978), 44; http://www.navweaps.com/Weapons/WNUS_5-38_mk12.php.
[13] Chesneau, ed., *Conway's 1922–1946*, 193; Anthony J Watts and Brian G Gordon, *The Imperial Japanese Navy* (Garden City, NY: Doubleday & Co, Inc, 1971), 274; Whitley, *Destroyers of World War Two*, 196.
[14] Alan Raven, 'Aspects of Japanese Warship Design, 1918-1945: No 2: Destroyers', *Warship* No 3 (July 1977), 22.
[15] —, *Japanese Naval Guns and Mounts, Article 1 – Mounts under 18"*, O-47(N)-1 (San Francisco, CA: U.S. Naval Technical Mission to Japan, 16 February 1946), 14–17; http://www.navweaps.com/Weapons/WNJAP_5-50_3ns.php#mountnote6.
[16] Jacques Mordal, 'The French Fantasque Class 1930', in *Super Destroyers: The big destroyers built in the 1930s for Britain, France, Germany, Italy, Japan and the United States*, edited by Antony Preston (London, England: Conway Maritime Press, 1978), 27; Jordan and Moulin, *French Destroyers*, 137–39.
[17] http://www.navweaps.com/Weapons/WNIT_6-53_m1926.php. This range figure must be understood within the context of all the Italian gun data from this period; guns in twin mounts, as was the case here, were commonly mounted too close together and demonstrated interference during salvo fire that led to inaccuracy due to excess dispersion. The problem was eventually solved by reducing the powder

charge and shell weight, which in turn reduced the maximum range of this gun to 24,716yds.

[18] Chesneau, ed., *Conway's 1922–1946*, 268; Jordan and Moulin, *French Destroyers*, 140; Whitley, *Destroyers of World War Two*, 42.

[19] Mordal, 'The French Fantasque Class 1930', in *Super Destroyers*, 28, claims that one of the ships of this class, *Le Terrible*, reached 45.029kt on trials, which he states stood as a speed record for ships of this size for at least forty years. This author has no means of verifying that claim.

[20] Ibid; Jordan and Moulin, *French Destroyers*, 144.

[21] Mordal, 'The French Fantasque Class 1930', in *Super Destroyers*, 27.

[22] Jordan and Moulin, *French Destroyers*, 160–61.

[23] 'Rome Joins Naval Accord, Ends Rivalry with Paris; To Complete 5-Power Pact', *New York Times*, Vol. LXXX, No. 26,699, 1 March 1931, p 1, 5.

[24] 'Text of Franco-Italian Naval Accord', *New York Times*, Vol. LXXX, No. 26,710, 12 March 1931, p 19.

[25] 'Cruiser B Approved by Reichstag Body', *New York Times*, Vol. LXXX, No. 26,710, 12 March 1931, p 19.

[26] Jordan and Moulin, *French Destroyers*, 161.

[27] Chesneau, ed., *Conway's 1922–1946*, 269; Jordan and Moulin, *French Destroyers*, 164; Whitley, *Destroyers of World War Two*, 44.

[28] http://www.navweaps.com/Weapons/WNFR_55-50_m1929.php#mountnote3.2back.

[29] —, *The Treaty of Peace with Germany (Treaty of Versailles)*, 1919 For. Rel. (Paris Peace Conference, XIII), Art 181.

[30] Article 185 of the Versailles Treaty specified that the Germans were to surrender forty-two of their most modern destroyers, as chosen by the Allied Powers.

[31] Gerhard Koop and Klaus-Peter Schmolke, *German Destroyers of World War II*, trans. Geoffrey Brooks (London, England: Lionel Leventhal Ltd, 2003), 9, 13–14.

[32] —, *Exchange of Notes between His Majesty's Government in the United Kingdom and the German Government regarding the Limitation of Naval Armaments, 18 June 1935*.

[33] The German navy officially changed its name from the Reichsmarine to the Kriegsmarine on 21 May 1935.

[34] Chesneau, ed., *Conway's 1922–1946*, 232; Koop and Schmolke, *German Destroyers of World War II*, 26; Whitley, *Destroyers of World War Two*, 59.

[35] The various sources give widely divergent figures for the standard displacement of these ships, all the way from 1625 tons (*Conway's*) to 2223 tons (Koop and Schmolke). I have chosen to give an intermediate figure since the true displacement is undoubtedly impossible to ascertain at this remove.

[36] https://destroyerhistory.org/goldplater/1500ton/; Antony Preston, 'The German Narvik Type 1936', in *Super Destroyers: The big destroyers built in the 1930s for Britain, France, Germany, Italy, Japan and the United States*, edited by Antony Preston (London, England: Conway Maritime Press, 1978), 64.

[37] Koop and Schmolke, *German Destroyers of World War II*, 16.

[38] Ibid, 17.

[39] Chesneau, ed., *Conway's 1922–1946*, 329; Whitley, *Destroyers of World War Two*, 229.

Chapter 8: Let slip the dogs of war – the floodgates open: 1935–1939

[1] —, *Treaty for the Limitation of Naval Armament (Second London Naval Treaty)*, London, March 25, 1936, Part I, Article 1, Section B Categories, Subsection (3) Light Surface Vessels, Sub-subsection (c).

[2] Ibid, Section C Over Age, Subsection (d) Light Surface Vessels, subcategory (c).

[3] Much of the content in this section comes from Norman Friedman, *British Destroyers & Frigates: The Second World War and After* (Barnsley, S Yorkshire: Seaforth, 2008), 81–95* and Whitley, *Destroyers of World War Two*, 117–21.

[4] Chesneau, ed., *Conway's 1922–1946*, 41; Whitley, *Destroyers of World War Two*, 117.

[5] Friedman, *British Destroyers & Frigates*, 98*.

[6] Whitley, *Destroyers of World War Two*, 118.

[7] Friedman, British Destroyers & Frigates, 115*.

[8] Chesneau, ed., *Conway's 1922–1946*, 42; Whitley, *Destroyers of World War Two*, 124.

[9] Chesneau, ed., *Conway's 1922–1946*, 299; Whitley, *Destroyers of World War Two*, 165.

[10] Chesneau, ed., *Conway's 1922–1946*, 300–1; Whitley, *Destroyers of World War Two*, 169.

[11] The reported standard displacement of these ships varied from 1620 tons to 1830 tons depending on the builder and who was doing the reporting.

[12] Koop and Schmolke, *German Destroyers of World War II*, 17–18.

[13] Whitley, *Destroyers of World War Two*, 67, 69.

[14] Koop and Schmolke, *German Destroyers of World War II*, 19–20.

[15] Whitley, *Destroyers of World War Two*, 236.

[16] Ibid, 239; Chesneau, ed., *Conway's 1922–1946*, 332.

[17] Chesneau, ed., *Conway's 1922–1946*, 194; Whitley, *Destroyers of World War Two*, 199; Chihaya Masataka and Abe Yasuo, *IJN Yukikaze/Destroyer/1939–1970*, Warship Profile 22 (Windsor, Berkshire: Profile Publications Ltd, 1972), 227.

[18] With the renunciation of treaty limits, the IJN ceased using standard displacement to measure the mass of their warships, instead using a system they described as 'trials displacement', which measured a ship with a full load of ammunition and two-thirds of all other consumables, such as feed water and fuel. This information comes from Fukui Shizuo and Robert Dulin, via Randy Stone.

[19] It is a known 'quirk' of hydrodynamics that a properly shaped transom stern can act like a much longer stern in terms of reducing drag, particularly at higher speeds.

[20] Masataka and Abe, *IJN Yukikaze*, 227–28.

[21] There is disagreement among sources as to whether the destroyer *Akigumo* was the last (and nineteenth) of the Kagero class or the second unit of the follow-on Yugumo class. The latest analysis seems to point to the former. Given the minimal differences between the classes, it is more of a book-keeping issue than anything else.

[22] Thanks to Randy Stone for this information about when the Type 93 was introduced and the retrofitting programme.

[23] Friedman, *U.S. Destroyers*, 94–96.

[24] Ibid, 410–11; Chesneau, ed., *Conway's 1922–1946*, 128; Whitley, *Destroyers of World War Two*, 273.

[25] Those four were DesDiv22, which comprised USS *Gwin*, *Meredith*,

Grayson, and *Monssen* (DD 433–36). They would see active service in the South Pacific, accounting for three of the five losses among the original twenty-four of the class. Cf., Robert C Stern 'Perilous Mission to Guadalcanal', *Naval History* Vol 36, No 5 (October 2022), pp 34–41, relates the story of a supply convoy that involved three of the four of this group and saw *Meredith*'s loss.

[26] Robert C Stern, *The US Navy and the War in Europe* (Barnsley, S Yorkshire: Seaforth, 2012), 25–30.

[27] https://destroyerhistory.org/benson-gleavesclass/desron12/, last accessed 8 May 2023.

[28] Friedman, *U.S. Destroyers*, 111–16.

[29] Ibid, 412; Chesneau, ed., *Conway's 1922–1946*, 130; Whitley, *Destroyers of World War Two*, 279.

Chapter 9: The test of battle, Part 2: 1942–1945

[1] Vincent P O'Hara, *Struggle for the Middle Sea: The Great Navies at war in the Mediterranean Theater, 1940–1945* (Annapolis, MD: Naval Institute Press, 2009), 108–12; Jack Greene and Alessandro Massignani, *The Naval War in the Mediterranean 1940–1943* (London, England: Frontline Books, 2011), 161–64.

[2] Derek Howse, *Radar at Sea: The Royal Navy in World War 2* (Annapolis, MD: Naval Institute Press, 1993), 73–74. A number of these sets, with their fixed antenna, were sent to Alexandria, along with technicians to perform the installations, at Cunningham's request at the beginning of 1941. According to Friedman, Norman, *Naval Radar* (Annapolis, MD: Naval Institute Press, 1981), 195–96, while this radar was not highly regarded for a number of reasons, including the non-rotating antenna and inability to distinguish between 'returns' ahead or astern of the antenna, it was reliable, lightweight, and available in quantity – sharing electronics with the RAF's ASV Mk I radar already in production – and could be installed at any port with a dockside crane in under a week.

[3] Howse, *Radar at Sea*, 80; Nino Arena, *Il Radar, La Guerra sui Mari*, Vol 1 (Modena, Italy: STEM Mucchi,1976), 121.

[4] The Navigatori class destroyers, described in Chapter 6, had initially been typed as *esploratori leggeri*, but had been re-typed as *cacciatorpe-diniere* in 1938. Cf., Chesneau, ed., *Conway's 1922–1946*, 299.

[5] —, 'Report of an Action against an Italian Convoy on the Night of the 15th/16th April, 1941', in *Supplement to the London Gazette* of Tuesday, the 11th of May, 1948, No 38287 (London, England: Her Majesty's Stationery Office, 1948). The sighting is recorded as being logged at 1157Z; the action in this account all took place in the Central European Time zone, which is Z+1, but pushed ahead another hour because of the adoption by the British of year-round Summer Time in 1940. The RAF Maryland was an American-built twin-engine medium bomber developed in response to a 1937 USAAC competition. The Martin Model 167, while losing the competition, found a willing buyer in the French, who ordered 290 aircraft. Two hundred fifteen aircraft were delivered to the French before June 1940; the seventy-five aircraft remaining on the French order were delivered to the British, and an additional 150 with up-rated engines were ordered by the RAF, which they dubbed the Maryland, using them as bombers and fast reconnaissance aircraft. The British liked the aircraft, because it was fast, had good range, and was durable.

[6] Cristiano D'Adamo, *Regia Marina Italiana: Sinking of the Tarigo Convoy, April 16th, 1941*, https://web.archive.org/web/20060910011054/http://regiamarina.net/engagements/tarigo/tarigo_us.htm.

[7] This is from Capt Mack's report for 14th Flotilla included as part of —, *Report of an Action against an Italian Convoy*, 2913–14.

[8] O'Hara, *Struggle for the Middle Sea*, 109–10.

[9] —, *Report of an Action against an Italian Convoy*, 2914.

[10] —, *Attacco da Parte di Forze Leggere Nemiche al Convoglio "Tarigo" Presso le Secche di Kerkenah (16 Aprile 1941/XIX°)*, Notizia 5^ (Rome, Italy: Ufficio di Stato Maggiore della Regia Marina, undated), 2.

[11] Norman Friedman, *Naval Radar* (Annapolis, MD: Naval Institute Press, 1981), 196.

[12] Alastair Mitchell, 'The Development of Radar in the Royal Navy 1939–1945, Part 1', *Warship* No 13 (January 1980), 11.

[13] —, *Attacco da Parte di Forze Leggere*, 1.

[14] Ibid, 2. The reader who may be wondering why there would be lighted buoys marking sea lanes in a war zone is reminded that Tunisia at this time was a French colony, and France was, in April 1941, technically a neutral, non-belligerent country, and so maintained navigational aids as in peacetime.

[15] —, *Report of an Action against an Italian Convoy*, 2914.

[16] The record here is confused. It is not surprising that four different reports by four officers on four ships (one of which sank) with four different clocks compiled after the fact from notes scribbled by yeomen in real time under stressful circumstances share few common details. The two best historical accounts, in O'Hara, *Struggle for the Middle Sea*, 110–11 and Greene and Massignani, *The Naval War in the Mediterranean*, 162–64, give accounts that differ considerably in details, and the account to be offered here by this author will differ yet again, as his analysis of the original reports has led him to construct some of the details of this engagement somewhat differently. For example, Greene and Massignani state that the convoy was first sighted by *Nubian*'s radar at 0158; O'Hara states that lookouts in *Jervis* made visual sighting at that same time, with no mention made of radar. The four individual narratives in —, *Report of an Action against an Italian Convoy*, 2914–16 tell yet another story, which I follow here.

[17] —, *Report of an Action against an Italian Convoy*, 2914. This is from *Jervis*' narrative; alone among these narratives, it is uncredited, but since an earlier footnote lists Mack as *Jervis*' captain, it must be presumed he wrote this narrative as well.

[18] Ibid.

[19] Ibid, 2915. The description of the target as a 'small one funnel destroyer' immediately rules out *Tarigo* as the target, which was larger with two funnels. That left *Baleno* and *Lampo*, which fit the description. Of the two, *Baleno* was much closer and plainly visible, while *Lampo* would by now have been aft of *Jervis* and probably obscured by the intervening transports of the convoy.

[20] Ibid. This was a case of the British use of 'RDF' as a synonym for radar. It should be remembered that the Type 286 radar was never intended to be a fire control system; any use to control gunfire was improvisational.

[21] Ibid, 2914.

[22] —, *Attacco da Parte di Forze Leggere*, 10.
[23] —, *Report of an Action against an Italian Convoy*, 2915.
[24] —, *Attacco da Parte di Forze Leggere*, 3.
[25] Ibid, 7–8.
[26] —, *Report of an Action against an Italian Convoy*, 2916.
[27] Ibid, 2915.
[28] LCdr Enrico Marano, RM, *Rapporto Relativo al Combattimento Sostenuto dai CC.TT. "TARIGO," "BALENO" e "LAMPO" la Notte sul 16 Aprile 1941/XIX°*, 4.
[29] While these ships had been completed with 13.2mm Breda MGs as light AA mounts, as the war progressed, these were replaced by 20mm/65 single and twin mounts, which fired a heavier round and were better liked.
[30] —, *Attacco da Parte di Forze Leggere*, 8. This translation has been 'augmented' by some details from LCdr Marano's attached report not included in the main document.
[31] Marano, *Rapporto*, 4.
[32] —, *Attacco da Parte di Forze Leggere*, 8–10.
[33] —, 'Report of an Action against an Italian Convoy', 2914.
[34] Ibid, 2915.
[35] Ibid, 2914–16.
[36] —, *Attacco da Parte di Forze Leggere*, 3.
[37] Ibid, 5.
[38] —, *Report of an Action against an Italian Convoy*, 2915.
[39] Ibid.
[40] Ibid.
[41] Ibid, 2914.
[42] —, *Attacco da Parte di Forze Leggere*, 10.
[43] —, *Report of an Action against an Italian Convoy*, 2914.
[44] Ibid. The first two sentences in this quote are from the end of a longer entry marked with the time 0227; the next entry is marked 0240. It must be presumed that this part of the action took place closer to 0240 than 0227.
[45] 'Fixed sight range' is also known as 'point blank range' or 'firing over open sights'. It simply means that all the gunner had to do was point the gun at the target and fire; no range calculation was necessary.
[46] Ibid, 2915.
[47] —, *Attacco da Parte di Forze Leggere*, 5.
[48] Ibid, 5–6.
[49] Ibid, 6.
[50] —, *Report of an Action against an Italian Convoy*, 2914.
[51] Ibid, 2916.
[52] Ibid.
[53] http://www.navweaps.com/Weapons/WTIT_List.php#World_War_II. The primary destroyer torpedo used by the Regia Marina during the Second World War was the 53.3cm Si270/533.4×7.2M, which, at ranges less than 2nm, could reach speeds up to 48kt.
[54] —, *Report of an Action against an Italian Convoy*, 2916.
[55] Ibid.
[56] Ibid.
[57] These casualty figures come from http://www.naval-history.net/xGM-Chrono-10DD-34Tribal-HMS_Mohawk.htm.
[58] At the risk of blatant self-promotion, this author can recommend no better introduction to the kamikaze campaign and the efforts to combat it than Robert C Stern, *Fire from the Sky: Surviving the Kamikaze Threat* (Barnsley, S Yorkshire: Seaforth, 2010).
[59] It is, in fact, more complex than stated here. The kanji characters that can be vocalized 'kamikaze' 神 風 were used to describe these attacks, but were vocalized 'shinbu' in Japan at the time. Both are valid vocalizations, but the latter was considered more appropriate; multiple possible vocalizations of kanji characters, and the use of different vocalizations in different linguistic situations, is a common feature of modern Japanese. The term kamikaze was first commonly used by Americans during the war, and only later did it become common usage in Japan.
[60] Ugaki Matome was appointed chief-of-staff of the Combined Fleet under Adm Yamamoto Isoroku in August 1941, which posting he retained until that officer's death in April 1943. While not an early advocate or organizer of *tokko* operations, in February 1945, Ugaki was appointed commander of the 5th Air Fleet in Kyushu, which put him in charge of the kamikaze operations in defence of Okinawa.
[61] Robin L Rielly, *Kamikazes, Corsairs, and Picket Ships* (Drexel Hill, PA: Casemate, 2008), 5. It was never the intention of VAdm Richmond K Turner, in charge of TF51, the amphibious task force for the Okinawa invasion, that all designated radar picket stations would be occupied simultaneously, or even at all, and some were designated as stations to be occupied only during daylight hours, with ships assigned there to be moved to other stations after dark to confuse the enemy.
[62] These numbers are derived from Samuel Eliot Morison, *History of United States Naval Operations in World War II, Vol. XIV, Victory in the Pacific, 1945* (Boston, MA: Little, Brown & Co., 1960), 373–85. They are 'nominal' in the sense that they represent the numbers reported at the beginning of the campaign; operational losses and subsequent replacements would have meant that on any given day, those numbers may have varied. Also, not all of those fighters could be airborne at the same time, nor were all assigned CAP duty; the actual number flying CAP patrol at any one time would likely have been less than a quarter of those totals.
[63] The US Navy handled the need to rest its fleet and carrier task force commanders, something it had been unable to do in 1942, and allow for the meticulous planning for the carrier support for its rapid drive across the central Pacific by means of a very clever expedient. After each major operation, the entire high-level command would switch out between Adm William F 'Bill' Halsey and his staff and Adm Raymond A Spruance and his. When commanded by Halsey, the main body of the Pacific Fleet was known as the Third Fleet and the Fast Carrier Force was known as TF38, commanded by VAdm John S 'Slew' McCain; when commanded by Spruance, it was known as the Fifth Fleet and the carrier force was called TF58, commanded by VAdm Marc A Mitscher.
[64] Rielly, *Kamikazes, Corsairs, and Picket Ships*, 85.
[65] Robert C Stern, *Fire from the Sky: Surviving the Kamikaze Threat* (Barnsley, S Yorkshire: Seaforth, 2010), 339. The term 'constructive total loss' is used in the maritime insurance industry to refer to a ship that, while technically still seaworthy, is damaged to a point that it is considered too costly to repair and is therefore written off.
[66] 'Val' was the name given, in the Allied reporting system developed

in Australian in 1942 to simplify identification of Japanese aircraft, to the Aichi D3A Type 99 Carrier Bomber. A low-wing, single-engine aircraft with fixed landing gear, the Val had been, at the outbreak of war, the IJN's standard carrier dive bomber, but by 1945, the Val was obsolescent; still available in large numbers and relatively docile to fly, it was an ideal candidate as a kamikaze aircraft.

[67] *DD793*, 7.

[68] Ibid.

[69] Ibid. *Cassin Young*'s ship's company was able to effect sufficient repairs to her powerplant to the extent that on 16 April she was able to head towards Ulithi atoll in the Carolines for permanent repairs, which were completed at the end of May 1945, after which she returned to combat off Okinawa.

[70] Ibid; *DD724–2*, 19.

[71] *DD724–2*, 19.

[72] Ibid, 21.

[73] Ibid, 21–22.

[74] Ibid, 22.

[75] Ibid, 23, 63–65. The second of those citations is a standard AA reporting form.

[76] Ibid, 23.

[77] Ibid.

[78] Ibid.

[79] It was US Navy practice to designate gun mounts with one number indicating gun size and another number (or numbers) indicating sequence from the bow aft. Thus the after main-battery mount on Laffey, the third 5in mount from the bow, was mount 53. Groups of machine guns, particularly 20mm guns, were sometimes given a single-mount number. When mounts (or groups) were paired port and starboard, the starboard side mount was given an odd number and the port side mount the next even number; this was even true when the 'paired' mounts were not exactly abreast of each other, so that in the case of *Laffey*, group 23 was actually aft of group 24 and the same was true with mounts 43 and 44, but group 25 was right aft on the centreline.

[80] This whole section is from Ibid, 24–26, with additional reference to the individual AA reports on pages 63–131

[81] *Laffey*'s action report only notes that eight men were taken ashore for burial, cf., Ibid, 27–28. It must be presumed that the remaining dead were either blown overboard during the attacks or their bodies were rendered unrecoverable by the explosions in which they died.

Chapter 10: Building from experience: 1942–1945

[1] Chesneau, ed., *Conway's 1922–1946*, 195; Whitley, *Destroyers of World War Two*, 202; http://www.combinedfleet.com/shimak_n.htm.

[2] Mark R Peattie, *Sunburst: The Rise of Japanese Naval Air Power, 1909–1941* (Annapolis, MD: Naval Institute Press, 2001), 151.

[3] http://www.navweaps.com/Weapons/WNJAP_39-65_t98.php, last accessed 30 January 2023. Although it was described as a dual-purpose weapon, the only rounds developed for this gun were timed-fuse HE rounds, suitable for anti-aircraft use.

[4] Chesneau, ed., *Conway's 1922–1946*, 195; Whitley, *Destroyers of World War Two*, 204–5; http://www.combinedfleet.com/akizuk_n.htm.

[5] Friedman, *British Destroyers & Frigates*, 278*.

[6] http://www.navweaps.com/Weapons/WNBR_47-45_mk9.php.

[7] Friedman, *British Destroyers & Frigates*, 279*. The Hazemeyer Bofors was a mount for a twin 40mm/56 Bofors machine gun that initially incorporated a rangefinder and an analog fire control computer. Later versions supplemented, and then replaced, the rangefinder with Type 282 or Type 285 radar. The prototype for this mount had been brought to Great Britain on a Dutch destroyer fleeing the fall of that country in 1940.

[8] Ibid, 282*.

[9] Chesneau, ed., *Conway's 1922–1946*, 44; Whitley, *Destroyers of World War Two*, 138–40; Ibid, 909–10*.

[10] Friedman, *U.S. Destroyers*, 123.

[11] Ibid, 124.

[12] Ibid, 413; Chesneau, ed., *Conway's 1922–1946*, 132–33; Whitley, *Destroyers of World War Two*, 288–90.

Sources

[1] In the United States, there are four privately-owned newspapers that are generally considered 'newspapers of record', meaning they took upon themselves the responsibility to report national as well as local news and to reproduce official documents released by the federal government. Besides the *New York Times*, these are the *Washington Post*, the *Wall Street Journal* and the *Los Angeles Times*.

Index

Ships or persons shown in images or mentioned in captions are indicated in *italics*. For navy abbreviations, see the 'Abbreviations/Acronyms' section at the beginning of the book; for those nations whose navies are not so defined, nationality codes are used, see https://en.wikipedia.org/wiki/ISO_3166-1_alpha-2; for ship type designators, see https://www.history.navy.mil/research/histories/ship-histories/abbreviations.html. (It should be noted that, despite being quite comprehensive, the list of ship type designators at the above-mentioned site does not include all the ship types described here, and some 'invention' was required. The reader should also note that the navy abbreviation 'KM' is used to designate all the successive German navies covered here. When two ships or classes in a navy share the same name, they are distinguished by the year in which they entered service.) Items mentioned in ship's specifications in the text are not indexed.